# ZYMURGY

### FOR THE HOMEBREWER AND BEER LOVER

The Best Articles and Advice
from America's
#1 Home Brewing Magazine

*Other Avon Books by*
**Charlie Papazian**

THE HOME BREWER'S COMPANION
HOME BREWER'S GOLD
THE NEW COMPLETE JOY OF HOME BREWING

---

Avon Books are available at special quantity discounts for bulk purchases for sales promotions, premiums, fund raising or educational use. Special books, or book excerpts, can also be created to fit specific needs.

For details write or telephone the office of the Director of Special Markets, Avon Books, Dept. FP, 1350 Avenue of the Americas, New York, New York 10019, 1-800-238-0658.

# ZYMURGY®
## FOR THE HOMEBREWER AND BEER LOVER

The Best Articles and Advice
from America's
#1 Home Brewing Magazine

Edited by
## Charlie Papazian
and the American Homebrewer's Association®

AVON BOOKS • NEW YORK

This book is published in cooperation with The American Homebrewers Association and Brewers Publications, both of which are divisions of the Association of Brewers, P.O. Box 1679, Boulder, CO 80306, U.S.A., (303) 447-0816 or (888) U CAN BREW; website address http://beertown.org.

AVON BOOKS
A division of
The Hearst Corporation
1350 Avenue of the Americas
New York, New York 10019

Copyright © 1998 by American Homebrewers Association
Published by arrangement with American Homebrewers Association
Visit our website at http://**www.AvonBooks.com**
ISBN: 0-380-79399-7

All rights reserved, which includes the right to reproduce this book or portions thereof in any form whatsoever except as provided by the U.S. Copyright Law. For information address Avon Books.

Zymurgy : the best articles and advice from America's #1 home brewing
 magazine / edited by Charlie Papazian and the American Homebrewer's
 Association
    p. cm.
  Includes index.
  1. Brewing—Amateurs' manuals.  I. Papazian, Charlie.
 II. American Homebrewers Association.
 TP570.Z96   1998
 641.8'73—dc21                                          97-49055
                                                          CIP

First Avon Books Trade Printing: May 1998

AVON TRADEMARK REG. U.S. PAT. OFF. AND IN OTHER COUNTRIES, MARCA REGISTRADA, HECHO EN U.S.A.

Printed in the U.S.A.

OPM  10  9  8  7  6  5  4  3  2  1

---

If you purchased this book without a cover, you should be aware that this book is stolen property. It was reported as ''unsold and destroyed'' to the publisher, and neither the author nor the publisher has received any payment for this ''stripped book.''

This book is dedicated to homebrewers everywhere who take the time to share their knowledge, experience, and ideas, especially the dedicated writers whose articles in homebrewing magazines, brewspapers, club newsletters, and books continue to enrich the hobby and beer culture throughout the world.

# ACKNOWLEDGMENTS

Very special thanks to Cathy Ewing, Vice President of the Association of Brewers; Dena Nishek, past Editor of *Zymurgy* magazine; Tyra Segars, Graphic/Production Director for the Association of Brewers; and Carolyn Robertson, Graphic Designer for the Association of Brewers, for their essential and tireless effort to help compile this book.

Also, grateful acknowledgment to past Editors of *Zymurgy* Tracy Loysen and Elizabeth Gold, and to John and Lois Canaday, who copyedited *Zymurgy* for over fourteen years.

And a very extra special salute to Kathy McClurg, who has been Associate Editor for every issue of *Zymurgy* since volume 1, number 1 (December 1978).

# CONTENTS

**INTRODUCTION**   xv

A GRAND AND MYSTERIOUS PHENOMENON   xv
*Charlie Papazian*

A SATISFYING SIP OF HOMEBREW HISTORY   xvii
*Dena Nishek*

**CHAPTER 1 • BREWING HISTORY IN THE AMERICAS**   1

"NATIVE" BREWING IN AMERICA   1
*William Litzinger*

BRINGING COLONIAL BREWING AND MALTING TO LIFE   5
*Rich Wagner*

| | |
|---|---|
| THE OLDEST BREWERY IN AMERICA<br>  Charlie Papazian | 8 |
| **CHAPTER 2 • BEER STYLES** | **13** |
| I'M A MILD MAN MYSELF, BUT . . .<br>  Howard Browne, M.D. | 13 |
| FLAMING STONE: BREWING TRADITIONAL Steinbiere<br>  Phil Rahn and Chuck Skypeck | 17 |
| GOTLANDSDRICKA, THE ANCIENT BREW OF GOTLAND<br>  Håkan Lundgren | 21 |
| LEANN FRAOCH—SCOTTISH HEATHER ALE<br>  Bruce Williams | 23 |
| THE REGAL ALTBIERS OF DÜSSELDORF<br>  Roger Deschner | 28 |
| BREWING BETTER BELGIAN ALES<br>  Phillip Seitz | 35 |
| PERFECT YOUR PORTER<br>  Terry Foster | 48 |
| CONFESSIONS OF TWO BITTER MEN<br>  Tony Babinec and Steve Hamburg | 60 |
| **CHAPTER 3 • EQUIPMENT AND GADGETS** | **75** |
| HOLY HYDROMETER, BATMAN!<br>  David A. Weisberg | 75 |
| WORT CHILLERS: THREE STYLES TO IMPROVE YOUR BREW<br>  Mindy and Ross Goeres | 81 |
| ALL-GRAIN ON A SHOESTRING<br>  Mark Moylan | 84 |

A BOTTLER'S GUIDE TO KEGGING ........................................ 89
  Ed Westemeier

THE COUNTERPRESSURE CONNECTION—
  WHERE BOTTLES AND KEGS UNITE ................................. 110
  David Ruggiero, Jonathan Spillane, and Doug Snyder

HOW TO OXYGENATE YOUR WORT—CHEAPLY .......................... 131
  Russ Klisch

THE JOCKEYBOX ........................................................ 135
  Teri Fahrendorf

ON THE SURPLUS TRAIL ................................................ 139
  Randy Mosher

## CHAPTER 4 • BREWING THEORY ........................... 142

AN EASY GUIDE TO RECIPE FORMULATION ........................... 142
  Monica Favre and Tracy Loysen

BREW BEER YEAR-ROUND ............................................. 148
  Byron Burch

BREW BY THE NUMBERS—ADD UP WHAT'S IN
  YOUR BEER .......................................................... 150
  Michael L. Hall, Ph.D.

## CHAPTER 5 • MALT ......................................... 168

THE ENCHANTING WORLD OF MALT EXTRACT—
  MAKE THE MOST OF IT .............................................. 168
  Norman Farrell

MALT—A SPECTRUM OF COLORS AND FLAVORS ..................... 182
  Neil C. Gudmestad and Raymond J. Taylor

## CHAPTER 6 • HOPS — 198

HOP VARIETIES AND QUALITIES — 198
  Bert Grant

CALCULATING HOP BITTERNESS IN BEER — 203
  Jackie Rager

## CHAPTER 7 • YEAST — 209

ACTIVE DRY YEAST FOR THE HOMEBREWER — 209
  Rodney Morris

BECOME SACCHAROMYCES SAVVY — 214
  Patrick Weix

YEAST STOCK MAINTENANCE AND STARTER
  CULTURE PRODUCTION — 219
  Paul Farnsworth

## CHAPTER 8 • WATER — 247

WATER TREATMENT: HOW TO CALCULATE SALT ADJUSTMENT — 247
  Darryl Richman

BEER FROM WATER—MODIFY MINERALS TO MATCH
  BEER STYLES — 252
  Jon Rodin and Glenn Colon-Bonet

## CHAPTER 9 • UNUSUAL INGREDIENTS — 259

OPTIONS FOR ADDING FRUIT — 259
  Al Korzonas

POTIONS! — 261
  Randy Mosher

A BREWER'S HERBAL—RETURNING TO FORGOTTEN FLAVORS — 267
  Gary Carlin

## CHAPTER 10 • BETTER BREWING — 274

Steeping: The Easy Step — 274
  Shawn Bosch

Start Mashing! — 279
  Charlie Papazian

Decoction Mashing — 283
  Gregory J. Noonan

The Detriments of Hot-Side Aeration — 301
  George Fix

pH and the Brewing Process — 310
  Eric Warner

## CHAPTER 11 • BEER EVALUATION — 314

The Sensory Aspects of Zymological Evaluation — 314
  David W. Eby, Ph.D.

## CHAPTER 12 • WORLD OF WORTS — 327
  Charlie Papazian

Thistle Do It Export Pilsener — 327

Here to Heaven Oktoberfestwine Ale — 330

Slumgullion Amber Ale — 332

## CHAPTER 13 • DEAR PROFESSOR SURFEIT — 336

## CHAPTER 14 • HOMEBREW COOKING — 358

Homebrew Cooking with the BrewGal
  Gourmet: A Feastive Fall Menu — 358
  Candy Schermerhorn

## CHAPTER 15 • HOMEBREWER OF THE YEAR WINNING RECIPES — 364

## CHAPTER 16 • JACKSON ON BEER — 378
Michael Jackson

### Drinkability: What Is It? — 378

### When a Beer Goes to Sleep — 382

## CHAPTER 17 • HOMEBREWING LORE — 386

### The Lost Art of Homebrewing — 386
Karl F. Ziesler

### Secret Satisfaction of Brewing — 391
John Goldfine

### Joy of Brewing — 393
Fritz Maytag

### Traveling with Homebrew — 396
Charles Matzen and Charlie Papazian

## CHAPTER 18 • THE LAST DROP — 402

### Last Drop—A Bit Overcarbonated — 402
John Isenhour

### Last Drop—You Know You're a Homebrewer If... — 403
Dean Booth

Appendix • *Zymurgy* Index — 407

Index — 433

# INTRODUCTION

# A Grand and Mysterious Phenomenon
## BY CHARLIE PAPAZIAN

"Beer does not make itself properly by itself. It takes an element of mystery and of things no one can understand. As a brewer you concern yourself with all the stuff you can understand, everywhere." These words were offered by Fritz Maytag, President and Master Brewer of Anchor Brewing Company, at the 1983 National Homebrewers Conference. I've been brewing since 1970 and I don't believe there is any beer wisdom better expressed.

Like Fritz and perhaps yourself, I feel the same way about brewing and beer. Beer is a wonderfully grand, enigmatic phenomenon transcending art, science, industry, culture, and humanness. As hobbyists or professional brewers, we strive to perfect our beer quality. I know of few endeavors receiving such grand passions as ours. I've come to understand and appreciate just how special beer and brewing are not only for me, but for all the hundreds of millions of people throughout the world whose lives are touched by them. There's a lot of magic in what we make. There's also a lot of magic in how we make it.

I must admit, in 1978 when I first published *Zymurgy* with my friend Charlie Matzen, I really had no business plan or glimmer of the magazine's future and the significant role it would play in championing beer enthusiasm throughout the world. Simply put, it was begun with a gut feeling (no pun intended) that there would be a never-ending discovery of information about beer that would be of interest and benefit to homebrewers in their quest for quality homebrew. I'm pleased to note the original premise still holds true.

For the last twenty years, *Zymurgy* has strived to offer a balance of traditional brewing science and techniques, cutting-edge research and ideas, people, history, style appreciation, news, events and activities, product analysis and information, equipment evaluation, literary reviews, exploration of the world's greatest or little-known beer traditions, championing innovations, tips, and gadgets, and a forum for communication between homebrewers and beer enthusiasts throughout the world.

Selecting the articles for *The Best of Zymurgy* was not easy, but it certainly was fun. For me, a lot of great memories came to mind of having worked with pioneer beer enthusiasts around the country in search of the holy ale and legendary lagers. Twenty years of *Zymurgy* magazines is a lot of American homebrew history. It was fun reliving some of the moments that eventually evolved into better malt, hops, yeast, equipment, and techniques that helped us all make the best possible beer and cultivate the spirit with which to enjoy it.

I am no longer the Editor of *Zymurgy*, though as founder and President of the Association of Brewers, I serve as an adviser to the staff that works its magic 10 feet across the hallway from me. With their expert insight and advice, and that of the American Homebrewers Association staff, we've managed to portray in this little book the spirit *Zymurgy* has championed for two decades, while offering many of the practical articles and classic information that have made *Zymurgy the* magazine at the forefront of homebrewing culture.

The *Zymurgy* editors, American Homebrewers Association staff, and I hope you enjoy this glimpse of homebrew history and wealth of information. If you haven't already, perhaps you'll pour yourself a homebrew and join the American Homebrewers Association in continuing the great brewing tradition that has helped make our "homes sweet homebrew." Homebrewing is forever. I look forward to continuing to be a part of it. I hope you will, too. And remember:

Relax. Don't worry. Have a homebrew.

*Charlie Papazian*

# A Satisfying Sip of Homebrew History
BY DENA NISHEK, EDITOR, ZYMURGY

As Editor of *Zymurgy*, I've had the privilege of representing the American Homebrewers Association (AHA) in its enviable position of talking beer with enthusiasts, teaching homebrewers how to brew better beer, furthering their appreciation of beer styles, refining their tasting skills, promoting a fun hobby, and enjoying with them the pleasures of a satisfying pint. The homebrewers I've met along the way have made it very clear that this hobby is something special. The community of homebrewers is a proud, friendly, and serious group in search of quality beer and quality information. Many consider homebrewing a lifestyle; the community, a family. I'm proud to be part of both, as well as a part of the Association that has been instrumental in developing the culture of homebrewing.

As in any community, homebrewers have diverse backgrounds and areas of expertise, and their interests range from technical/scientific and practical how-to information to history, beer styles, and breweriana (although any homebrewer you speak to can turn a pint of beer into a poem of a recipe). The contributors to *Zymurgy* represent this community perfectly. Homebrewers and beer lovers all have brought to the craft the expertise of their chosen work—plumbing, engineering, and mathematics, for example—and a flair for experimentation that have thus enabled them to contribute a wealth of information to other enthusiasts and to unite brewers, both amateur and professional, around the world.

In the beginning, the first homebrewers, who faced practical challenges in their kitchens and garages, shared their solutions and successes, and showed other brewers they were not alone in the quest for quality homebrew. And as these homebrewers gained notoriety through their articles in *Zymurgy*, many became regular speakers at AHA conferences and other beer events. They published books and became professional brewers at award-winning breweries. But their memory of those days when information was hard to come by remains clear. Thus these graduates of homebrewing's youth are apt to apply their professional techniques to homebrewing, expanding the base from which homebrewers can draw information.

I was reminded of the broad and changing interests and needs of homebrewers as we chose articles for this book. This collection represents the work the AHA has done in the last 20 years educating and uniting the international homebrewing community. Looking back through *Zymurgy*'s history—nearly ninety issues—it is easy to see how much the information and products have evolved. The homebrewing tools have become better engineered for their tasks. The ingredients are of higher quality and greater diversity. The articles are more technical and in-depth. Picking classic and landmark features from such a large pool of articles was not an easy task.

Throughout the process we had to remind ourselves not to select too many new articles, too many old articles, too many beer-style features, or too much from any one subject. The collection had to represent then and now; it had to touch on ingredients, brewing methods, and styles. With so many great choices, how could the subjective task of compiling the best of *Zymurgy* ever be perfect? We hope you find the collection a satisfying sip of homebrewing history, the American Homebrewers Association, and *Zymurgy*.

# one

# Brewing History in the Americas

## "NATIVE" BREWING IN AMERICA

BY WILLIAM LITZINGER

*This article originally appeared in* Zymurgy, *Fall 1980 (vol. 3, no. 3)*

Taking my first drink of "batari," as the Tarahumara people of the Sierra Madre of Mexico call their maize (corn) kernel beer, I examined its qualities with great anticipation. The beverage was delightfully effervescent with natural carbonation, giving off a strong sweet aroma, but tasting pleasantly tart and definitely alcoholic. The appearance of the beverage was not equally pleasing. It was a dull yellowish color, with a consistency that reminded me of buttermilk. Aside from the abundance of fibers and other plant parts that had to be discreetly strained through the teeth while drinking, it was thoroughly enjoyable.

If you have been an ardent reader of past issues of *Zymurgy*, you probably already know a little about native homebrews in other parts of the world ("Vale Vakaviti-Fiji Homebrew," vol. 2, no.4, page 10). But I'll bet that most of you American homebrewers do not know much about "Native" American homebrews.

Actually, not much is known about Native American brewing processes. After considerable reading and searching, I found only a few discussions of native brews and how they are made. However, there is quite a bit of incidental information about native alcoholic beverages.

Christopher Columbus and several members of his crews were the first non-natives to try American homebrew and write about their experience. It seems that nearly everywhere that Columbus visited on his voyages to the New World, he was offered local beverages like cassava beer, maize (corn) kernel beer, and pineapple wine. Ever since then, hundreds of accounts from Spanish soldiers and priests, early explorers, colonial records, historians, and anthropologists have added to our knowledge of native American homebrews. Unfortunately, most of these accounts contain only brief mention of alcoholic beverages, and are often given with negative connotations.

In recent years I have been fortunate to have had some firsthand experience with a number of Native American homebrewers. One such group are the Tarahumara of the Sierra Madre of Mexico. I will go into some detail about how the Tarahumara make their maize kernel beer, but before I do, I would like to present to you a brief description of Native American brews and brewing in general.

Native American alcoholic beverages can generally be classed in the same categories as our wines and beers, depending on whether starch conversion to sugar was used in preparing the fermentable substrate. There are some unique beverages.

## NATIVE AMERICAN WINES

Native wines were made from many types of fruits including pineapples, cactus fruits (sahuaro, organpipe, prickly pear, and others), chokecherries, guavas, mesquite pods, and many lesser-known tropical fruits. Some types of wines were made from roots and barks, such as mesquite bark wine, pine bark wine, and the better-known sarsaparilla root wine (actually a mildly alcoholic wine and not a "root beer"). Mescal wine is a well-known wine made from the expressed juice of the baked stem and heart of agave (the century plant). Today in Mexico the popular distilled drinks such as tequila and mescal are the distilled products of mescal wine.

Wines from honey are also made by native brewers. The Mayan people are known to produce only one alcoholic beverage: a special ritual wine called Balche, made from honey and the bark of a tree.

Many of you have probably heard of the alcoholic drink called pulque, from Central Mexico. Pulque is made from the sap of special varieties of the agave. Unlike mescal wine wherein the plants are killed and hearts and stems baked, pulque is made with sap collected from living plants. The plants are tapped for sap by special procedures, just before they send up their flower stalk. As much as 400 liters (about 100 gal.) may be tapped from a single plant over the course of nine to twelve months. Pulque is definitely a unique beverage, involving bacteria and yeasts in several stages of fermentation. Pulque's flavor is derived from various bark and other plant additives. Its texture is derived from the growth of a special form of "slimy" bacteria. Its inebriating qualities are unlike any other homebrewed beverage I have ever tried.

Native American beers are made from maize (corn) kernels and from several types of root crops such as cassava (tapioca).

## NATIVE AMERICAN BEERS

Maize kernel beer is an important dietary staple for most Native Americans from Northern Mexico to Peru. In many areas people consume as much as 30 percent of their corn harvest as beer. (Less than 50 percent may be consumed as tortillas, tamales, and other food products.) In the Tarahumara country, which I am most familiar with, corn beer may be consumed on more than 200 days out of the year. Much communal work and social interaction takes place at these "beer-drinking parties." Often drinking events last several days, and from what I can tell, the Tarahumara must spend the other 165 days of the year recovering from these drinking feasts. I do not mean to imply that the Tarahumara lead a laid-back lifestyle. It is no easy task to be a farmer in the remote rugged country where they live. But their homebrew has a great deal of social and cultural importance for them. The Tarahumara actually consider their "beer" a special kind of food. (It is, in fact, highly nutritious!) Also, the Tarahumaran view of inebriation is quite different from our own. Drinking patterns are generally restricted to when and where, rather than how much. The main rule of consumption is that whatever quantity is produced must be consumed before it spoils, thus often leading to several days of drinking.

## HOW THE TARAHUMARA MAKE THEIR SPROUTED MAIZE KERNEL BEER

"Sunu batari," as sprouted maize (corn) kernel beer is properly known to the Tarahumara, is made by a very rudimentary process. The following description of the maize-malting process was written by Wendell C. Bennet and Robert M. Zingg (The Tarahumara, an Indian Tribe of Northern Mexico, Chicago,1935, page 46), and contains some particularly interesting comments:

> Three to five decaliters of corn are shelled into a basket or sack, thoroughly moistened and kept in an olla for two or three days in a warm place. Then a hole 12–18 inches deep is dug where the sun strikes the ground. It is lined with pine needles or grass. The wet corn is put in this hole and covered with pine needles and a layer of small stones. The corn is kept moist in this hole until it sprouts. When the sprouting has advanced to a satisfactory stage, the corn is taken out and ground on the metate two or three times. Then, with plenty of water, it is placed in a large olla and boiled all day until it becomes yellow. White beer is not acceptable.
>
> It is now ready to take off the fire and be cooled. When it is cool, the liquid is strained through a special basket, wari (basket) pagera (from page, "to strain"). It comes out looking like esquiate, and has a not unpleasant sweet taste. It is at this stage that the process is complicated. From his [sic] cornhouse the Indians take about a pint of brome grass, which resembles oats, Basiaowi (Bromus sp.), and which has been gathered and saved for this purpose. This furnishes the ferment and is an essential ingredient of tesquino. It is ground on the metate, but is not cooked, as it would not ferment.
>
> Wild oats are sometimes ground with a quantity of lichen (Usnea sp.), which they think stimulates fermentation and makes the liquor sweeter. Other methods of sweetening the liquor include the four mosslike plants mangora (Selaginella cuspidata), and which are often ground up in the drink, too. Other plants are added to change the flavor: dowisawa (Chimaphila maculata), to make it stronger, and the root of the dolinawa (Stevia sp.), which also makes it stronger, especially when the tesquino goes flat. It sweetens it as well.
>
> We drank tesquino wherever we went, and it showed a great difference in taste. This was partly due, no doubt, to the various mixtures of plants. Several bunches of the sedge (Fibristylis sp.) are often ground with the corn to prevent colds.
>
> The combination of ingredients for fermentation are put in a special small olla (sikoli doela, "boiling pot"), which has "learned to boil well," i.e., to ferment. These special pots are carefully kept for this use alone and are never washed or used for cooking, as their facilities might be impaired. The Indians say that these pots learn how to "boil" from each other. The "boiling pot" thus is filled, and placed in a warm place near the fire, where it ferments all night. The next morning its contents are mixed with the strained corn liquid, and it ferments within three or four days. It is then ready to drink.

The Tarahumara are still making their "sunu batari" in almost the exact same way that Bennet and Zingg described, and from all that I can gather, the process is very old. The use of all the plant additives might in some ways be analogous to the use of hops in barley beer. The Tarahumara are also known to add narcotic and hallucinogenic plants such as jimsonweed (*Datura*) and peyote (*Lophophora*) to their corn beer in order to "fortify" it. Apparently the main fermenting organism involved in the brewing of "sunu batari" is the brewer's yeast, *Saccharomyces cervisiae*. But native brews are seldom as pure or uncontaminated with other organisms as one might think. In fact, it may be that the creamy texture of typical "sunu batari" is the result of the growth of a bacteria similar to the "slimy" bacteria found in pulque.

With a little experimentation and modification it might be possible to brew up a batch of sprouted-maize beer from ingredients available in any grocery store. I must admit, though, that I have not had much luck. Perhaps my fermentation pot has not yet learned to boil well.

*William Litzinger is a biologist and ethnobotanist who has studied Native American brewing processes in various parts of Mexico and Central America.*

# BRINGING COLONIAL BREWING AND MALTING TO LIFE

### BY RICH WAGNER

*This article originally appeared in* Zymurgy, Summer 1989 *(vol. 12., no. 2)*

Even before the state's founder, William Penn, arrived in his colony in 1882, brewing was fast becoming a prominent endeavor there. Penn's visit lasted less than two years, but in his "greene country towne" of Philadelphia, Pennsylvania, Penn reported there was "an able man who . . . set up a large brewhouse, in order to furnish the people with good drink, both there and up and down the (Delaware) River. . . ."

Penn erected a mansion at Pennsbury Manor, preferring a country estate for his family. He had a brew and bake house built adjacent to the home. Beer was brewed with malt purchased from Philadelphia, and it is reported that hops were grown in his garden.

Penn's love of beer influenced the laws governing the production of beer in the colony. Beer was preferred over distilled spirits, and domestic production was promoted through tax incentives. The government encouraged farmers to grow barley and hops by imposing tariffs on imported raw materials. From the brewing industry's infancy with that "able man" who knew how to brew good beer, Philadelphia would go on to become famous for its porter, preferred by none other than George Washington. Eventually Pennsylvania would become a world-renowned brewing center.

In the city of Philadelphia as well as the outlying areas there are examples of colonial architecture and clues to life in the past. There are even a few sites that provide glimpses into the brewing process of colonial times.

## GRAEME PARK

Graeme Park is a preserved country estate that was home to the proprietary governor of Pennsylvania, Sir William Keith. Recent attempts to discern the original intent of the structure have been a source of controversy. Was the house built as a malt house or as a dwelling?

Architectural consultants were hired to do the authoritative work on the estate to guide future restoration and development of the property. Mark Reinberger's assessment indicates that a small malt house was built behind the house. There were plans to build a larger malt house to process the locally grown barley, but due to economic forces they were dashed.

I toured the grounds in May 1987 and arranged to give a malting demonstration. Preparing it was a lot more work than I expected. I had brewed full-mash batches of beer before, and had even demonstrated brewing outdoors, but this malting business was completely different. I consulted *Zymurgy*, some obscure seventeenth-century references supplied by Clare Lise Cavicchi, Pennsbury Manor's curator, and other sources.

Preparations began with a trip to the Farm Bureau for some untreated feed barley. The smallest quantity available was 40 pounds. I proceeded to spread about 20 pounds of soaked grain over a cloth supported by a screen, propped up on blocks in the backyard.

A sample of the grain was taken every 24 hours for four days and placed in a jar to show the sprouting process. The grain was turned regularly, and when most of it had sprouted, it was heated on cookie sheets in the oven. The result was a caramel malt, of sorts.

The Friends of Graeme Park who attended my demonstration were amazed at the complexity of the malting process.

## HECKLERFEST

The Heckler Plains Folklife Society is dedicated to bringing colonial crafts alive and to making people aware of what life was like in the past. I was asked to brew beer at their annual Hecklerfest in September 1987 and 1988. With my homebrewing partner, Dan Brosious, we wowed and zowed countless Hecklerfesters with a full-mash brew from start to finish. The recipe included 25 pounds of six-row fancy and about 5 pounds of homemade caramel malt from the Graeme Park demonstration, and some hops grown in Montgomery County, Pennsylvania.

For the uninitiated, it is always astounding to see how involved brewing a 10-gallon batch of beer really is. To many, beer is just something that comes out of a tap or bottle. Along with the blacksmith, butter churners, spinners, weavers and quiltmakers, and a wine-maker across the way, the brewers fit right into the act. As an added attraction, Charlie Brem from the Homebrewers of Philadelphia and Suburbs contributed his expertise and some heavy-handed hop additions.

## PENNSBURY MANOR

Pennsbury Manor has beautifully landscaped grounds surrounding buildings reconstructed by the Works Project Administration in the 1930s. The gardens are attended by Charles Thomforde and include flowers, grape arbors, corn, flax, herbs, spices, and other plants.

On a visit to Pennsbury Manor in 1987 I began talking to Thomforde about hops in Penn's garden. He explained that several years ago he had transplanted some hops he found growing on a fence to the garden. I observed they would yield better if they were trained to climb a trellis, which he made of sapling poles to be as authentic as possible.

By mid-July the plants were struggling from the heat and dry weather, but they rebounded by late August, and that year the volunteer staff harvested about a pound of hops from the six or eight vines.

Visitors can now see some of the raw materials used in beer-making before they enter the bake- and brewhouse. The volunteers received hop pillows for their efforts. In colonial times hops were placed in pillows for their calming or sedative qualities to promote sound sleep.

Fortunately for the Pennsbury Society, Cavicchi has researched beer brewing on Penn's estate. She has begun working out the details for outfitting the brewhouse, so that future visitors may actually see beer being brewed as it was during Penn's time.

Plans stress authenticity, so the barrels will have wooden hoops. Imagine homebrewing without plastic tubing and liquid bleach! Thinking about the flavor of a dry-hopped ale made from molasses and fermented at room temperature for 24 hours staggers the imagination.

Cavicchi brewed brown ale using a colonial recipe that called for yeast to be added to the wort when it was "blood-warm" (a description used prior to the invention of the thermometer). Her first batch was really good.

Cavicchi has contracted with a craftsman in New Hampshire who specializes in re-creating artifacts from the past. Based on her research, she has ordered a scoop, funnel, and tun dish, three barrels, three coolers, one working tun, mash tun, and underback, as well as some small utensils. She hopes that brewing will begin sometime soon.

*Rich Wagner has conducted research on Pennsylvania's brewing history with partner Rich Dochter since 1980. They have amassed an inventory of more than 400 sites throughout the state. In addition, Wagner has been homebrewing since 1983. He has given homebrewing and malting demonstrations and grows hops for brewing.*

# THE OLDEST BREWERY IN AMERICA

## BY CHARLIE PAPAZIAN

*This article originally appeared in* Zymurgy, *Fall 1992 (vol. 15, no. 3)*

America's oldest existing brewery stands where it is eternally spring, at an elevation of nearly 10,000 feet surrounded by volcanoes towering over 20,000 feet, only 50 miles from the hot, humid tangle of the world's largest jungle. As though these surroundings were not spectacular enough, the brewery's history inspires the

CHARLIE PAPAZIAN

mind when contemplating the origins, wisdom, and tenacity of the founders nearly 450 years ago.

The high Andean air in the city of Quito, Ecuador, is bright, clean, and easily calmed by the country's readily available Pilsener and Club Premium lager beers. A country the size of Colorado, situated on the equator, Ecuador has a lot to offer—the towering Andes, tropical beaches, Amazon Basin jungle, brilliant deserts, Indian, African, and Spanish culture, and America's oldest brewery.

Quito was founded in 1534 during the Spanish conquest. It was only a matter of days before work on a church and monastery was begun by seven monks who had traveled to the South American continent from Flanders (now part of Belgium). With them they brought their yearning for beer as it was brewed in the old country. Eleven years later, wheat was brought over and cultivated. Then perhaps America's first brewery began malting, brewing, and fermenting wheat beer.

In the garden of the monastery one of the padres gave a short history of the brewery, which has been restored as a museum. Through an interpreter he explained that cultivation of wheat was followed in later years with the introduction of barley. The small, simple brewery, popular with the growing population, expanded into a more "serious" brewery in 1595.

Up until 1957 the 5-to-6-U.S.-barrel brewhouse malted its own barley and sun-dried it. The brewery continued operation until 1967 when, according to the padre, the Pope issued an order halting brewing operations. The tradition of the Franciscan Order embraced a vow of poverty and humility, and as the Vatican evidently saw it, the brewing of beer did not fit into the modern interpretation of these values.

However, until the brewery ceased operations, the monks were brewing twelve batches per week, half kept for the monastery church itself and the rest distributed to other churches and monasteries in the area. Wheat beers, pale beers, and very dark beers were brewed regularly, according to the padre.

These beers are now only a memory in the minds of a select few residents of Quito. The manager at the small hostel where I stayed related how his father used to work in the brewery. He couldn't recall very much except that a locally produced proteinous substance was used in the brewing process. I figured it must have been some sort of clarifying agent similar to isinglass, which is derived from the swim bladders of sturgeon.

With the help of two Ecuadorean breweries, Cervezas Nacionales of Guayaquil and Cerveceria Andina of Quito, the brewery has been preserved as a museum. The museum has a beautifully preserved small beer hall next to it. The flavor of it all seems to linger as a cross between a Flanders and a Germanic brewing tradition. A visit to the monastery brewery is well worth the effort if you are ever in Quito.

And I know the padre appreciated the bottle of my own homebrew I gave him after our brief encounter at the Brewery of the Monastery of San Francisco.

*Author's Note: Many thanks to Guillermo Moscoso for introducing to me the tale of the Brewery of the Monastery of San Francisco; to Dori Whitney, editor of* The Brewers Digest, *who uncovered the January 1966 issue that featured the brewery; and to Lilian Bejarano, who is working on the monastery restoration and graciously arranged a brewery tour.*

*AHA founder Charlie Papazian visited the brewery of the San Francisco Monastery on a vacation trip to Ecuador.*

Charlie Papazian

*The following is excerpted from an article by Paul V. Grano about the monastery brewery and one type of beer brewed there. "The Brewery of the Monastery of San Francisco" originally appeared in the January 1966 issue of* The Brewers Digest.

The brewhouse consists of a copperlined concrete mash kettle, a wooden lauter tub with a bronze false bottom and a hot water tank made of an old German hop cylinder. All heating is by direct wood fire. Thirty-three pounds of Pilsener malt plus 50 pounds of caramel malt and 20 pounds of black malt (milled in a coffee grinder in the monastery) are mashed in and the usual infusion method is followed. Since the agitator that was installed for the fathers is inadequate they have to take turns stirring the mash in order to avoid burning on to the bottom. This is quite a hot and heavy job.

The mash then runs to the lauter tub by gravity and the wort is pumped back to the mash kettle by hand. Here 100 pounds of brown sugar dissolved in hot water are added. The hop rate for the batch is 2 pounds. When the boiling of the wort is finished the wort once again runs by gravity to the surface cooler and is left there overnight. The next morning, the yeast that we (La Victoria Malting/Brewing Company) provide them is added and fermentation goes on at about 60 degrees F (15.5 degrees C). The original extract is about 10.5 degrees Balling and when the beer has fermented down to 1 degree Balling above the end fermentation the beer is bottled immediately. The bottles are filled through four siphons and crowned by hand. The empty bottles as well as the crown corks were (recently) given to the monastery by our brewery when this type of bottle was discontinued for commercial purposes some years ago.

The fermentable extract still in the beer produces about 0.5 percent $CO_2$ by weight or 2.7 percent by volume after 10 to 12 days of storage when the beer is ready for consumption. The alcohol is about 3.4 percent by weight and the taste is very pleasant. The fathers are served one glass for lunch and one for supper. About one hour before serving, the crown corks are lifted slightly, otherwise the beer would pour like Champagne.

To conclude, I would mention a small episode that took place a few weeks after we had first agreed to assist the fathers. I went over to the monastery to ask the "Brewmaster" how the beer had turned out, to which he answered with a smile that where he had previously used one padlock to close his storeroom, he now had to use two!

# two

# Beer Styles

## I'M A MILD MAN MYSELF, BUT . . .
### BY HOWARD BROWNE, M.D.

*This article originally appeared in* Zymurgy, Fall 1990 *(vol. 13, no. 3)*

Philip Davia is my kind of guy. In his "I'm a Bitter Man," *Zymurgy*, Winter 1989 (vol. 12, no. 5), he reflects on the meaning of life and the importance of real ale. I'd like to extend his thesis and throw in a few cautions for the newcomer.

I lived in London in the fifties. Actually, we lived in Hendon, northwest London. My parish church was noted for several things, but the one that concerns us is that the parish was landlord of a pub. Landlord in the real estate sense, not the publican sense. Years ago a triangular strip was detached from the churchyard and "The Greyhound" built. The name comes from being at the top of Greyhound Hill. Or is it the reverse?

In those days there was no "real ale movement." It *was* real ale. It came from pulls, or taps, back of the bar, usually about four in number. One was the "mild" that Mr. Davia discusses (known as beer, or "pig's ear"), another would be "ordinary" bitter, and a third "best" bitter. I think keg stout and porter had gone by that time. There might be another tap or two with another ordinary bitter.

Old-timers used to ask for traditional mixtures of which you may have heard: "half-and-half" was half mild and half bitter, and a "mother-in-law" was stout and bitter. In the summer the men of the parish co-opted me, the token Yank, to help clean the graves in the churchyard. Nettles encumbered the place, and it was hot, thirsty work. One of the men said after the first session, "You'll want something light and refreshing; try a shandy." "What's that?" I said. "There's ginger beer shandy and lemonade shandy. You'll probably like the ginger beer one better." I gulped and held my opinion to myself as I watched the barman draw a half-pint of mild and fill it up with ginger beer from a bottle. I was delighted with the taste; it wasn't nauseating at all. You can't make a decent shandy in this country unless you can get ginger beer. Even so, it's not the same without old-fashioned mild.

Another concoction prescribed for me once when I had a cold and was feeling seedy was Guinness and port—the less said about that the better.

I learned that Britain's reputation for serving "warm beer," so widespread by Americans who served in the U.K. during World War II, was undeserved. I needn't explain to this audience, but I learned that British ales and beers taste best at "cellar" temperature—about 50 degrees F (10 degrees C). The slurs were perhaps somewhat justified in that during the war publicans couldn't get skilled help who knew the proper keeping techniques of the cellar.

I learned the basic anatomy of a pub. The "public bar" was large and occupied one part of the bar. In it beer was maybe a penny a pint cheaper (when a pint of best bitter was tenpence, ordinary ninepence, and mild eightpence, or eightpence ha'penny). The "private bar" was smaller and upscale. Sometimes there was a "lounge bar," or "snug," which was much smaller and more intimate. These divisions seem to have gone with the times. You paid on delivery and didn't leave your change lying about—it was considered ostentatious. The publican's word was law, his clock was five minutes fast, and when he said, "Last orders please" and "Time, gentlemen, please," you obeyed. I noted with interest that you hardly ever saw anyone drunk in the streets, very few fat people, and no young men with beer bellies. But then they walked or rode bicycles everywhere.

Over the intervening years when I made trips back, the individuality and quality of the beer began to decline. The lads began drinking lager out of a bottle. Before, this had been a ladies' drink, with lime added. The triumph of industry over artistry was almost complete when the world's most successful consumer revolution, the Campaign for Real Ale (CAMRA), rescued traditional brews and brewers from destruction. But don't get cocky; real ale pubs seem to come and go with great rapidity. But it is possible to get a good glass of beer. You just have to know the hallmarks of a good pub, and spend a little time finding one.

First, the industrial giants have caught on. You see pubs with old-fashioned signboards advertising "traditional ales." Inside, they are dishing up mass-produced fizzwater from a pressurized faucet. So don't be misled. The key sign is "free house." However, this is not a guarantee but a sine qua non. It's what's inside the free house that counts. The words "real ale" with or without the CAMRA logo are a pretty good guarantee. The label will tell, so get to know names, as Davia suggests. Make sure the beer gushes forth from real beer engines or pulls, not fake pulls that are cleverly attached to a pressurized tap.

On my trip this year I found a few pubs new to me: the Yorkshire Grey, Theobalks Road at Grey's Inn Road (Holborn), is a brewpub and worth going out of your way for. The " . . . and Firkins" can still be counted on, and the Pig and Parrot in the old Kew Gardens railway station pleases me with their brews. The latter Pig and Parrot is not to be confused with the Hog in the Pound just off Oxford Street and next to the Bond Street tube station. Once outside the latter I saw a large, green tank truck draw up, disgorging through hoses the contents of several tanks bearing an industrial name. This was not an oil delivery but beer. Does anyone remember Think-a-Drink Hoffman, an old vaudevillian who could pour different requested drinks out of the same cocktail shaker? There is still a real push by certain American industrial beers to invade the British market, or "try to teach their grandmother how to suck eggs." You'd better believe that the CAMRA people haven't lost the need for eternal vigilance and public relations.

Now that the "real ale" movement is a qualified success, I wish someone would start a "real food" movement. One of the joys of pubs back along was pub food. A "ploughman's lunch" of crusty bread, cheddar off a wheel, and a pickled onion, or cold pies of pork, ham, chicken, and veal, or sandwiches and "pickle" (we'd call it relish) were the staples. One or more of the above with a pint was the best simple lunch you could buy. Now you have to look long and hard, as the Brits want to serve you *hot* lunches, kept for hours under an infrared bulb. There's stuff they call "lasagne" and, my God, "chile." You're lucky if they include a decent traditional steak and kidney pie. So unless you're very lucky or look long and hard to find a pub with both real ale *and* real food, you may have to drink your pint without lunch, unless you have cast-iron digestion.

Speaking of lucky, in 1988 I was touring the West Country and stopped for the night in a bed-and-breakfast in the village of Wyveliscombe, Somerset. The village had a good country pub and I was standing at the bar talking to the locals about my favorite subject. A man on a stool next to me said, "You like beer, do you? How'd you like to come round tomorrow and visit my brewery?" Does a dog have fleas? It turned out he was a partner in the Golden Hill Brewery, which does the work of sixty men with less than a dozen thanks to modern technology. He gave me a tour of the large traditional one-room operation, where they brew Exmoor Ale, Exmoor Dark, and other special brews. I see occasionally that one of Golden Hill's brews has won some prize or other. I feel slightly proprietary, and grateful that for once my big mouth got me a rare experience instead of into trouble.

Here are the makings of your own shandy, kindly provided by Russ Schehrer of Wynkoop Brewing Company in Denver, Colorado.

*Ingredients for 5 gallons of mild*:
- 4 lbs. light dry malt extract
- 1 lb. crystal malt
- ¼ lb. chocolate malt
- 1 oz. Cascade hops (boiling)
- ½ oz. Cascade hops (finishing)
- ale yeast(Muntona is suggested)

*Brewer's Specifics*
Crush specialty malts, put in cold water, bring just to a boil, strain out husks.

*Ingredients for 5 gallons of ginger beer*:
- 3 lbs. light dry malt extract
- 3 lbs. amber dry malt extract
- 3 oz. freshly grated ginger root
- 1 oz. Northern Brewer hops (boiling)
- ½ oz. Willamette hops (finishing)
- ale yeast(Muntona is suggested)

*Brewer's Specifics*:
Add half the grated ginger at the ½-hour boiling point. Add the remainder as a finishing spice. Strain it out.

Howard Browne, M.D., *credits* Charlie Papazian's *book* The Complete Joy of Home Brewing (*and others*) *and the education he receives from*

Zymurgy. Says Howard, "Years ago I tried making wine. I found that the best wine I could make was worse than the worst wine I could buy. However, the worst beer I can brew is better than the best (standard American industrial) beer I can buy. So keep it up, and count me as a loyal colleague. Finally, relax, yes, but be aggressive in spreading the faith."

# FLAMING STONE: BREWING TRADITIONAL STEINBIERE

BY PHIL RAHN AND CHUCK SKYPECK

*This article originally appeared in* Zymurgy, Winter 1992 *(vol. 15, no. 5)*

Memphis, Tennessee, is not known as the beer capital of the world. Want a fresh ale on tap? Forget it. Only two brands are available, and they are in bottles. The selection of lagers is not much better. Name the top fifteen imports and you have it. Microbrewed beers? Don't waste your time looking. With these sad facts in mind, beer connoisseurs in Memphis can face long trips to St. Louis or New Orleans for a better selection, or they can brew their own. We take the homebrewing approach more often than not. Because examples of beers we would like to brew are not available, research into the specifics of styles is part of our brewing routine. When we decided to brew a German "steinbiere" we found ourselves doing more research than we bargained for.

Most homebrewers are familiar with the general precepts behind steinbiere. *Stein* is the German word for "stone." The style supposedly dates back to the time before metal kettles. Because wood kettles could not be set on an open fire, red-hot stones were added to the wort to bring it to a boil. Besides imparting a smoky, phenolic flavor to the beer from the ash and soot on the rocks, the intense heat in the stones caramelized sugars in the wort. Often the stones would be added back into the beer after fermentation, when a secondary fermentation would occur as the caramelized sugars dissolved into solution.

This general information was in Michael Jackson's *The New World Guide to Beer* (Running Press, 1988). We found additional information on the style, such as original gravity, hop rates, and color, in Fred Eckhardt's *The Essentials of Beer Styles* (Abis, 1989), but these sources hardly answered all the questions we had about the style. What types of rocks are best? What is the ratio of stones to wort? Can the addition of stones affect the pH? While seeking the answers to these questions, we came across information we believed was worth sharing.

Steinbiere is brewed commercially in Germany by Rauchenfels at Neustadt. While we waited for a reply from the brewery concerning their production methods, we took matters into our own hands. Our first attempt at brewing stone beer was more an experiment to test rock samples and a heating system. We found out quickly what would not work. Memphis, along the Mississippi River Delta, has just as poor a selection of surface rocks as it does of fine beers. Our two choices were soft limestone and quartzite, which is metamorphic sandstone. The limestone was a poor choice. It cracked and popped in the fire. When added to the wort, it dissolved, elevating the pH and ruining the beer's flavor. (Speaking from experience, bricks and beer are not a good combination.) The quartzite, on the other hand, held up well when heated and was coated with a layer of delicious caramelized malt when cooled. We had our stones.

We had yet to hear from Rauchenfels, so we made some decisions and pressed ahead. We selected stones that weighed about 2 pounds each and were the size of average oranges for ease in handling. They also fit inside the opening of a 5-gallon soda canister used during secondary fermentation. The stones were scrubbed, boiled in water, and soaked overnight in acid to ease our concerns about contamination and lime content. Not knowing the ratio of stones to wort, we decided to brew a 6-gallon split batch with two stones in "A" and three stones in "B." "A" contained 0.77 pounds of stones per gallon of wort; "B" had 1.53 pounds per gallon. (Subsequent research turned up the fact that Rauchenfels brews with 0.64 pounds of stones per gallon, so we were in the ballpark.

For our experimental batch we heated the stones in the coals of an oak-wood fire. This method produced very little smoke flavor in the beer and caused a lot of ashes to stick to the stones. We decided to use a different technique in our split batch. We constructed a fire chamber of cinder blocks and used grates from Phil's fireplace to hold the stones above the flames. We placed an electric fan in front of the opening to make a hotter fire and to direct the flames up past the stones. Although this method melted the plastic fan housing, it made the stones hot enough to pop a few flakes off.

While the stones were heating we began our mash. In formulating our recipe, we decided to use traditional German ingredients. Because the stone beer style is an ale, we chose to use one of

Phil's award-winning Altbier recipes. The combination of 8 pounds of German Pilsener malt (Ireks), 2 pounds of light Munich malt, and 1 pound of wheat malt became the backbone of our steinbiere, which we named "Flaming Stone."

We mashed the grains at 126 degrees F (52 degrees C) for 30 minutes (protein rest), and at 154 degrees F (68 degrees C) for 90 minutes (starch conversion). After a negative iodine test, the mash temperature was raised to 165 degrees F (74 degrees C) and the mash sparged with 168-degree F (75.5-degree C) water. We boiled the sweet wort for 30 minutes and then split it into two 3-gallon batches.

At this point we lowered two hot stones weighing 2 pounds 5 ounces altogether into the wort of "A," and three stones weighing 4 pounds 9 ounces altogether into "B." Large stainless steel tea strainers cradled the stones. As the stones contacted the wort, great clouds of steam were released. The stones seemed angry as they hissed, growled, and rumbled during the first few minutes. The worts, which we had removed from their burners, began to boil again. We removed the stones while they were still too hot to touch. They were coated with a shiny layer of caramelized sugars. We cooled the stones, sealed them in containers, and put them in the freezer until secondary fermentation.

We maintained the boils over propane burners and added one half ounce of compressed German Northern Brewers hops to both "A" and "B." After 20 minutes, we added 11.4 grams of fresh Tettnanger hops to "A" and 17 grams to "B." After 40 more minutes of boiling, the batches were each finished with three quarters ounce of Tettnanger hops and given ice baths to cool them quickly. During the boil we both noticed something different in the air. The brewery, which used to be Phil's garage, was filled with a pleasant soft aroma of caramelized malt.

After cooling, we strained the worts from the hops into 3-gallon carboys to settle, then siphoned them into 5-gallon carboys for fermentation. During the split of the sweet wort after the initial boil, "A" received less wort. When topped with water to 3 gallons, the original gravity of "A" was 1.045 and "B" was 1.050. (Rauchenfels Steinbiere has an original gravity of 1.045.) We pitched both batches with 1 pint of a Wyeast No. 1338 European ale yeast starter, then fermented at 65 degrees F (18 degrees C).

After 12 days we racked the batches onto the stones in 3-gallon stainless steel soda canisters. After two days we racked both batches into 3-gallon carboys to finish secondary fermentation. One week later the beers were chilled, fined with three ounces of isinglass solution, kegged, carbonated, and bottled.

The results? Both beers were fantastic, but "B" had a more pronounced caramel flavor and slightly more body. "A" was lighter in flavor with a hint of caramelized malt, light hop character, and a clean finish. Missing was a light smoky flavor, which may have been hidden by the distinctive caramel character. Both beers produced a thick, creamy head. In "B" this creamy note in the aroma was carried into the flavor. Both beers were great and distinctly different from one another. The obvious difference was the number of stones boiled in the wort. Rauchenfels uses metamorphic stones from their own private quarry that supposedly "bloom," or expand their surface area, when heated. This in turn adds more caramel and smoky flavors to the beer. This information left us wishing for some authentic stones to experiment with.

Our research convinced us that we produced a reasonable example of an authentic steinbeire. The color of our homebrewed stone beer was a deep, rich gold, and was identical to color photos sent to us by Rauchenfels. The rock-to-wort ratio, hop rate, and original gravity were all confirmed by the official analysis of Rauchenfels Steinbiere performed by the Versuchsanstalt für Bierbrauerei in Nürnberg. We also received a document from the above institute declaring that Rauchenfels Steinbiere has no harmful effect and is good for human consumption.

This would have been the end of our experiment with stone beers until Brewmaster Franz-Joseph Sailer of Rauchenfels spurred our curiosity. Knowing of our interest, Sailer sent us an article from the Austrian brewing and hops trade publication, *Gambrinus*, published in 1910, that told of stone brewing techniques used in 1906. The process did not involve the use of stones to boil the wort, but to heat and control temperature in the mash and to roast the hops. The hops were roasted on top of a layer of hot stones in the mash tun before the brewing liquor and malt were added. Consequently the hops were present in the mash tun throughout the process. The mash was eventually boiled after it was gradually warmed by the addition of more and more stones, but the wort was not boiled. The yeast was pitched immediately after sparging. Half the beer was generally kegged after 7 to 10 hours of fermentation and consumed after about 6 days. The other half was allowed to ferment for two days, kegged, then consumed after the first half was gone, usually in two weeks.

These techniques point to practices of brewers in ancient times. Another clue is the fact that juniper branches were used in the bottom of the mash tun, which doubled as the lauter tun. Juniper was a popular bittering agent before the use of hops. It appears that the evergreens also served as a filtering medium during sparging. Boiling the mash with the stones may have been a predecessor to decoction mashing, the German method wherein portions of the mash are removed, boiled, then returned. The mash, in the 1906 Steinbiere version, was a combination of equal parts of barley malt, wheat malt, and oat malt.

In the presence of so many interesting facts and unanswered questions, it appears that our stone brewing experiments may have just begun.

*Phil, a homebrewer for almost twenty years, has covered his office wall with ribbons and brewing awards. Chuck is a National Judge in the Beer Judge Certification Program. This article is available in Library 13-AHA/Zymurgy/Clubs on CompuServe's Beer and Wine Forum as STNEBR.W92.*

# GOTLANDSDRICKA, THE ANCIENT BREW OF GOTLAND

## BY HÅKAN LUNDGREN

*This article originally appeared in* Zymurgy, Special 1994 (vol. 17, no. 4)

Homebrewing in Scandinavia originated on Gotland, the largest island east of Sweden in the Baltic Sea. The traditional "gotlandsdricka," a malt beverage with flavors of smoked malt and juniper, can be traced to evidence of brewing on the neighboring island of Oland, where malt from the thirteenth century was found.

The ancient tradition has survived. Today an estimated 500 active homebrewers are among the island's population of 60,000. Each year they produce about 92,400 gallons (350,000 l.) of gotlandsdricka, which probably was the everyday drink of the people of Gotland during the Viking era (A.D. 800 to 1066). Mead was their celebration drink.

The flavor profile of gotlandsdricka is complex. When young, it can be very sweet and is called "woman-dricka." The background is smoked malt, sometimes with a salty touch and, of course, juniper, bringing the freshness and spiciness to the "dricka." There is only minimal hop flavor evident. Aged gotlandsdricka will be drier with a taste of tar and a sourness similar to that found in Belgian lambics.

**Gotland**

The ingredients used depend on the recipe, but typically include smoked barley malt as the major extract source. Juniper and nowadays even hops are added to most recipes. It is common to use other extracts and spices as well as honey, rye, wheat, and different kinds of sugar. The amounts of these ingredients generally are not large enough to dominate the flavor except for the taste of alcohol in strong versions of gotlandsdricka.

The first step is malting, which takes place on Gotland from September to April. Barley is steeped for three days, drained for one day, and germinated on the floor in layers about four inches deep. After a week of growth the barley is kilned in woodsmoke, often from birch. This step may take as long as one week.

Before mashing, the mash liquor is prepared by boiling water with juniper branches and berries for about one hour. At the same time the combined mash and sparge tun, called *rostbunn*, is prepared with a kind of false bottom built up with juniper sticks arranged around the runoff hole in the bottom. Some brewers use very old sticks from previous generations, believing they have a magic force. Some lay juniper and straw layers on top of the sticks before the crushed malt and other ingredients are added to the tun.

After sparging, the sweet wort is sampled through the hole in the bottom of the tun and is boiled with hops and sometimes honey or sugar. The wort is cooled, and yeast pitched.

After one week of fermentation the gotlandsdricka can be enjoyed. Typically the yeast is still working—a guarantee of freshness and preservation. In the following weeks the brewer feeds yeast with a sugar cube every day until the keg or cask is emptied.

If we translate the old methods to modern homebrewing technology, the following instructions will give a nice result.

### King Gorm's Gotlandsdricka

*Ingredients for 5 gallons*
- 5 lbs. home-smoked barley malt
- 1 lb. honey
- 4 lbs. juniper (branches with berries)
- ⅓ oz. Perle hops, 6% to 7% alpha acid (60 min.)
- baker's yeast, cake the size of a sugar cube, dissolved in water

- Original specific gravity: about 1.060,
- Final specific gravity: depends on when it is consumed

Boil 8 gallons (30 l.) of liquor with juniper for 1 hour. Mash the smoked malt for 90 minutes at 154 degrees F (68 degrees C), then sparge with the juniper-infused liquor. Boil 60 minutes with hops. Primary ferment one week at 65 to 68 degrees F (18 to 20 degrees C). You may want to try a batch with beer yeast and compare. Secondary fermentation is optional.

*Håkan Lundgren is a member of the Swedish Homebrewers Association and has been a homebrewer for ten years.*

# LEANN FRAOCH—SCOTTISH HEATHER ALE
## BY BRUCE WILLIAMS

*This article originally appeared in Zymurgy, Special 1994 (vol. 17, no. 4)*

*Leann fraoch* (pronounced "lyan fray oogh" with a soft *oogh*) is Gaelic for heather (*fraoch*) ale. Heather is a low-growing shrub common in

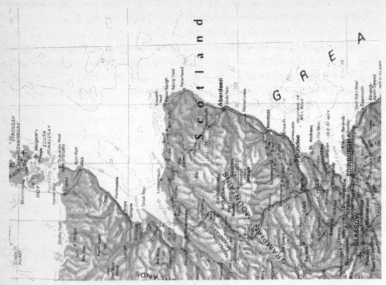

Scotland

the peat hills of Scotland. Bell heather, also called bonnie bells (E*rica tetralix* and *E. cinerea*), has bell-shaped white to purple flowers from April to June. Ling heather or broom heather (*Calluna vulgaris*) has small budlike flowers that are white, red, or purple and flower from August to September.

For brewing, use only the top 5 centimeters of the plant. Heather flowers should be used within 36 hours of picking or stored at temperatures lower than 38 degrees F (3 degrees C) because they lose their aroma value. Moss (*fog*) grows on the woody stem inside the heather plant, not near the flowers, and contains wild yeast. F*og* has some narcotic effects that have been omitted from commercial recipes. The moss grows deep within the stalks but does fly around when the flowers are picked. It is a lightweight white powder that can be easily removed by rinsing in cold water.

Heather ale is undoubtedly Scotland's oldest brewing heritage. The brew was made by first mashing Scotch ale malt, boiling the wort with flowering heather tips, then covering the surface with fresh heather flowers, leaving it to cool and ferment for 12 days until the heather blackened. It was drunk straight from the *cran* (barrel) with the tap hole one quarter of the way up. It is an amber-colored lightly conditioned ale with a soft bitterness, firm oily body, and winelike finish—just ask Michael Jackson. This is the drink the French called Scottish burgundy during the eighteenth century Auld Alliance and the English referred to as Scottish malmsey.

The brewing of heather ale predates history. An archaeological dig on the Scottish Isle of Rhum discovered a Neolithic shard, circa 2000 B.C., in which were found traces of a fermented beverage containing heather. Much later, circa 100 B.C., the definitive Europeans (the Celts) were known to produce an intoxicating decoction from heather flowers and honey, but the most powerful heritage of heather ale comes from the time when Scotland was Pictland.

The famous fourth century B.C. navigator Pytheas noted that the Picts were accomplished brewers, and the Scottish dictionary notes, "The Picts brewed some awful grand drink they ca't heather ale out of heather and some unknown kind of fogg." The kings of Pictland fought off many invasions from Europeans, Anglo-Saxons, and Irishmen, and they even turned away the mighty Roman Empire. For this they became known as a ferocious race with a secret magical potion called heather ale. One legend still told today is recorded by Neil Munro, Sir Herbert Maxwell, and is the subject of a poem by Robert Louis Stevenson—the accounts differ only in detail.

In A.D. 400, an Irish king invaded Dalriada in southwest Scotland and began to wipe out the local Picts. In his frenzy to win the battle he was thought to have killed them all before remembering the heather ale. He sent his army to find survivors and they returned with a Pictish chief and his son. The Irish king was going to torture

**heather**

them to gain the secret of the heather ale when the chief agreed to tell the secret if they would kill his son quickly. Once his son was dead the old chief took the Irish king up to a crop of heather by a cliff and, according to Robert Louis Stevenson, said, "But now in vain is the torture, fire shall never avail, here dies in my bosom, the secret of the heather ale." The chief threw himself at the king and they both fell to their death over the cliff.

Although this legend is based on fact, the Picts were not totally wiped out. They were cleared from the Dalriadic area and the custom of brewing heather ale continued, particularly in the Highlands.

By the twelfth century the Pict and Dalriadic lands had united to become Scotland, "Alba" to the Gaels, and heather ale became the common drink of the clans. One Ceilidh legend told of a cold winter inside a cave in the Highlands where a Gaelic clan was gathered. They sat around a pot of heather ale warming over the fire, telling stories, singing and drinking. Meanwhile, the steam from the heather ale was condensing on the ceiling and dripping straight into a bowl on the ground. The clansmen drank from the bowl and experienced sensations of euphoria, warmth, and coolness they had never known. "Uisge-beatha!" they exclaimed, and water of life was discovered that night. Its name was soon shortened to "uisge," uis-ge, which was bastardized by the English language to whiskey.

In the eighteenth century Scotland was at its darkest. After the massacre at Glencoe and fifty years of fighting the British Empire, Bonnie Prince Charlie's 1745 rebellion was defeated by overwhelming odds and Scotland lost its independence. To prevent any further uprising, the British government tried to destroy the entire clan system by banning the wearing of tartan or any traditional Highland clothing, the carrying of arms, outlawing the Gaelic language, and generally repressing Highland communities. This and the Highland clearances resulted in the loss of many crafts and skills; in fact, a whole cultural way of life was under threat. Heather ale was soon reduced to legend. Legislation prevented the use of anything but hops, malt, and water in the brewing of ale. This ethnic cleansing by the British caused thousands of Scots to be transported to the West Indies, New Zealand, Maryland, or South Carolina, which in turn created a mass exodus. Scottish society began to emigrate and follow their clans.

The twentieth century saw the recognition of Scotch whiskey as the definitive distillate, and Scotch ale was being shipped to destinations all around the world. Heather ale is known to have been produced, perhaps in defiance, in the remote Highlands and islands from ancient Gaelic recipes recited by clan alewives to their descendants. In Glasgow in 1986, a Gaelic-speaking islander translated one such recipe for me, and like the Bruce before me, I tried and tried again before successfully reviving the *leann fraoch*.

I began by brewing the ale in July 1986 at the small West Highland Brewery in Argyll and selected three pubs in Edinburgh and three in Glasgow to sell the naturally conditioned cask heather ale. The publicity was good and the product took off. The six pubs I chose were ordering more than the capacity of the plant and by September I was renting 120-barrel facilities at the Thistle Brewery in Alloa (heather ale being made at the Thistle Brewery!). By October the heather flowers were gone and only one batch of 18,000 17-ounce (50-cl.) bottles had been produced. I kept 1,000 bottles and the rest were sold within ten weeks. *Leann fraoch* became a cult product within the year.

Brewing heather ale is dependent on the flowering season. The heather pickers begin in May and the first pubs to receive stock will have it on sale in Scotland by the end of June. This could fast become the Beaujolais Nouveau race for cask ale.

For technical buffs the numerical specifications for the bottled ale are: alcohol by volume, 4.9 percent; original gravity, 1.048; pH, 4.1; color, 9 SRM (23 EBC); and bitterness, 21 IBUs.

## Heather Ale

*Ingredients for 5 gallons*

    6⅔ lbs. (3 kg.) crushed Scotch ale malt, or 6 lbs.. (22.7 kg.) U.S. two-row malted barley and 10½ oz. (300 g.) amber malt (crystal or Cara-type)
    12⅔ c. (3 l.) lightly pressed flowering heather tips
    ³⁄₁₀ oz. (8 g.) Irish moss (10 min.)
    2⅗ gal. (10 l.) soft water
    lager yeast
½ to ¾ c. corn sugar (to prime)

- Original specific gravity: 1.048
- Final specific gravity: 1.011

Mash the malt at 153 degrees F (67 degrees C) for 90 minutes. Sparge to collect up to 5⅓ gallons (20 l.). Add about ½ gallon (2 l.) of lightly pressed heather tips and boil vigorously for 90 minutes.

Run hot wort through a sieve filled with 2 cups (½ l.) of heather tips into the fermenting vessel. Allow to cool and ferment at 61 degrees F (16 degrees C) for 7 to 10 days. I recommend a lager-type yeast. My original yeast was a Scotch ale yeast, but years of cold slow fermentation have evolved a strain with a bottom-fermenting bias. When the gravity reaches 1.015, usually the fifth day, remove ½ gallon (2 l.) of ale, add 2 cups (½ l.) of heather flowers, and warm to 158 degrees F (70 degrees C). Cover and steep for 15 minutes, then return to the fermenter.

Condition the ale as usual. For those needing a hop fix, add 1 ⅘ ounces (50 g.) of 6 percent alpha acid hops for the 90-minute boil to provide bitterness that will not unbalance the flavors. Late-addition aroma hops would compete with the delicate heather.

## SOURCES

Glenbrew, Bruce Williams, 736 Dumbarton Rd., Glasgow G116RD United Kingdom; phone (041)3393479 or fax (041)3376298.
Northeast Heather Society, Walter Wornick, P.O. Box 101, Alstead, NH 03602; phone (603) 835-6165.
Speyside Heather and Heather Craft, Fran Rowley, Dulnain Bridge, Iverness-Shire, PH26 3PA, United Kingdom; phone (047)985359 or fax (047)985396.

# THE REGAL ALTBIERS OF DÜSSELDORF

## BY ROGER DESCHNER

*This article originally appeared in Zymurgy, Winter 1994 (vol. 17, no. 5)*

It was late afternoon as I drove in heavy traffic into Düsseldorf. I found a drab but inexpensive hotel room in The Altstadt (old city) and parked the car. Düsseldorf is not frequently mentioned as a tourist destination because it is primarily a medium-sized financial center. It might be comparable to a city like Hartford, Connecticut.

The guidebook that led me here was Michael Jackson's *Pocket Guide to Beer* (Simon & Schuster, 1992). Jackson had already led me to such highlights as the beer halls of Munich and the fabulous Hirsch Brewery-Hotel in Ottoburen, where a swimming pool is heated by the brewkettles.

My destination was the Altbier breweries of old Düsseldorf,

## The Regal Altbiers of Düsseldorf

Germany

particularly Zum Uerige. The brewpub is very large, covering the first floor of several adjoining buildings. On your way into the dimly lit rooms to find a table you must constantly dodge wooden kegs of beer being rolled to the serving stations, where two people lift them to high shelves for dispensing by gravity. The tall, narrow glasses are filled by the usual German method of filling halfway, waiting until the head settles a bit, then filling completely. Once you order one beer, it is a standing order and you are brought glass after glass until you persuade your waiter to stop and settle the bill, kept by tick marks on your coaster.

The Zum Uerige brewery, visible through windows in a back room, is all gleaming polished copper with antique equipment still in use, such as an original Baudelot ammonia-based refrigeration system. The place is always crowded, although the food menu is minimal with only (delicious!) sausage and cheese plates. The large crowd spilled out onto the sidewalk of a side street that evening, soaking in the warmth of a pleasant fall evening along with the warmth of great beer.

But Zum Uerige the place is overshadowed by Zum Uerige Altbier. It has a beautiful deep copper color and a nice white head, a sturdy, complex German maltiness with a wonderful, huge hoppiness that brings everything into balance. It is a very big beer with a big, intense flavor profile, but neither sweet nor unusually alcoholic. Zum Uerige Altbier is truly one of the world's magically special beers. (Michael Jackson gives it four stars.) I will always remember my evening at Zum Uerige, drinking perhaps the best beer I have ever had, and watching the staff roll out the wooden barrels. It made Düsseldorf, an otherwise unspectacular city, the high point of the trip.

Unfortunately, my visit did not coincide with the seasonal specialty called *Sticke* (secret), which has a higher gravity and is dry-hopped and even more intense. Zum Uerige brews one batch of *Sticke* to serve in September and one in January.

I also visited Zum Schlussel (The Key) brewery, which had a characteristic romantic, very German setting, a full food menu, and very good Altbier. However, it was not quite as intense as Zum Uerige's beer.

Where did this mystical elixir, Düsseldorfer Altbier, come from? Why doesn't anybody know about it? To answer the second question first, perhaps the fame of Düsseldorfer Altbier has not spread farther because most of the breweries making it in Düsseldorf are brewpubs (called house breweries). While Altbier is not an appellation like Kölsch is for Cologne, this kind of Altbier is not made outside of Düsseldorf's half dozen or so brewpubs, not even elsewhere in Germany. Very little of the Düsseldorf product is exported anywhere, much less across the Atlantic. (This situation is changing. See Commercial Examples at the end of this article.) The major problem is simply that most people have never tasted this gem of the brewing world. Even though the American Homebrewers Association wisely gives "Düssledorf-style Altbier" its own subcategory, it may be the least understood classification. Well-meaning American homebrewers or microbrewers usually miss the mark because they have not tasted it, since traveling to Düsseldorf is about the only way to gain the experience.

Other beers called Altbier are brewed in other regions of Germany. The most distinct of these is the Münster style, a remarkable brew in its own right, having a slight lactic character somewhat like a Berliner Weiss. It is made with as much as 40 percent wheat malt, making it fairly light-colored. Pinkus Münster Alt is imported to the United States, but it is clearly distinct from the Düsseldorf style. Many German brewers, such as the larger breweries of Dortmund, call their ordinary dark beer "Alt." These tend to be darker, less hoppy, sweeter, and unremarkable compared to the Düsseldorf style. Many of the American efforts appear to be imitating the Dortmund-style Altbiers.

To understand its history it helps to know how real Düsseldorf-style Altbier is made. Altbier brewing starts with a lengthy upward step infusion mash, or a decoction mash, using a mixture of traditional German malts in relatively soft water. Hops are primarily Spalt with some Tettnanger and Saaz. Primary fermentation is at a relatively warm 60 to 65 degrees F (16 to 18 degrees C) using specific clean single-strain yeast followed by cold (41 degrees F or 5 degrees C) conditioning like a lager beer. Esters produced in the warm primary phase are subdued in the cold secondary. This process is the hallmark of an Altbier.

Altbiers should fall in the initial gravity range of 1.045 to 1.051, but can vary greatly in final gravity. There is a lively debate among brewers about which of the so-called Altbier yeasts is the real one, and the answer probably is that they all are, even though they differ greatly in attenuation. One might suspect that the low-attenuating Wyeast No. 1338, which leaves a big malty profile, could be more like Zum Uerige, while the drier Wyeast No. 1007 could be similar to the strains used by some of the other Düsseldorf brewhouses. Although the exact hop bill varies from one brewery to another, Altbier is a hoppy beer—from 35 to 50 IBUs. Randy Mosher's table in *The Brewer's Companion* (Alephenalia Publications, 1994) shows that Düsseldorfer Altbier has the highest rate of hop bitterness per unit of gravity of all the world's beer styles. A typical malt bill would contain mostly German two-row Pilsener malt, Munich and Vienna specialty malts, and no more than a small amount of caramel malt. A touch of black malt is used to adjust the color to a deep copper, but not brown. Some brewers use a portion of wheat malt as well. Too many brewers make the mistake of using too much crystal malt and too few hops, the result being a sweet brown ale they call Altbier but that misses the mark completely.

Altbier brewing predates the invention of refrigeration and the Pilsener style that refrigeration fostered. *Alt* is German for "old," meaning simply top-fermented the old way. But the Düsseldorf style also is "old" in that it has a big flavor profile, in contrast to the increasingly light and clean Pilseners pioneered in the nineteenth century. We are indeed lucky that the city of Düsseldorf continues this unique brewing style today. My visit was one of the reasons I began to brew at home. I searched in vain for anything like the beer I had in Düsseldorf, so I resolved to make it myself. My homebrewed versions have yet to hit the mark, although I am getting closer with each revision, and the research I did in preparing this article should improve them another step.

If you want to try brewing a Düsseldorfer Altbier, start with the right ingredients. It is important to use German-type malts and hops to give this beer a German accent; otherwise it could be an India pale ale or a California common. But the most frequently overlooked ingredient is hops. Think "medium-dark, malty, German beer for hop-heads." To be truly in style, use "noble-type" hop varieties for bittering, as well as for flavor and aroma. This does make a difference and the cost is insignificant at home-brewer quantities. If you have the facilities, try kegging Düsseldorfer Altbier in a wooden barrel, as is done at Zum Uerige. Researchers studying the effect on beer flavor have investigated certain bacteria that live only in wooden beer vessels.

This recipe has several alternatives, such as choice of mashing procedures or making the higher-gravity *Sticke* Altbier. To make *Sticke*, follow the general recipe below, but raise the original gravity to about 1.058, add a little more roasted malt, raise the hopping to about 60 IBUs, and dry-hop when racking to the secondary. (*Sticke* is an intense brew!)

## Düsseldorfer Altbier

*Ingredients for 5 gallons*
- 1 lb. Munich malt
- 1 lb. Vienna malt
- 1 lb. caramel malt
- 1 to 4 oz. black patent malt (Vary this to achieve SRM color in the mid-teens—a deep copper but not brown.)
- 6 lbs. German two-row Pilsener malt (Vary this if your mashing efficiency is not about 85% to achieve original gravity between 1.045 and 1.050.)
- 3 oz. Spalt hops (Saaz, Tettnanger, or Hallertauer can also be used. Avoid Cascade or other distinctively fruity American hop varieties.) Adjust the amount of hops based on your boil gravity and the alpha acid levels of your hops—try to achieve 35 to 50 IBUs.
- moderately soft water
- Wyeast No. 1338 (My preference, but whatever you use, make a starter! Alternate alt yeast selections can be found in the "Table of Available Yeast Strains," *Zymurgy*, Summer 1994 [vol. 17, no. 2].)

Dr. Bob Technical's wheels of malt and hops can help you get close to the target gravity and hop level here. The right mashing procedure is important. Some Düsseldorf brewers use a standard decoction mash, while others use this 2½-hour upward step infusion mash:

- 122 degrees F (50 degrees C), hold for 45 minutes
- raise temperature to 144 degrees F (62 degrees C) at 1 degree C per minute, hold for 20 minutes
- raise temperature to 158 degrees F (70 degrees C) at 1 degree C per minute, hold for conversion
- raise temperature to 169 degrees F (76 degrees C) for mash-out
  Boil for 1½ to 2 hours with the following hop additions:
- 40 percent of hops after 15 minutes
- 20 percent after 1 hour
- 20 percent 5 minutes before end
- 20 percent at end of boil

Do not skip the later flavor and aroma hop additions, as some might suggest.

Ferment at 60 to 65 degrees F (16 to 18 degrees C) until fermentation stops. Rack and condition at 41 degrees F (5 degrees C) for three weeks or longer. Residual gravity may be high and this is OK. Prime or force carbonate. Serve relatively fresh while the hop character is still evident.

## COMMERCIAL EXAMPLES

One Düsseldorfer Altbier is now being imported to the United States in limited quantities. While none of the other examples available in the United States are "it," if you want to taste an approximation of the Düsseldorf-style Altbier, you might try the following:

**Schlösser Alt** ("Locksmith"—not to be confused with Zum Schlussel, "The Key") Made in Düsseldorf and now available for the first time in a few U.S. cities, on draft only.

**Alaskan Amber** (Alaskan Brewing Company, Juneau, Alaska) Revives an old tradition of brewing altlike beers in Alaska. Probably not an accident that it goes especially well with smoked salmon.

**Widmer Alt** (Widmer Brewing Company, Portland, Oregon) Though unfortunately toned down from its earlier aggressive imitation of Zum Uerige, is still brewed using the unique Düsseldorf process.

**Anchor Steam** (Anchor Brewing Company, San Francisco, California) The California Common beer style is superficially similar, and Anchor's process is close to the Düsseldorf process except in the use of lager yeast instead of ale yeast. Anchor Steam is moderately close to Düsseldorfer Altbier in color, texture, hoppiness, and a certain element of "wonderfulness." It is plausible to suppose that German immigrant brewers in old San Francisco, faced with the near impossibility of obtaining ice, turned to the old alt process they remembered from Düsseldorf because it worked and produced drinkable beer under the circumstances.

Go to Düsseldorf—why not taste the real thing? There is no U.S.-made beer that compares with Zum Uerige Altbier fresh from the source. Watch for transatlantic airline fare wars. Düsseldorf is a frequent landing point for flights from the United States, and fares have occasionally dipped to the price of a flight across the United States. Düsseldorf is easy to reach by car or train from Belgium, The Netherlands, or Luxembourg. Bring your copy of *Pocket Guide to Beer* as your tour guide!

American Altbiers not in the Düsseldorf style—Düsseldorf Ale (Indianapolis Brewing Company, Indianapolis, Indiana), Schmaltz' Alt (August Schell Brewing Company, New Ulm, Minnesota), Old Detroit Ale (contract-brewed by Frankenmuth Brewery, Inc., Frankenmuth, Michigan), St. Stan's Alt (St. Stan's Brewery, Modesto, California).

## REFERENCES

Eckhardt, Fred, "German Style Ale," *Zymurgy*, Traditional Beer Styles 1991 Special Issue (vol. 14, no. 4)

Eckhardt, Fred, *The Essentials of Beer Style*, Fred Eckhardt Associates, Inc., 1989

Jackson, Michael, *Pocket Guide to Beer*, Simon & Schuster, 1986, 1988, 1991

Mosher, Randy, *The Brewer's Companion*, Alephenalia Publications, 1994

*Roger Deschner is a homebrewer, member of the Chicago Beer Society, Recognized BJCP Judge, amateur photographer, and professional computer systems programmer from Chicago, Illinois. This article is available on Library 13-AHA/zymurgy/Clubs on CompuServe's Beer and Wine Forum as ALTBER.W94.*

# BREWING BETTER BELGIAN ALES

## by Phillip Seitz

*This article originally appeared in* Zymurgy, *Spring 1995 (vol. 18, no. 1)*

Interest in Belgian beer has blossomed in the past few years. At the same time many of the yeasts and specialty ingredients needed to make them have appeared on the homebrew market, and more people are deciding to brew Belgian-style beer every day. Like a lot of people, I've been trying for some time to capture a taste of Belgium in my own home. In the process I've poured my share of beer down the drain, but with luck, this article will spare you some of the same trauma and start you on the road to brewing quality Belgian-style beers.

I'll focus on four popular styles: doubles, triples, strong ales, and white beers. The first three have a lot in common and make wonderful cool-weather drinking; the last will get you through the summer in style.

## DOUBLES

Doubles (dubbels) originated as a Trappist style but now are made in secular breweries as well. A good double will be malty and sweet with a noticeable plum character, range in color from dark amber to brown, and have an original gravity from 1.060 to about 1.070. This is not a light beer, so modest alcohol flavor is permissible, as are low levels of fruity-tasting esters. But the sweet, plumlike character should predominate in both the palate and the nose.

These beers are usually full-bodied with fairly mousselike carbonation that produces a very nice head. Bitterness levels vary from nonexistent to low (about 15 IBU at most). Commercial examples available in the United States include Westmalle Dubbel (5.1 percent alcohol by weight, 6.5 percent by volume), Grimbergen Dubbel (4.9 percent alcohol by weight, 6.2 percent by volume), Steenbrugge

Dubbel (5.1 percent alcohol by weight, 6.5 percent by volume), and Affligem Dubbel (5.5 percent alcohol by weight, 7 percent by volume).

Almost all Belgian brewers use Pilsener malt as the base malt for their beers. For doubles a substantial amount of caramel malt is added for color, sweetness, and flavor. The Belgian caramel malts now on the market work especially well here, particularly the CaraMunich and Special "B" because they provide some of the plum and raisin flavors necessary for the style. One pound (0.45 kg.) of CaraMunich and ½ pound (0.23 kg.) Special "B" make a good starting point for a 5-gallon (19-l.) batch, and you can adjust the quantities to your liking in subsequent batches.

Candi sugar is added to keep the body comparatively light for the strength and to provide some alcoholic warmth. Start with ½ to 1 pound (0.23 to 0.45 kg.) in the kettle for a 5-gallon (19-l.) batch. Roasted malts can be used in very small amounts to deepen the color, but there shouldn't be any roasty flavors in the finished beer. Toasted malts like Biscuit, aromatic, and Munich contribute pleasantly malty or nutty flavors, and can be used in fairly high quantity. Try 2 pounds (0.91 kg.) of toasted malts for a 5-gallon (19-l.) batch. Their use, however, requires mashing.

Hop selection offers plenty of flexibility, but you should lean toward the "noble type" rather than varieties like Cascade. Keep the bitterness low and be sparing with late additions so any hop flavors blend in and don't stand out too much.

Yeast choice offers some flexibility, though strains with a smooth, fruity character complement the raisin/plum flavors of the caramel malts better than yeasts yielding spicy flavors. None of the readily available yeasts are perfect for this, but either of the Belgian Wyeast strains will work. Ferment at temperatures below 65 degrees F (18 degrees C), preferably around 60 degrees F (16 degrees C).

Extract brewers can't use the toasted malts, but otherwise should be able to produce a nice malty brew. Start with pale extract and a hefty infusion of Belgian caramel malts, add sugar to the kettle, and choose your yeast with care.

Because doubles are very effervescent, brewers should prime with ⅞ to 1 cup (125 to 150 g.) of corn sugar for a 5-gallon (19-l.) batch.

## TRIPLES

Triples (tripels) also come from Trappist roots. These are strong, pale beers with a neutral aroma and comparatively light body and flavor for their strength. Frequently they are slightly sweet. With original gravities of 1.080 to 1.095, the alcoholic strength will be evident but smooth, followed by a subtle mix of yeast, hop, and malt flavors. High carbonation levels are the norm, sometimes reaching 3.5 volumes of $CO_2$. Commercial examples available in the United States include Brugse Tripel (7.5 percent alcohol by weight, 9.5 percent by volume), Affligem Tripel (7.1 percent alcohol by weight, 9 percent by volume), Grimbergen Tripel (6.5 percent alcohol by weight, 8.31 percent by volume), Westmalle Tripel (6.4 percent alcohol by weight, 8 percent by volume), and Steenbrugge Tripel (7.1 percent alcohol by weight, 9 percent by volume).

To keep the color light, you'll need to use Pilsener malt or light malt extract, sugar, and nothing else. These beers also have a very light body for their strength, which is directly attributable to the use of large quantities of highly fermentable adjuncts. Sugar and corn are the most common and allow the beer to attain higher strength without adding heaviness to the body. With a triple you can count on using 2 or more pounds (0.91 kg. or more) of a light-colored sugar—corn sugar, for instance—for every 5 gallons (19 l.) of beer. Hopping levels should be kept low (from 18 to 25 IBU), with classic varieties such as Saaz and Hallertauer preferred. Because this beer doesn't usually have much hop aroma, most of your hops should be added at the beginning of the boil. Pick a Belgian yeast like Wyeast 3944 that ferments well but has a relatively low aromatic profile. Keep in mind that you want a yeast that can handle high gravities without producing too many fusel alcohols, or your beer will end up a headache in a bottle. [Editor's Note: Individual reactions to fusel alcohols vary, as does scientific evidence confirming or denying the connection of fusels to headaches.]

Triples are highly carbonated, and 1 cup (150 g.) of sugar should be used to prime a 5-gallon (19-l.) batch. It's a good idea to add some fresh yeast at bottling time to help with carbonation. Just add 1 pint of yeast starter to your bottling bucket along with the sugar solution and finished beer.

## STRONG ALE

The Belgian strong ale category covers a lot of ground. According to Belgian tax law, this includes everything with an original gravity greater than 1.062 (including doubles and triples), but most Belgians refer to the strong ales as "specials." Strong ales can be light or dark, and sweet or dry. Only rarely do they have hoppy or roasted flavors, which don't have much appeal for the Belgian drinking public. Usually they're wonderfully aromatic and derive most of their enticing flavor and aroma from the yeast and from a restrained blend of spices, hops, and malts. Coriander and orange peel are frequently included, and sometimes anise is used. Original gravities range from 1.062 to 1.120 (though usually closer to 1.080 or 1.090), and the best examples may be noticeably strong but still have no alcohol flavor. Flemish examples tend toward higher terminal gravities (1.025 to 1.050) and are therefore sweeter, while Walloon versions are usually more attenuated and drier.

Fortunately there are a lot of commercial examples available in this country. Included among them are Corsendonk Blond (6.4 percent alcohol by weight, 8 percent by volume), Corsendonk Brown (6.4 percent alcohol by weight, 8 percent by volume), Saison DuPont (5.1 percent alcohol by weight, 6.5 percent by volume), Gouden Carolus (5.5 percent alcohol by weight, 7 percent by volume), Scaldis (9.4 percent alcohol by weight, 12 percent by volume), Duvel (6.7 percent alcohol by weight, 8.5 percent alcohol by volume), Brigand (7.1 percent alcohol by weight, 9 percent by volume), Pauwel Kwak (6.4 percent alcohol by weight, 8 percent by volume), Celis Grand Cru (7 percent alcohol by weight, 8.9 by volume), Mateen (7.1 percent alcohol by weight, 9 percent by volume), Artevelde Grand Cru (5.3 percent alcohol by weight, 6.7 percent by volume), and Chouffe (6.3 percent alcohol by weight, 8 percent by volume).

Yeast should provide the foundation flavors for these beers, and all other ingredients should be added to support or accentuate this. Tasting other beers made with the yeast you are considering using should help with recipe formulation.

Use Pilsener malt as a base, but you may also want to try adding substantial quantities of sugar or flaked corn as an adjunct; count on using at least 1 pound (0.45 kg.) of one or the other in a 5-gallon (19-l.) batch, and 1 pound of each isn't too much. Caramel, Munich, and toasted malts often are used in small quantities to add color, fullness, or flavor accents; roasted malts are sometimes used in very small amounts for coloring only. All classic hop varieties are common, but are used in small and judicious quantities. Sugars are added in the kettle, as are spices. Because many spices have delicate aromas, they should be boiled only for a few minutes.

Extract brewers will do fine in this category. Steep judicious quantities of caramel malts, then add pale extract and sugar (1 to 2 pounds, 0.45 to 0.91 kg.) to the kettle.

The secret for either brewing style is to choose the right yeast and keep your ferment as clean as possible. Prime with ⅞ cup (125 g.) of sugar for 5 gallons (19 l.). Adding fresh yeast at bottling time is a good idea because of the high alcohol content of the beer.

## WHITE BEERS

White beers are another matter. These are wheat beers intended for summer consumption with original gravities ranging from 1.044 to 1.055. The grist typically is 50 percent barley malt and 50 percent unmalted wheat, although sometimes a small quantity of oats is added. The finished beer should be golden when warm and a very cloudy yellow when chilled. Coriander usually is added in substantial amounts, and most commercial examples are mildly acidic—both contributing refreshing qualities.

As with most Belgian styles, white beers are rarely bitter or hoppy, and unlike the styles listed previously, you shouldn't taste any alcohol. Carbonation is usually about average or a little higher, and some versions show a bit of yeast character. Commercial examples include Celis White (3.7 percent alcohol by weight, 4.7 percent by volume), Riva Blanche (3.9 percent alcohol by weight, 5 percent by volume), and Blanche de Bruges (3.9 percent alcohol by weight, 5 percent by volume).

Extract brewers will have a hard time getting the traditional yellow color and won't be able to add oats, which require mashing. But if you use 50 percent wheat extract and follow the guidelines below, you will still have a very distinctive and satisfying beer.

For all-grain brewers, making a white beer is the ultimate adventure. Use 50 percent raw unmalted wheat (by weight), and if you're interested, you can add 5 percent oats for some silkiness. Rolled oats work fine as an alternative to whole oats. Unmalted wheat or wheat berries are available in many health-food stores and food co-ops. There's debate whether the soft white or hard red variety is preferable, but both seem to work. One thing is indisputable: the stuff is a nightmare to grind by hand. Grinding unmalted wheat in my Corona mill is like grinding rubber bullets. Find someone with a two-roller mill or a mechanized grinder to help you out.

The secret to making white beers is an extended protein rest. Museums use wheat starch as a glue, and once you mash in, you'll understand why. Start with a loose mash using 2 quarts of water per pound of grist (4.2 l. per kg.), and plan on using a protein rest of 45 minutes to 1 hour at temperatures between 117 and 126 degrees F (47 and 52 degrees C). You'll be amazed how well the proteolytic enzymes work, turning a mass of wallpaper paste into a light, workable mash. Never has the miracle of mashing been better demonstrated.

This extended protein rest includes a trade-off. Keep in mind that wheat has no husk and is filled with proteins and gluey starches. If you run the protein rest longer—1 hour—you'll break down more of these and get an easily spargeable mash. But it also produces a clearer beer that may be less hazy and lack the color you want. Rests of 45 minutes or less give wonderful color, but can be sticky to lauter. I use a 45-minute rest and watch the lauter tun carefully.

If you're a nervous brewer and are willing to sacrifice some authenticity, you also can substitute several pounds of malted wheat for a portion of the unmalted variety. Some people have used decoction mashes to handle the wheat. This will certainly work, but is not the traditional method and involves additional labor.

For hopping I prefer floral varieties such as East Kent Goldings or Styrian Goldings, but nearly all the "noble type" hop varieties work, including Hallertauer and Saaz. Keep your hop levels around 18 to 20 IBU, and since this can be an aromatic beer, you may want to include a late hop addition.

You'll also need bitter orange peels and ground coriander. These are described in more detail under Special Ingredients, but you'll want to start by boiling ⅓ ounce (9 g.) of bitter orange, for about 20 minutes, and 1 ounce (28 g.) of ground coriander, for about 5 minutes, then you're ready to chill and ferment.

Almost any yeast seems to work, ranging from neutral American ale yeasts to German wheat beer strains and the more adventurous Belgian cultures. Creativity counts, so if you have an interesting idea, you should give it a try.

Mild acidity is a classic feature of a good white beer. The brave can attempt a fermentation that includes lactic acid bacteria, but there's an easy shortcut: add a very small quantity of 88 percent lactic acid to your beer at bottling time. Amounts between 1 and 3 teaspoons (5 and 15 ml. per 5 gallons (19 l.) work well. Be aware that the acid will need some time to blend with the other flavors, usually about two months.

## BREWING AND FERMENTING BELGIAN ALE

All-grain brewers may wonder which mashing techniques to use. Different Belgian breweries use them all: infusions, step, and decoction mashes. You're welcome to experiment, but you shouldn't have any problems using a simple infusion or step mash. I suggest including a 15-minute protein rest at 120 degrees F (49 degrees C) and a 10-minute mash-out at 170 degrees F (77 degrees C). Even with fully modified grains, this appears to improve yield, clarity, and fermentation. Some Belgian breweries use multiple saccharification rests, starting with one at about 145 degrees F (63 degrees C) and then moving to one at about 160 degrees F (71 degrees C). This territory is wide-open for experimentation.

To ferment high-gravity worts your yeast will need a lot of help or you will suffer the indignity of a stuck ferment. Believe me, it's very sad to watch your beer drop from 1.092 to 1.050, then have it sit there as the yeast goes on strike. Even if your ferments don't stick, yeast needs to be in good health to get the clean flavors you want in a Belgian-style beer.

To avoid headaches—literally—keep a few things in mind:

**Pitch a lot of yeast.** Count on using at least 1½ quarts (1.42 l.) of yeast starter in anything you brew. For beers in the 1.080 range you can double that. To avoid diluting your beer you can make the starter well ahead of brewing time and let the yeast settle out. Pour off the clear liquid above the settled yeast and add a pint of fresh wort just before you start to brew. You'll have lots of active yeast ready to go by pitching time, and your total fluid addition volume will be quite small.

**Aerate.** When fermenting strong beers, the health of the yeast is as important as quantity. Splashing wort as you transfer it into the fermenter is a start, but it won't solve your problems. Many homebrew supply shops sell aquarium pumps with in-line air filters, and if you're serious about brewing Belgian beers, you should invest the $15 or $20. With 30 minutes of aeration just after pitching, your lag times will be reduced and ferments will improve dramatically no matter what kind of beer you're brewing.

**Watch your fermentation temperatures**. Fermenting high-gravity worts at elevated temperatures is a prescription for headaches, the kind that come in a bottle. If you're brewing beers with original gravities higher than 1.060, you definitely need to keep the ambient temperature below 65 degrees F (18 degrees C). Remember that fermentation creates heat all by itself, so your beer is going to be warmer than the place you put it. The stronger the beer you make, the cooler you want to keep it while it's fermenting. I use a thermostat-controlled refrigerator for fermentation, and for beers like these I usually set it to 60 degrees F (16 degrees C). If you've chosen a good yeast, you'll still get plenty of flavor when you ferment at this temperature.

In Pierre Rajotte's book, *Belgian Ale* (Brewers Publications, 1992), he mentions that some Belgian breweries use warm and even hot fermentations. This is true, but trust me, if you try it at home, you're going to be sorry.

White beer fermentations don't require any unusual attention, although some of the commercial white beer yeasts get a bit sluggish when fermentation temperatures drop below 65 degrees F (18 degrees C).

## SPECIAL INGREDIENTS

Sugar plays an essential role in Belgian brewing. Because it ferments almost completely, it allows you to brew strong beers without the heavy body typical of barley wines. For all-grain brewers, using sugar also lets you brew at original gravities that exceed your mashing and lautering capacity. My Zapap lauter tun fills up with 15 pounds of grain, but by adding sugar to the kettle, I can increase either the gravity or the quantity of the finished wort. Candi sugar is the most common form used in Belgian brewing. Basically it is just rock candy. You can get light or dark candi sugar in liquid solution or various-sized chunks. It has little flavor other than a clean sweetness and provides the same number of gravity points per pound as corn sugar. Light candi sugar adds little or no color, and the dark stuff—at least what I've bought from Belgian supermarkets—doesn't provide much color either. When you make up your recipe you can assume that dark candi sugar delivers about 20 °L per pound. If you can't find any, go ahead and substitute any other neutral-tasting sugar, and definitely relax, don't worry, have a homebrew.

Coriander is a Belgian staple. Just about any kind should give a very pleasant orangelike flavor and aroma. For best results you should buy fresh coriander seeds and grind them to a powder with a mortar and pestle (food processors grind too coarsely). Whole seeds are available at health-food and ethnic grocery stores.

If you're using coriander in a strong ale, you're probably trying to add a relatively subtle extra flavor. Half an ounce (14 g.) works well for 5 gallons (19 l.); add it for the last 5 minutes of the boil. If you want lots of coriander flavor and aroma, use 1 ounce (28 g.), particularly for white beers. Boiling coriander too long (more than 15 minutes) or grinding it too coarsely will reduce the flavor and aroma.

Bitter orange peels come in greenish gray chips or quarter-of-an-orange slices. They also are known as Curaçao oranges, and despite their name, they are not very bitter and do not taste much like orange. Instead, they give a nice herb-tea flavor resembling chamomile.

Bitter orange is used in white beers. Start with 0.1 ounce of peel per gallon of beer you plan to make (0.5 g. per l.). This is about ⅓ ounce (9 g.) for a 5-gallon (19-l.) batch. If you want more, you can go up to a full gram per liter. I usually boil the peels for about 20 minutes. Don't substitute ordinary orange peel or your beer will have a peculiar hamlike aroma. The bitter peels can sometimes cause this aroma. To see what the peel will smell like, boil some dried peel in a small pot of water.

Finally there is sweet orange peel, which is not the supermarket variety, either. Usually it comes in strips or ribbons, as if someone peeled an orange in one piece, and is thinner and more orange in color than the bitter peels. Use it in strong ales in roughly the same quantities as bitter orange in white beers and it will produce a heavenly orange flavor very similar to Cointreau or Grand Marnier. This goes well with some of the Belgian yeasts and particularly well with coriander. Again, boil it for 20 minutes.

Unfortunately, at the time I'm writing this, sweet orange is only available in Belgium and cannot be purchased in the United States. A number of people are looking into importing it, so tell your homebrew dealer you want it. This is the last important Belgian brewing ingredient that is not available to homebrewers in America. (Some people have had success substituting tangerine peel.)

## RECIPES

### Andy Anderson's Aaron's Abbey Ale (slightly revised)

Ingredients for 5 gallons (19 l.)
- 9 lbs. Belgian Pilsener malt (4.08 kg.)
- 2 lbs. Belgian biscuit malt (0.91 kg.)
- 1 lb. Belgian CaraMunich (0.45 kg.)
- 4 oz. Special "B" malt (113 g.)
- 1 lb. dark candi sugar (0.45 kg.)
- 1⅖ oz. Tettnanger hop pellets, 4.4% alpha acid; goal is 25 IBUs (39.7 g.) (60 min.)

½ oz. Hallertauer Hersbrucker hop plug, 2.9% alpha acid (14 g.) (5 min.)
1 tbsp. Irish moss (14.8 ml.) (15 min.)
1 qt. Chouffe yeast starter (0.95 l.)
1 pt. Chouffe yeast (0.47 l.) (add with priming medium)
⅘ c. dextrose (120 g.) (to prime)

- Original specific gravity: 1.065
- Final specific gravity: 1.014

The malt bill assumes an extraction rate of 25 points per pound, so adjust to fit your brewing setup. Mash with a protein rest for 30 minutes at 120 degrees F (49 degrees C). Boost temperature straight to 158 degrees F (70 degrees C) for saccharification. Hold until conversion is complete.

Mash out at 170 degrees F (77 degrees C) for 10 minutes and sparge with 170-degree F (77-degree C) water. Ferment at 60 degrees F (16 degrees C).

Note: The original recipe (of which this is a variation) took first place at the Home Wine and Beer Trade Association competition in 1994.

## Delano Dugarm's Batch #28 Triple

*Ingredients for 5 gallons (19 l.)*
9 9/10 lbs. (3 boxes) Northwest Gold liquid malt extract (4.49 kg.)
1½ lbs. corn sugar (0.68 kg.)
1 3/10 ozs. Hallertauer hops, 4% alpha acid (36.8 g.) (60 min.)
3/10 oz. Saaz hops, 3% alpha acid (8.5 g.) (60 min.)
3/10 oz. Saaz hops, 3% alpha acid (8.5 g.) (2 min.)
Wyeast No. 1214 Belgian yeast
¾ c. corn sugar (113 g.) (to prime)

- Original specific gravity: 1.080 (est.)
- Final specific gravity: not taken

Boil for 60 minutes, cool, and pitch slurry from 1½-quart (1.42-l.) culture. Ferment very cool (60 degrees F or 16 degrees C). Rack to secondary and bottle when ready.

## Jeff Frane's Strong Ale

Ingredients for 5 ¾ gallons (22 l.)
- 9 lbs. DeWolf-Cosyns Pilsener malt (4.08 kg.)
- ⅗ lb. DeWolf-Cosyns aromatic malt (0.3 kg.)
- 1 lb. DeWolf-Cosyns CaraMunich malt (0.45 kg.)
- 1 lb. flaked maize (0.45 kg.)
- 1½ lbs. light candi sugar (0.68 kg.) (75 min.)
- ¼ oz. Saaz hops, 3% alpha acid (7 g.) (60 min.)
- 1 oz. British Columbian Golding hops, 4.5% alpha acid (28 g.) (15 min.)
- 1 oz. Mount Hood hops, 3.5% alpha acid (28 g.) (15 min.)
- ½ tbsp. rehydrated Irish moss (7.4 ml.) (75 min.)
- Wyeast White yeast No. 3944
- 1 c. corn sugar (150 g.) (to prime)

- Original specific gravity: 1.062
- Final specific gravity: 1.012 (est.)

Mash in the malts (not the maize) at 98 degrees F (37 degrees C) in 3½ gallons (13.25 l.) water and adjust pH. Raise to 120 degrees F (49 degrees C) and hold for 30 minutes. Raise to 153 degrees F (67 degrees C), add maize, and hold until conversion, about 45 minutes. Raise to 175 degrees F (79 degrees C) for 15 minutes for mash-out.


## Rick Garvin's Cherry Blossom Wit

Ingredients for 5 gallons (19 l.)
- 4 lbs. Pilsener malt (1.81 kg.)
- 3⅗ lbs. unmalted wheat (1.63 kg.)
- ⅖ lb. rolled oats (181.4 g.)
- 9/10 oz. Styrian Goldings hops, 6.2% alpha acid (25.2 g.) (60 min.)
- ½ oz. bitter orange peel (14.5 g.) (20 min.)
- ⅖ oz. Saaz hops, 3.2% alpha acid (10.2 g.) (5 min.)
- 1¼ ozs. ground coriander (5 min.) (35 g.)
- Wyeast White yeast No. 3944
- ¾ c. corn sugar (120 g.) (to prime)

- Original Specific Gravity: 1.048
- Final specific gravity: 1.008

Mash in at 117 degrees F (47 degrees C). Rest 20 minutes at 117 degrees F (47 degrees C) and 122 degrees F (50 degrees C). Rest 60 minutes at 146 degrees F (63 degrees C). Mash out at 160 degrees F (71 degrees C). Boil 30 minutes before the first hop addition.
Note: This uses a lot of coriander. Use less if you are faint of heart.

In June 1994 Brewers United for Real Potables (BURP) of Virginia sponsored a monthlong advanced course on judging Belgian beers during which the following guidelines were developed. Draft versions of the guidelines were then circulated to beer judges via Internet's JudgeNet Digest, and the comments of many participating beer judges were incorporated. These are based on the current AHA categories, but we feel they provide more detail than the current guidelines. These descriptions provide additional information for people who may not be familiar with these types of beer. The guidelines were used in BURP's Spirit of Belgium Homebrew Competition.

## DOUBLES

Original gravity (Balling/Plato): 1.060–1.070 (15–17.5)

Percent alc./wt. (alc./vol.): 4.7–5.9 (6–7.5)

International Bitterness Units: 18–25

Color SRM: 10–14

Dark amber to brown. Sweet malty aroma. Faint hop aroma OK. Medium to full body. Malty, plumlike flavor. Very low bitterness, no hop flavor. Medium to high carbonation. Low esters OK. No roasted flavors or diacetyl.

Description: Doubles should be malty and sweet with a noticeable plum character. Modest alcohol flavor is OK, as are low levels of esters, but the malt flavors should predominate. Doubles are usually lighter in body than their maltiness would suggest, with a fairly moussy carbonation that produces a very nice head.

## TRIPLES

Original gravity (Balling/Plato): 1.080–1.095 (20–23.75)

Percent alc./wt. (alc./vol.): 5.5–7.9 (7–10)

International Bitterness Units: 18–25

Color SRM: 3.5–5.5

Light or pale color. Low ester, malt, or hop aroma OK. Medium to full body. High carbonation. No diacetyl. Strength should be evident; alcohol flavor OK.

Description: This is a strong, very pale beer with a relatively neutral character. These beers should have low esters (by Belgian standards) and comparatively light body and flavor for their strength. Frequently they are somewhat sweet. Alcoholic strength should be evident, followed by a subtle mix of yeast products and hop and malt flavors. Some commercial examples are well hopped, but most are not. Some spicy (phenol) character is OK. High carbonation levels are the norm.

## STRONG ALE

Original gravity (Balling/Plato): 1.062–1.120 (15.5–30)

Percent alc./wt. (alc./vol.): 4.7–9.4 (6–12)

International Bitterness Units: 16–30

Color SRM: 3.5–20

Pale to dark brown. Low hop bitterness and aroma OK, but should blend with other flavors. Medium to high esters in flavor and aroma. Phenols OK. Often highly aromatic. Spices or orange OK. Strength evident, but alcohol flavor subdued or absent. Medium to full body, sometimes with a high terminal gravity. Medium to high carbonation. No roasted flavors or diacetyl.

Description: Should be formulated to show off yeast character, with all other ingredients playing a supporting role. The flavor may be subtly complex, but should not be crowded. Body is comparatively light for beers of this strength, due to use of brewing adjuncts or of Pilsener malt only. High carbonation also helps; these beers should feel like mousse on the palate and have an impressive head. The best examples may be noticeably strong but still have no alcohol flavor. Flemish examples tend toward higher terminal gravities (1.025–1.050, 6.25–12.5°P). Trappist and Saison clones should be submitted in this category.

## WHITE BEERS

Original gravity (Balling/Plato): 1.044–1.055 (11–13.75)

Percent alc./wt. (alc./vol.): 3.5–4.3 (4.5–5.5)

International Bitterness Units: 15–22

Color SRM: 2–4

Cloudy yellow color, coriander flavor, and mild acidity essential. Wheat and bitter orange peel flavors desirable. Mild hop flavor and aroma OK. Low to medium bitterness. Low to medium body, medium or higher carbonation. No diacetyl. Low to medium esters.

Description: These beers should be average in gravity with a definitely hazy yellow color and a dense, rich, dazzlingly white head. May or may not have a slightly orangy aroma (due primarily to the coriander) or mild hop aroma (preferably floral rather than spicy). Body should be medium or a bit lighter, and the carbonation should be reasonably aggressive. Bitterness should be low, mild acidity is essential, no alcohol flavor. Esters are OK, but shouldn't predominate. Should be very drinkable.

*Phillip Seitz is the curator/historian/archivist/librarian of a small medical museum in Alexandria, Virginia, an all-grain brewer, Recognized BJCP Judge, and member of BURP (Brewers United for Real Potables). He has visited Belgium eight times since 1987 and can't wait to go back.*

# PERFECT YOUR PORTER

## BY TERRY FOSTER

*This article originally appeared in Zymurgy, Summer 1996 (vol. 19, no. 2)*

Porter probably is the most confusing beer style in the homebrewer's portfolio. The original London version certainly was the first definitive beer style, in that it was the first beer to be recognized by its drinkers as a definite type of beer, but we cannot be sure exactly what that was because we cannot taste the original. Commercial brewing of porter ceased entirely in London earlier this century. Keith Thomas has presented porter's dubious origins and pointed out that there were two different styles, the original and what he refers to as Victorian porter. In the last fifteen years or so the style has been revived in the United States and England mainly by homebrewers, microbreweries, and brewpubs, and some English regional brewers also have reintroduced porter.

One of England's largest modern breweries, Whitbread, started out as a London porter brewery in 1742. The company has seen fit to bring out a cask-conditioned draft porter in recent years that one might expect to be a definitive porter. Although the beer is based on one of their original recipes, it has been adjusted "to suit modern tastes," brewspeak for watered down. Further, the recipe they used is a nineteenth-century one, and in the Victorian rather than the original style.

The Association of Brewers addressed the confusion about the style's origins by distinguishing between two types of porter in its competition categories, "brown" and "robust." Some of the porters offered by the new brewers are idiosyncratic, to say the least, with at least one highly respected micro offering a porter that is clearly a stout. In addition, Yuenglings of Pennsylvania, in business since 1829, brews the one example of an American porter with a genuinely long pedigree but uses a bottom-fermenting yeast not authentic for the style. Sierra Nevada Porter is another widely available commercial example that Michael Jackson describes as firmly dry with a gentle toffee-coffee finish.

As Thomas points out, few of the new porters match the original, even in such basics as original gravity and alcohol content. In fact, the same could be said for modern versions of some other long-standing beer styles, notably pale ales and stouts. Drinkers in general, conscious of health and other issues, just do not want their beer to be as strong as it was in the eighteenth and nineteenth centuries.

Of course, it is impossible to match the original porter because we have never tasted it. Although recipes are available from the last century, nothing similar has been uncovered for porters of the eighteenth century. Some generalized accounts from the period give us a fairly broad grasp of the nature of the style, but brewers apparently kept a close guard on their recipes and did not commit them to paper.

Brewing techniques have changed and methods of storage and

dispensing are different. Even more important, there have been great changes in raw materials, particularly malt and hops. Perhaps modern knowledge of hop chemistry can permit us to approximate the hop flavors of eighteenth-century porters, but the same is certainly not true of malt. It is clear that the first porters were brewed from brown malt, but the nature of that malt is not clear. Some writers, including Graham Wheeler and Thomas, suggest it may have been brown on the outside but actually pale on the inside, resulting in a beer with a much paler color than we've assumed. Brown malt may have had a marked smoked character, which Wheeler believes carried through to the beer, while H. S. Corran thinks it probably was not noticeable in the finished product. Since I wrote *Porter* (Brewers Publications, 1992), English brown malt has become available in the United States but is no longer produced by the wood-fired methods used some 200 years ago, and cannot be regarded as a direct match for the original.

Nevertheless, there does seem to be some consensus about the modern styles of porter. Perhaps this is because stout, a derivative of porter, has continued to be brewed, and has developed into a style of its own, which limits porter at the top end of its flavor spectrum. Overall, porter should be a balanced beer with no outstanding single flavor characteristics. That does not mean it is in any way bland, or lacks complexity, or that individual flavor components cannot be detected by careful tasting. Rather, it means the complexities blend together well, complementing one another to give a smooth-drinking beer. Such a description certainly fits the brown porter designation. The robust variety, using black rather than chocolate malt, will have a more definite roast malt character, putting it closer to stout in style. In the early days of stout, brewers often called their beers "stout porter."

Following is a summary of what I would look for in a modern porter.

## MODERN PORTER CHARACTERISTICS

Original gravity 1.045 to 1.060 (11 to 15 °P).

Finishing gravity 1.010 to 1.015 (2.5 to 3.8 °P).

Translucent ruby red color, not black or opaque.

Slight roasted malt flavor, can be more assertive in robust style.

Some nuttiness from crystal and/or brown or chocolate malt.

Moderate but definite hop bitterness.

Full-bodied, an all-malt beer.

Fairly dry, but some residual sweetness from malt.

Some estery, fruity character—it must be top-fermented.

Hop aromatic character not essential, but certainly not out of place. Choose the classic English hops or their American versions like Fuggles or Willamette.

These are the characteristics describing most modern commercial examples of porter. Many brewers tend to use lower-temperature single-infusion mashes, resulting in a beer lower in final gravity and lacking residual dextrins. I have come around to the opinion that this results in too low a finishing gravity, and that porters benefit from finishing a little higher, at 1.015 to 1.020 (3.8 to 5 °P). This extra sweetness gives the beer more depth and puts it a little closer to the original.

## BREWING INGREDIENTS

The preceding characteristics mean porter is a style that lends itself nicely to brewing from either malt extract or all grain. There are porter kits and dark malt extracts on the market, but I prefer to use unhopped pale ale malt extract as the main source of fermentables and add roasted and crystal malts, as well as hops, for the color and main flavor notes. This approach simply offers the extract brewer better control over the end result. Some malt extracts can give somewhat low finishing gravities, so I generally add maltodextrin powder to my formulations. About ½ pound of maltodextrin in a 5-gallon batch (225 g. in 19 l.) will ensure sufficient residual malt sweetness to round off the beer.

For all-grain brewing, use pale malt (preferably English two-row) as the base, which is just what the London porter brewers did when roasted malts first became available in the 1820s. You should note that I have taken a yield of 1.032 for 1 pound of pale malt per U.S. gallon (454 g. per 3.8 l.) in the all-grain recipes that follow. This is not the highest you can expect to achieve, and the yield you get may be lower or higher depending on your brewing setup. You may want to adjust the amount of pale malt given according to your own home-brewery performance to obtain the target gravities noted.

Crystal malts were not part of the grist for the early versions of porter, but are well suited to the balanced complexity of the modern style. They are available to the homebrewer in a variety of colors, with the darker ones reflecting a higher level of caramelization. For that reason, and because of the reddish tinge they give to the beer, the darkest-colored crystal malts are generally preferred. An interesting range of Belgian crystal malts is now on the market. I have experimented only with biscuit malt in porter brewing and found it to be excellent. As the name implies, biscuit malt adds a nutty-biscuity flavor, nicely complementing the roasted malt.

The main malt flavorings come from brown, chocolate, and black malts. Brown malt is the least heavily roasted and is not the same as the brown malt originally used to produce porter. Therefore, if you were to brew only from brown malt, the flavor would be quite overpowering. It can be used in combination with other roasted malts, up to a maximum of 1½ pounds for 5 gallons (680 g. in 19 l.), giving an excellent nutty flavor and increasing the mouth feel. You can make brown malt easily at home. John Harrison and Robert Grossman each provide instructions (see References).

Chocolate malt is more roasted than brown, and is my favorite for both color and flavor in porter. Used in reasonable proportions, up to a maximum of ½ pound for 5 gallons (227 g. for 19 l.), it gives a definite but smooth, roasted coffeelike flavor, and that warm red hue that so delights the eye in a well-made porter.

Black malt is the most roasted of the three, meaning it is roasted longer and hotter than brown or chocolate malt, and pound for pound, black malt will contribute stronger roasted flavors. Its use really is what distinguishes a robust porter from a brown porter. It has a harsher flavor than chocolate malt and I find it can easily throw a porter out of balance, so should be used sparingly—not more than 4 to 6 ounces per 5 gallons (113 to 170 g. per 19 l.). However, it was the principal malt flavoring component of Victorian porters, and warrants attention in re-creating historical brews, as you will see in one of the following recipes.

German Rauch malt or a home-smoked malt provides a way to achieve the subtle smoked flavor that may have been characteristic of the original porters resulting from the brown malt, which reportedly was wood-smoked. For the less style-conscious or creative homebrewer, an adventurous alternative is Scottish peated malt. Peated malt should not be overdone. I suggest a maximum of 1 pound per 5 gallons (454 g. per 19 l.), and it requires mashing. It is useful for malt extract beers only if a partial mash is used. An alternative is to use liquid smoke if you want to add a hint of smoke flavor. I have not experimented with it, but Wheeler addresses this and malt smoking in his book, *Home Brewing, The CAMRA Guide* (CAMRA,

1993). The smoke character achieved from peat malt is really quite different from that produced from wood-smoked malt.

No beer worth the name can be discussed without mentioning hops. There are few restrictions on the varieties that may be used for brewing porter. For bittering, standard porters normally fall toward the lower end of the range, 25 to 45 IBU or 7 to 12 HBU in 5 gallons (19 l.). Almost any variety can be used for this purpose, including the high-alpha-acid types such as Chinook and Nugget, although the slightly less aggressive Perle and Northern Brewer are my preferences. However, both English Goldings and Fuggles have a better claim to authenticity in porter brewing, and the American Fuggle derivative, Willamette, is a good alternative. Roger Protz reports that Challenger hops are often used in U.K. porter examples. English and American Fuggles as well as Willamette will serve well as aroma hops, but should not be overdone in this style of beer. Probably no more than ½ ounce for a 5-gallon batch (14 g. for 19 l.) is enough when added at the end of the boil. One hop I recommend avoiding for aroma purposes in porter is Cascade, which I believe is a little overpowering and unbalancing for the style, but is often found in West Coast examples.

Finally, yeast plays an essential role in determining the flavor of the finished beer. What we want for porter is a good top-fermenting ale yeast. Almost any strain will do, but some strains might just bring your porter closer to perfection. Among the dry types, Whitbread is an excellent choice, although it probably is far removed from the original porter strain. My own choice would be one of the liquid strains sold by most suppliers, including British ale, London ale, and Irish ale, with the latter particularly suited to the robust style of porter. London ale yeast with its relatively low attenuation will tend to give a higher finishing gravity and slightly higher residual sweetness in the finished beer. As indicated earlier, this is a desirable characteristic in porter. You may want to experiment with other low-attenuating ale yeasts on the market.

You may have noticed that the recipes provided encompass almost all aspects of porter character: chocolate malt as the definitive roasted malt, brown malt as the sole source of roasted character, a black-malt-based "robust" porter, a genuine Victorian brew, a smoked version, and a highly hopped style. As far as a recipe for the "three threads" that reportedly inspired Ralph Harwood to brew the first porter, I cannot provide that because we do not really know what these beers were like.

I have never attempted to brew anything like that, but if we assume that the three threads were pale ale, mild brown ale, and stale brown ale, as some versions have it, I can suggest an approach. Brew an old pale ale at around 1.090 original gravity, 100 IBUs, and mature it for a year. Make a porter at an original gravity of 1.080, 70 IBUs, using a sour mash technique (after mash-in, leave the mash at around 122 degrees F or 50 degrees C for several hours), and mature the finished beer for at least three months. Finally, prepare a similar brew by normal infusion mashing, mature for one month, then mix with the other two brews in whatever proportions you find suitable. If you do give this a shot, I shall be very interested to hear the results. I am not convinced it would be worth the effort because, after all, porter is supposed to have originated to avoid such a complicated procedure!

**BROWN MALT'S ROLE IN BREWING PORTER**

From the beginning of porter brewing, brown malt was a defining ingredient of the style and once accounted for all of the grist. However, by the early 1800s, the more lightly roasted amber malt was a common ingredient. Through the 1800s, the use of brown and amber malt declined and eventually disappeared altogether early in this century. By 1900, pale and black malts (or roast barley) had become the defining ingredients of the porter style.

"To emulate the early porters, you can roast and/or smoke your own brown and amber malts. I recently made a porter that I believe closely emulates the eighteenth-century version:

*Ingredients for 4 gallons*:

 8½ lbs. brown malt, roasted/smoked in small batches over hardwood coals for 10 to 15 min. (3.86 kg.)
 6 lbs. pale malt (2.7 kg.)
 Progress hops for bittering (45 IBUs)

Original specific gravity: 1.068
Final specific gravity: 1.021
Mash at 153 degrees F (67 degrees C) for one hour.

—Ray Daniels, author of *Designing Great Beers* (Brewers Publications, 1996)

## BREWING METHODS

No special techniques are required for the malt extract recipes. Roasted and crystal grains should be crushed and, for a 5-gallon (19-l.) batch, mixed with 1 to 2 quarts (1 to 2 l.) of cold water, then brought to a boil, stirring occasionally. Do not actually boil the grain because you may extract some harsh flavors from the husk that will unbalance the beer. Strain off the liquid, add remaining water, dissolve the extract in the solution, add maltodextrin if required, and boil as indicated in the recipe.

For the all-grain recipes, crush all pale and roasted grains and mash with 1 quart of water per pound of malt (1 l. of water per 454 g. malt). Simple infusion mashing is all that is required, but you should aim for 90 minutes at relatively high temperatures, preferably 153 to 155 degrees F (67 to 68 degrees C) to ensure a high dextrin level in the wort. Sparge with 170- to 180-degree F (77- to 82-degree C) water and collect about 5½ gallons (21 l.), then boil. I do not have space here to address the complexities of water treatment, but the brewing liquor should have both permanent and temporary hardness. A soft water would require about 1 teaspoon (5 g.) each of gypsum and precipitated chalk.

Boil the wort vigorously for a full 1½ hours, adding the bittering hops 10 to 15 minutes after the start of the boil. Add Irish moss about 20 minutes before the end of the boil to ensure a good break. Turn off the heat and stir in any aroma hops. Use a wort chiller to cool the wort to fermentation temperature as rapidly as possible. If you are using a malt extract and can boil only part of the wort, cooling can be done by making up the required volume with cold water. However, for best hop utilization I recommend the total volume of liquid be boiled.

Fermentation should be carried out at 60 to 70 degrees F (16 to 21 degrees C) and no higher than 75 degrees F (24 degrees C). Primary fermentation, preferably in glass, should last 5 to 7 days. The beer should then be racked into the secondary fermenter and held there for another 7 days at temperatures similar to primary fermentation temperatures.

At the end of secondary fermentation the beer can be primed and bottled or kegged. The stronger versions will benefit from storing in a stainless steel keg for one to two months before kegging or bottling. You can use standard levels of priming (4 to 6 ounces or 113 to 170 g. of corn sugar for 5 gallons or 19 l.) to achieve the desired level of carbonation. I prefer lower carbonation rates, which result in a smoother, more satisfying porter flavor, so I prime with only 2 to 3 ounces (55 to 85 g.) of corn sugar per 5 gallons (19 l.). When force-carbonating in the keg, apply 10 psi of carbon dioxide over 2 to 3 hours. Then, after dispensing a few pints, reapply the gas at the same pressure for no more than a few minutes. This will result in a simulated English-style real ale with enough carbonation to dispense the beer with a good head but without excessive gassiness. Of course, if you want to chill the beer instead of drinking it at cellar temperature in the "correct" manner, you may want to carbonate to higher levels.

One of the features of original porter was storage in wooden vats for up to a year, during which time they developed a sour, acidic flavor. That is not necessary with most of the weaker modern porters. Yet for a beer that should be balanced, even the lower-gravity versions will benefit from conditioning for two to three months before drinking. Conditioning should be at cellar temperature, around 55 degrees F (13 degrees C). At warmer temperatures your porter will be ready sooner.

To avoid repetition, I will not give specific instructions for each recipe. Follow the preceding instructions according to whether the recipe indicates an extract or all grain. Recipes are for 5 U.S. gallons (19 l.) final batch size.

## **Not So Brown Porter (all-grain)**

This is a beer relying only on brown malt for color and flavor. The nutty-caramel nature of this malt is sufficient to carry the other flavor components into a balanced whole, and is all that is needed for a porter at the low end of the strength spectrum for this style. The result in terms of color is quite pale, perhaps paralleling the original version in this respect. This beer is dangerously drinkable while still young!

- 6 lbs. English two-row pale malt (2.7 kg.)
- ½ lb. wheat malt (227 g.)
- 1 lb. brown malt (454 g.)
- 2½ oz. English Fuggles hops, 3.4% alpha acid (71 g., 8.5 HBU, 32 IBU), for bittering
- London ale yeast

- Original specific gravity: 1.048 (12 °P)
- Final specific gravity: 1.014 (3.5 °P)

## Popeye Porter (malt extract)

A robust porter with the biscuit flavor of Belgian malt and a higher alcohol content to balance the bite of the black malt. This beer will benefit from a six-month or longer maturation.

- 7½ lbs. pale malt extract syrup (3.4 kg.)
- ¾ lb. Belgian biscuit malt (340 g.)
- 6 oz. black malt (170 g.)
- ½ lb. maltodextrin powder (227 g.)
- 1¼ oz. Northern Brewer hops, 7.5% alpha acid (35 g., 9.4 HBU, 35 IBU), for bittering
  Wyeast Irish ale No. 1084 liquid yeast

- Original specific gravity: 1.060 (15 °P)
- Final specific gravity: 1.018 (4.5 °P)


## Smoky the Beer (all grain)

A beer with a smoked flavor, but components contributed by the brown, chocolate, and crystal malts subdue the smoky element. This is true of the only two commercial examples of smoked porter that I know, from the Alaskan Brewing Company and the Vermont Pub and Brewery. If you are a smokehead, you may want to increase the proportion of peated Rauch or smoked malt. A slightly smoked character may well be truer to the original porter than most modern versions.

- 6½ lbs. English two-row pale malt (2.95 kg.)
- ½ lb. 40 °L pale crystal malt (227 g.)
- 6 oz. Scottish peated malt, German Rauch malt, or home-smoked malt (170 g.)
- ½ lb. brown malt (227 g.)
- ½ lb. chocolate malt (227 g.)
- 1½ oz. Perle hops, 7.0% alpha acid (43 g., 10.5 HBU, 40 IBU), for bittering
  Wyeast Special London ale No. 1968 liquid yeast

- Original specific gravity: 1.052 (13 °P)
- Final specific gravity: 1.015 (3.75 °P)

"Blackjack Porter (robust) is a fine, drinkable, well-balanced, ruby red beer. It is what I call a 'big beer,' with an original gravity of 1.064. We use the finest English malts (no black malts) and Kent Goldings hops. I remember the day we formulated this recipe. We bought every English and American porter we could find and took notes on body, hops, character, residual sweetness, and yeast—a masterpiece was born," according to Jim Martella, brewer, Left Hand Brewing Company, Longmont, Colorado. Blackjack Porter won a gold medal in the robust porter category at the 1995 Great American Beer Festival and a Bronze medal in 1994.

## London Porter (malt extract)

A malt-extract version of an 1850 Whitbread recipe, and an example of Victorian porter. It is almost black in color and has a high alcohol content to balance the high bitterness contributed by the generous proportion of black malt. The large amount of brown malt gives the beer enough body and mouth feel to make the addition of maltodextrin unnecessary. It becomes impressive with six months or more of maturation.

   7½ lbs. pale malt extract syrup (3.4 kg.)
   1¼ lbs. brown malt (567 g.)
   ½ lb. black malt (227 g.)
   2¼ oz. English Kent Goldings hops, 4.2% alpha acid (64 g., 9.5 HBU, 36 IBU), for bittering
   Whitbread ale yeast (what else?); try Wyeast British Ale No.1098 liquid yeast

- Original specific gravity: 1.063 (15.75 °P)
- Final specific gravity: 1.020 (5 °P)

## Chocolate Decadence (malt extract)

A straightforward brown porter, pleasant and satisfying.

   5 lbs. pale malt extract syrup (2.27 kg.)
   1 lb. 140 °L dark crystal malt (454 g.)
   ½ lb. chocolate malt (227 g.)
   ½ lb. maltodextrin powder (227 g.)

1¼ oz. Willamette hops, 5.4% alpha acid (35 g., 6.75 HBU, 25 IBU), for bittering
Wyeast Special London Ale No. 1968 liquid yeast

- Original specific gravity: 1.050 (12.5 °P)
- Final specific gravity: 1.015 (3.75 °P)

### **Kentish Porter (malt extract)**

This final recipe is my malt-extract version of a beer produced by a brewpub deep in the heart of England's Kent hop country. It breaks all the rules about hop rates for porter, but comes out balanced, especially when matured for three to four months or more.

7¼ lbs. pale malt extract syrup (3.3 kg.)
½ lb. 140 °L dark crystal malt (227 g.)
5 oz. chocolate malt (142 g.)
½ lb. maltodextrin powder (227 g.)
2 oz. English Kent Goldings hops, 7.8% alpha acid (57 g., 15.6 HBU, 59 IBU), for bittering
½ oz. (14 g.) Czech Saaz hops (75 min.)
1 oz. (28 g.) Czech Saaz hops (finish)
Whitbread ale yeast

- Original specific gravity: 1.057 (14.25 °P)
- Final specific gravity: 1.020 (5 °P)

## REFERENCES

Corran, H. S., *History of British Brewing*, David and Charles, 1975.
Foster, Terry, *Porter*, Classic Beer Style Series No. 5, Brewers Publications, 1992.
Grossman, Robert, "Home Grain Roasting," *Zymurgy*, Special Issue 1995 (vol. 18, no. 4).
Harrison, John, *An Introduction to Old British Beers and How to Make Them*, Durden Park Beer Circle, 1991.
Protz, Roger, *The Real Ale Drinker's Almanac*, Neil Wilson Publishing Ltd., 1993.
Thomas, Keith, "A Peek Into Porter's Past," *Zymurgy*, Summer 1996 (vol. 19, no. 2).
Wheeler, Graham, *Home Brewing, The Camra Guide*, CAMRA, 1993.

*Terry Foster has been homebrewing for more than thirty-five years, and studying and writing about brewing and breweries for more than twenty-five years. Terry is author of* Pale Ale *and* Porter (Brewers Publications, 1990, 1992).

# CONFESSIONS OF TWO BITTER MEN

## BY TONY BABINEC AND STEVE HAMBURG

*This article originally appeared in* Zymurgy, *Summer 1995 (vol. 18, no. 2)*

We're bitter.

All right, we confess. But can you really blame us? If your favorite style of beer were underappreciated, misunderstood, and almost completely unheard-of in your country, how would *you* feel? You see, our favorite style is cask-conditioned draft pale ale, known throughout Great Britain as "bitter." If you love beer and you've been lucky enough to spend time there—like us—you know that a pint of cask-conditioned bitter is one of the world's great drinking pleasures. Bitter is a wonderful social beverage, the perfect "session beer," packed with flavor yet low in gassiness and alcohol. Moreover, it is a testament to the wide range of flavors a brewer can evoke from the most basic ingredients of the craft.

Bitter is an inviting style for homebrewers of all stripes. Procedures are simple. Hop-heads are naturally attracted because, as the name implies, bitter really *is* hopped well beyond the taste threshold. And because of its moderate gravity, those hops don't have to fight their way through massive maltiness. Malt lovers also can find something to like, because the style is broadly defined enough to allow a wide range of malt expression.

So like many homebrewers, we returned from our trips to the Sceptered Isle raring to make and serve authentic bitter. After much research, we arrived at our own style definition and a consistent brewing process. Problem was, it wasn't long before we learned that our best, most on-target efforts performed poorly in competitions. This wasn't because of poor judging. On the contrary, judges did a great job evaluating our beers according to the then-current AHA guidelines. The problem was the guidelines themselves. By analyzing these more closely, it was easy to understand why even talented brewers and beer judges misunderstood the style. The most widely distributed standards did not accurately reflect current brewing practice in Britain, particularly when it came to original gravity and International Bitterness Unit ranges.

That was when we ceased being mild-mannered homebrewers. We became missionaries, pledged to train fellow brewers and judges in the proper characteristics of our favorite style. The following is what we discovered—they are the confessions of two Bitter Men!

## BRITISH OR ENGLISH?

Before we begin, let's be clear about our geography and language. The terms *British* and *Britain* encompass England, Scotland, and Wales, while *English* more narrowly applies to England. In our research we looked at beers from England and Wales, where bitter predominates. Although you can find bitter in Scotland, the Scots have their own distinctive brewing tradition, preferring maltier styles.

## IS IT BITTER OR PALE ALE?

Actually, it's both. The distinction between bitter and pale ale is quite fuzzy, and the styles overlap.

The term *bitter* is relatively modern in brewing parlance. It only shows up in the professional literature in the mid-twentieth century, in the Whitbread-published *The Brewer's Art* (1948). Here, for the first time, the distinction was made between bottled and draft pale ale. Clearly, bitter is the descendant of pale ale or, originally, India pale ale, Burton-upon-Trent's gift to the world. In Britain today, *bitter* remains the all-encompassing term for draft (draught) pale ale. A brewery might refer to its "premium" bitter as a pale ale or IPA, but when the beer is served in draft form, it's always bitter.

If pale ale and bitter are so inextricably linked, then why are they in separate competition categories? After all, if pale ale is simply bottled bitter, then there should be little or no difference between a bottled classic pale ale and bottled special bitter, right? We can only assume that bottled pale ale has taken on its own limited set of characteristics. Certainly there's higher carbonation in bottled pale ales, plus the color range (particularly for IPA) may be more narrowly defined and skewed toward lighter shades. But rather than worry about this, we prefer to relax and take comfort in the knowledge that there are more categories for more people to enter and win ribbons in.

# DEBUNKING MYTHS AND MISCONCEPTIONS

### Myth No. 1: *Bitter is served warm*

Bitter is not served warm, but at cellar temperature. It only seems warm if you're accustomed to icy cold lagers. A bitter certainly should not be iced down before serving because low temperatures will adversely affect the perceived flavor balance. In practice, cellar temperature is roughly in the range of 50 to 60 degrees F (10 to 16 degrees C). The Campaign for Real Ale (CAMRA) recommends maintaining 55 to 57 degrees F (13 to 14 degrees C) at all times. A pub that really cares about its beer will use cooling or heating equipment as necessary to protect it from temperature extremes. This is true even though the English climate is relatively cool and cellars are often well insulated.

### Myth No. 2: *Bitter is flat*

Bitter is not flat. It is simply not as fizzy or effervescent as most other styles. As conventionally prepared, most commercial and homebrewed beers in the United States have about 2.5 volumes of $CO_2$. In contrast, a proper draft bitter might have 1 volume. The lower carbonation is doubly beneficial: it allows the flavors to express themselves more and it makes for greater drinkability without the sensation of fullness, belching, and the like. Swirl a glass of fine bitter and you'll always see a nice profusion of bubbles.

### Myth No. 3: *Bitter lacks a head*

The simple generalization is this: Beers in northern England are served with a head; southern beers are not. The differences are because of slightly different brewing and fermentation methods, but the engineering of the beer engine also plays a role. In the North, a beer engine will have a "sparkler" attachment on the nozzle. Most also feature a long-reach "swan-neck" spout. Together they bring out the characteristic tight, creamy head. Northern brewers formulate their recipes expecting this form of dispense. Southern beers, in contrast, are almost always served from an engine with a standard spout and no sparkler. Thus, little or no head is not a fault. The beer still expresses its balance of flavors and aromas quite well. For the American traveler to London and environs, the latter will be the more common sight, smell, and taste. Gravity dispense reduces the distinction somewhat, but northern beers will still exhibit a more prominent collar of foam.

*Myth No. 4: Every English brewery has three bitters: an ordinary, a special, and an extra special bitter (ESB)*

Not so. Individual breweries have varying numbers of beers. Some are always available, some only seasonally. Although bitter is surely the dominant English style, brewers typically produce different versions as more suitably befit the occasion. Among these, some brews prove so popular that they, too, are served year-round. Fuller's ESB, deservedly renowned even here, actually originated as a strong winter specialty. It is definitely extra special, because few brewers serve anything comparable. Keep this in mind, because we'll return to this subject later.

*Myth No. 5: Bitter is light to pale in color*

In fact, bitter varies in color quite a lot. At the extreme light end you can find examples like Boddington's Bitter, which is essentially straw-colored. Bass Ale, at 10 °Lovibond, is a good reference point for amber color. On the dark end, bitter can be ruby-colored, reflecting a relatively high proportion of crystal malt, or perhaps a touch of highly roasted malt in the grain bill. Terry Foster suggests a target color range of 8 to 20 °Lovibond, which is closer to the mark than even the latest AHA guidelines, listing 8 to 14 °Lovibond as the range (Foster 1990). Michael Jackson describes the range as "straw to chestnut" (Jackson 1993). And no wonder: there are multiple commercial examples at both ends of the spectrum.

## BITTER AS THE ENGLISH BREW IT

We developed our own definition by studying how the English actually brew bitter. That's where the discrepancies in the old AHA guidelines really came to light.

Bitter roughly encompasses the starting gravity range of 1.034 to 1.046 (CAMRA). One end of this range admits Fuller's Chiswick Bitter; the high end includes Young's Special and Courage Directors Bitter. Beers with a gravity higher than 1.046 can be variously classified as strong bitter, pale ale, winter warmer, old ale, or the like, depending on various stylistic considerations. Remember that the boundaries are not hard and fast. A point or two outside either end of the range is not much to haggle about.

The term *bitter*, lacking any modifying prefix, is an "ordinary." The terms *best* or *special bitter* typically signify a higher-gravity "pre-

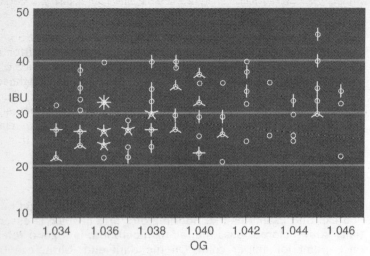

Figure 1: English Bitter (OG 1.034–1.046)
Original Gravity by International Bitterness Units

mium" bitter in the brewer's line. Note that these are ideal types. You'll find examples of commercial beers called "best bitter" that have an original gravity of less than 1.040. Similarly, you'll find so-called IPAs with gravities in the high 1.030s or low 1.040s. Brewers will use whatever name strikes their fancy.

What about extra special bitter? We contend that the AHA-style Extra Special Bitter—at least through 1994—has always been defined by one beer: Fuller's ESB. Few, if any, English brewers make a beer called an extra special bitter. In fact, they are so rare that Roger Protz's *The Real Ale Drinker's Almanac* includes only three ESBs among his 500 or so beer descriptions: Fuller's at 1.053, Mitchell's of Lancaster at 1.050, and Pitfield at 1.044 (Protz 1993). CAMRA's *Good Beer Guide* 1994 adds but two more: Big Lamp at 1.046 and Marston Moor at 1.050. There are, however, a fair number of original gravity 1.050+ English ales that don't use the ESB moniker. These can best be described as strong bitters. CAMRA calls them exactly that, even providing a separate category for them in the Great British Beer Festival. We should do likewise in our own competitions, defining the category with a gravity range of 1.046 to 1.060, IBUs between 30 and 50.

Anyone seriously interested in bitter and other British styles should have a copy of Protz's *The Real Ale Drinker's Almanac*. Through the years he has obtained detailed information from commercial

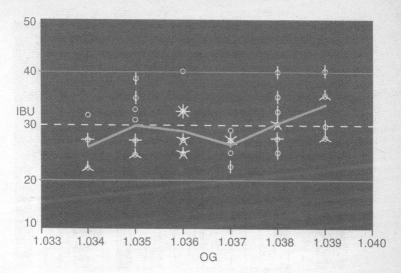

Figure 2: English Ordinary Bitter (OG 1.034–1.039)
Original Gravity by International Bitterness Units

brewers, including ingredients, original gravity, alcohol by volume, color rating (using the European scale), and bittering units (in IBUs).

We created a data base of every English and Welsh beer in Protz's book for which there was information on both original gravity and IBUs. While this information was missing for many entries, our sense is that these beers are not qualitatively different from the ones where data are available. We excluded all milds, porters, stouts, and old ales, then applied an original-gravity-based selection filter to the remaining sample.

The result was 129 commercial bitters in the original gravity range 1.034 to 1.046, with bitterness ranging from 19 to 45 IBUs. Figure 1 shows a sunflower scatterplot of IBU vs. original gravity for these beers.

A problem with standard scatterplots is that they do not show multiple hits in the same space. The "sunflower" plot rectifies this problem, because the "petals" count the number of beers in the vicinity of each sunflower center. This plot shows a number of things:

- The average bitterness of the beers is quite high, especially given the gravities.
- At any given starting gravity, there is a range of bitterness in beers at that gravity.

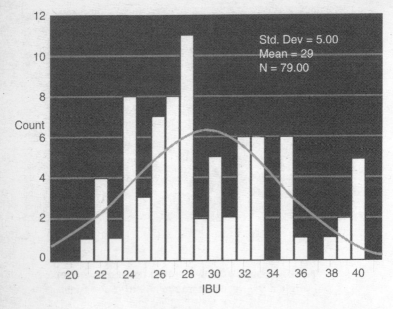

Figure 3: English Ordinary Bitter (OG 1.034–1.039)
Distribution by International Bitterness Units

- Some gravities have a lot more beers than some other gravities.
- While a higher-gravity category (1.045) has on average some highly bitter beers, there is no simple linear trending up of bitterness as gravity increases.

Figure 2 shows a sunflower scatterplot of IBU vs. original gravity for ordinary bitter by our definition.

The dotted reference line at 30 IBUs shows how the beers fall around this level of bitterness. The angular line is a lowess (locally weighted estimate) fit line that shows how bitterness trends slightly upward as gravity increases. The histogram in Figure 3 shows the number of beers at each IBU level. Note that the mean (average) is 29. The standard deviation of 5 indicates that nearly 70 percent of all beers will fall in the range of 29 plus or minus 5 (24 to 34). Compare this with the best bitter data displayed in Figures 4 and 5.

For whatever reason, the original gravity 1.045 beers are noticeably more bitter than the others. Yet in general, the other best bitters are really no more bitter than ordinary bitters in the high

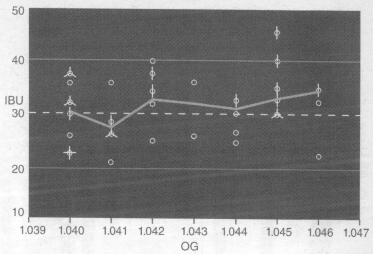

Figure 4: English Best Bitter (OG 1.040–1.046)
Original Gravity by International Bitterness Units

1.030s. Note the distribution in Figure 5. The average IBU level is 32, compared to 29 for ordinaries.

Compare these data with the old AHA guidelines. Ordinary bitter had a gravity range of 1.035 to 1.038, 20 to 25 IBUs; special bitter was 1.038 to 1.042 original gravity, 25 to 30 IBUs. Of the beers in our sample, sixty were within the old ordinary gravity range, fifty-six in the special. Of these, more than three fourths of the ordinaries (forty-six) and almost half (twenty-seven) of the specials would have been considered too bitter. As we have seen, the *average* bitterness of our sample beers was greater than the *maximum* IBU level in both subcategories. By following these guidelines, U.S. homebrewers were routinely making beers that weren't as bitter as the average commercial versions in England.

Partly as a result of this research and our subsequent presentation at the AHA 1994 National Homebrewers Conference in Denver, Colorado, the official guidelines were modified for 1995. They are now 1.033 to 1.038 original gravity, 20 to 35 IBU for ordinary; 1.038 to 1.045 original gravity, 28 to 46 IBU for special. Using the same data but the updated standards, now sixty-five of the seventy (93 percent) beers in the ordinary original gravity range meet the new IBU criteria (the five that don't are slightly higher). In the special original gravity range, more than 75 percent (fifty-six of seventy-four) are within the new IBU specifications, and none

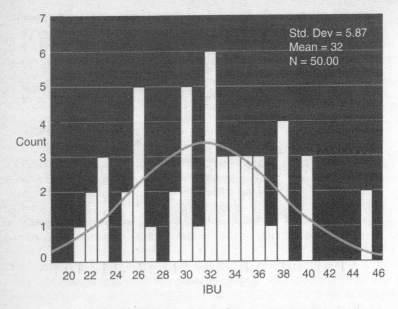

Figure 5: English Best Bitter (OG 1.040–1.046)
Distribution by International Bitterness Units

exceeds them. This is a major improvement that will help us all brew and judge bitter more accurately.

Interestingly, almost one fourth of the specials now fall below the new AHA IBU minimum. This would exclude such well-known English brands as Draught Bass, Badger Best Bitter, and Everard's Tiger (all 26 IBU). That's why we think a more relaxed 20 to 40 IBU range for both ordinaries and specials makes more sense. This would include 100 percent of the sample beers in the ordinary original gravity range, 97 percent (seventy-two of seventy-four) in the special. But in general, this is a minor point. We're thrilled with the adjustments that have been made. We may be *bitter*, but in light of these changes, it's in a much more positive sense.

## PROPERLY EVALUATING BITTER

With most any beer style you judge, it's nice to have a reference point. The problem with cask-conditioned bitter is that it is nearly impossible to find here. The commercial examples available from the United Kingdom are almost always premium bottled pale ales, bigger than your average bitter. Given this, and the old guidelines, it's not at all surprising that a spot-on ordinary bitter might be judged "too thin," "too bitter," or "poorly balanced." But now that we know better, what are the signature characteristics we should be looking for?

## AROMA

A fruity aroma is quite typical. With many traditional English yeast strains, these can even be quite citrusy. One thing is clear: authentic bitter tends to have much more of a fruity yeast profile than we're used to in the United States. If you are accustomed to the neutral cleanliness you get with Wyeast No. 1056, be prepared for a bit of a shock. Diacetyl may be present, but only slightly. A buttery note is a characteristic of some Yorkshire examples, particularly Samuel Smith, but in most bitter, especially in the 1.030 to 1.040 range, there's not a lot else in the beer to counteract diacetyl's effect. Hop aroma is always pleasant, but not required. Rich malt aromas are rarer, but not unheard-of. Finally, whiffs of grain can be common, even at the lowest gravities.

## APPEARANCE

The notion of clarity often seems foreign for a traditional ale. Yet as anyone who has had a proper pint can tell you, cask-conditioned bitter pulls bright and clear from the cask. And this is nothing new. Isinglass finings have been used to clear beer for centuries. (The oldest mention of this method was published in 1695!) When you evaluate a bitter, you can be tolerant of a slight haze, but it definitely shouldn't look like a Wit or Hefeweizen. As we've said, color can range widely, from straw gold to copper. And given its low conditioning, it's not a flaw if it doesn't have much of a head.

## FLAVOR AND BALANCE

Remember that bitter is dry, with low to medium maltiness. The watchword is the style's name: bitter. Yet until now, most American homebrewed versions have not been bitter enough. Especially in the lower gravity range, hop bitterness should be the dominant characteristic. Most ordinaries actually are more bitter than their stronger counterparts, with some examples more bitter than most pale ales. To quote Michael Jackson, " . . . the essential ingredient is the hearty smack of hops" (Jackson 1993). No proper bitter recipe ever skimps on the kettle hops. That is why, when evaluating the balance, you shouldn't base this on an equal perception of malt and hops. Within the target gravity range, beers at the high end will show more maltiness. But the expectation for balance should not be the same for an original gravity 1.035 ordinary and a 1.046 special. The differences between an ordinary and a 1.055 ESB will be even more pronounced.

Fight the temptation to judge every bitter as if it were Fuller's ESB. An ordinary or special just cannot compete with a beer of its complexity and "bigness." Other beers like Bateman's XXXB and Victory Ale, and Young's Special London Ale, also are too big to be proper models. All is not lost, however. Fuller's London Pride is an excellent benchmark when fresh. It is astonishingly flavorful for a 1.040 beer. Young's Ramrod also is decent, although the version we get is somewhat maltier than the brewery's dry-hopped cask-conditioned special. The best example of the current crop of imports may be Brakspear's Special Bitter (original gravity 1.043). It is a close cousin to Brakspear's Ordinary (original gravity 1.035, 38 IBU), which has, in our minds, the definitive dryness and massive hop character that turned us into *Bitter Men* in the first place.

## NOW BREW IT!

Now that you know more about the style, you should get out and start brewing it! We could easily do a separate article of equal length covering raw ingredients and brewing techniques, but we won't do it here. Besides the space limitations, why should we repeat what others have already said so well? Instead, we recommend you start with Terry Foster's *Pale Ale* (Brewers Publications, 1990) from the Classic Beer Style Series. There is little he doesn't cover, whether you're an extract or all-grain brewer. For more specific information from the British perspective, two CAMRA books—Graham Wheeler's *Home Brewing—The CAMRA Guide*, and Graham Wheeler and Roger Protz's *Brew Your Own Real Ale at Home*—are invaluable. The latter contains recipes that allow you to replicate many famous commercial cask-conditioned ales. Almost seventy bitter and pale ale recipes are included.

There has never been a better time to brew bitter. The highest-quality, most authentic raw materials can be purchased from homebrew suppliers all over the country. Maris Otter, Hugh Baird, and DeWolf-Cosyns pale ale and crystal malts are widely available. Traditional hop varieties like East Kent Goldings, Fuggles, Northern Brewer, and Styrian Goldings (actually a Fuggles variant) can be procured most anywhere. Perhaps best, there are now more suppliers of pure culture liquid yeast offering a growing variety of British commercial strains. If you experiment with them as much as we have, you'll discover striking differences in the flavor and perceived bitterness of your brews.

Obviously we hope you all will have the chance to go to England and savor bitter at the source. But until you do, a homebrewed version can come pretty close to the real stuff. Some say that bitter isn't for everybody. We think that is insane! In fact, we can't think of a better day-in-day-out drink. Sure, the lower alcohol means you can keep your wits about you, but you're not sacrificing anything in flavor profile. In a time when hop-head beers like Sierra Nevada Pale Ale and Liberty Ale are more popular than ever, it's easy to see that there should be a little bitter in everyone's future. The world could be better if everyone were a little more bitter.

## WATER TREATMENT: "BURTONIZING"

Much is said about the mineral content of water in Burton-upon-Trent and the desirability of some sulfates in your water to complement the hop bitterness. Here are levels of important minerals in Burton water, in parts per million:

| | | | |
|---|---|---|---|
| Calcium | 268 ppm | Chloride | 36 ppm |
| Magnesium | 62 ppm | Sulfate | 638 ppm |
| Sodium | 54 ppm | Alkalinity ($CO_3$) | 200 ppm |

Compare an analysis of your local brewing water to the description above. If you have low levels of minerals in your water, there are two things you should consider doing. First, take steps to rid your water of chlorine. You can either preboil the water to drive off the chlorine, or use a filter on your water tap. Second, "Burtonize" your water by adding some mineral salts. For a 5-gallon batch, use additions on the order of the following:

2–3 tsp. gypsum (calcium sulfate) (10–15 ml.)

½ tsp. Epsom salt (magnesium sulfate) (2.5 ml.)

¼ tsp. noniodized table salt (sodium chloride) (1.2 ml.)

Here are a few bitter recipes that come out mighty tasty:

### Ordinary Bitter

*Ingredients for 5 gallons (19 l.)*
- 5½ lbs. pale ale malt (Hugh Baird, Maris Otter, or DeWolf-Cosyns recommended) (2.5 kg.)
- ½ lb. 60 °L Maris Otter crystal malt (0.2 kg.)
- ½ lb. corn or cane sugar (0.2 kg.)
- 1 oz. DeWolf-Cosyns black malt (28 g.)
- 1 oz. Northern Brewer hops, 7% alpha acid (28 g.) (60 min.)
- ½ oz. East Kent Goldings hops, 5.2% alpha acid (14 g.) (15 min.)
- ½ oz. Styrian Goldings hops, 5% alpha acid (14 g.) (5 min.)
- Optional: dry-hop with ½ to 1 oz. of Kent Goldings or Styrian Goldings (14 to 28 g.)
- Yeast Lab YLA01 liquid Australian ale culture or a well-attenuating strain with good fruity notes. Pitch a good 1-quart (0.9-l.) starter.

- Original specific gravity: about 1.036
- IBUs: 33 to 34

Single infusion mash 90 minutes at 150 to 151 degrees F (66 degrees C). Raise to 168 degrees F (76 degrees C) for mash-out. Sparge with 170- to 175-degree F (77- to 79-degree C) water. Boil 90 minutes. Burtonize your water.

Ferment at 65 to 68 degrees F (18 to 20 degrees C) at least 7 days. Rack with priming sugar (1.5 to 2 oz. cane sugar as syrup) and optional finings to cask or keg.

Extract brewers can substitute 3 to 3.5 pounds (1.4 to 1.6 kg.) dry malt extract for the pale ale malt and steep the specialty grains.

### Flossmoor Best Bitter

Ingredients for 5 gallons (19 l.)
  7¼ lbs. Hugh Baird pale ale malt (3.3 kg.)
  ½ lb. 60 °L crystal malt (0.2 kg.)
  ¼ lb. flaked wheat (0.1 kg.)
  1⅓ oz. Northern Brewer hops, 7.1% alpha acid (38 g.) (60 min.)
  ½ oz. Styrian Goldings hop plug, 5% alpha acid (14 g.) (dry-hopped in keg) ale yeast (Wyeast No. 1968 and 1028, Brewer's Resource CL-160, Brewer's Resource CL-130, and Yeast Culture Kit Company NCYC 1187 are good choices)

- Original specific gravity: 1.044

Burtonize your brewing water. Mash grains at 150 degrees F (66 degrees C) for 90 minutes. Mash out at 170 degrees F (77 degrees C). Sparge and collect wort. Boil for 90 minutes.

### Bitter

Ingredients for 5 gallons (19 l.)
  5½ lbs. DeWolf-Cosyns pale ale malt (2.5 kg.)
  ¾ lb. 72 °L CaraMunich malt (0.3 kg.)
  1 lb. flaked maize (0.5 kg.)
  2 oz. DeWolf-Cosyns Special "B" malt (57 g.)
  ⅔ oz. Northern Brewers hops, 7% alpha acid (19 g.)
  ⅓ oz. Fuggles hops, 4% alpha acid (9 g.) (60 min.)
  ½ oz. Kent Goldings hops, 5% alpha acid (14 g.) (10 min.)
  ½ oz. Kent Goldings hops, 5% alpha acid (14 g.) (2 min.)
  ale yeast (See Flossmore Best Bitter examples)

- Original specific gravity: 1.039

Burtonize your brewing water. Mash grains at 152 degrees F (67 degrees C) for 90 minutes. Mash out at 170 degrees F (77 degrees C). Sparge and collect wort. Boil for 90 minutes.

## BITTER IN NORTH AMERICA

A short list of Two Bitter Men's favorite commercial examples in North America:

Goose Island Brewing, Chicago, Illinois: ordinary and best bitter, IPA, ESB, Yorkshire Bitter, and standard pale ale

Great Lakes Brewing, Cleveland, Ohio: Moondog Ale

Sherlock's Home, Minnetonka, Minnesota: Bishop's Bitter

Commonwealth Brewing Company, Boston, Massachusetts: Best Burton Bitter

## REFERENCES

Foster, Terry, *Pale Ale*, Classic Beer Style Series No. 1, Brewers Publications, 1990.
Jackson, Michael, *Beer Companion*, Running Press, 1993.
Protz, Roger, *The Real Ale Drinker's Almanac*, third edition, Neil Wilson Publishing Ltd./CAMRA, 1993.
Wheeler, Graham, and Roger Protz, *Brew Your Own Real Ale at Home*, Campaign for Real Ale, 1993.
Wheeler, Graham, *Home Brewing, The* CAMRA *Guide*, Campaign for Real Ale, 1993.
*Cellarmanship: Caring for Real Ale*, Campaign for Real Ale, 1992.
*Good Beer Guide 1994*, Campaign for Real Ale, 1994.

*Tony Babinec is an award-winning homebrewer, Certified BJCP Judge, and board member of the Chicago Beer Society.*

*A native of Washington, D.C., Steve Hamburg is a National BJCP Judge, homebrewer, and Chicago Beer Society board member. When not brewing, judging, drinking, traveling for beer, or watching sports, he occasionally works as a software quality consultant.*

*Although Tony and Steve occasionally brew separately, they usually collaborate on their shared half barrel.*

# three

# Equipment and Gadgets

## HOLY HYDROMETER, BATMAN!
### A For the Beginner column
#### BY DAVID A. WEISBERG

*This article originally appeared in* Zymurgy, Spring 1995 (vol. 18, no. 1)

If you are like I was, you probably didn't (or won't) use the hydrometer your brewing equipment kit came with until maybe your sixteenth batch. Either you couldn't be bothered or you were too excited (or nervous) about what ingredient to toss in next. When do I throw the yeast in? Is it ready to bottle yet?

After a while I learned how easy a hydrometer is to use and how important it is for brewing. I really could brew better beer with this weird thermometerlike instrument.

A hydrometer measures the density (thickness) of liquids compared to the density of water. Adding solids that dissolve (such as sugar) to water causes the specific gravity to rise from 1.000. It can help you know when to bottle, and will help you avoid those quiet time-bomb gushers that have an affinity for spraying the faces of any visiting relatives wanting a taste of fresh homebrew.

## THE BASICS

**1.** The scale most homebrewers use is called *specific gravity*. Most hydrometers measure specific gravity from 0.900 to 1.200 at a temperature of 60 degrees F (16 degrees C).

**2.** The Balling scale (measured in degrees Plato) may be found on your hydrometer and is used mostly by microbrewers and megabrewers. If you find a recipe showing gravities in degrees Plato, simply multiply this number by four, divide by 1,000, and add one to get the approximate specific gravity. For example, 12 °Plato × 4 = (48 ÷ 1,000) + 1 = 1.048 specific gravity. At high gravities, above 1.090, the conversion will be off by 1 °Plato.

**3.** Fact: The specific gravity of water is 1.000 at 60 degrees F (16 degrees C). The more dense a liquid is, the higher the hydrometer floats and the higher the specific gravity reading you will get.

**4.** To be cool, just say "ten-forty" (1.040) or "ten-fifty" (1.050), kinda like CB talk—no need to say "one point zero four zero," unless, of course, you are in the military.

## HOW DO YOU WORK THIS THING?

You have just boiled and cooled a delicious batch of porter. The wort (unfermented beer, pronounced "wert") has been transferred into the primary fermenter. (If you added concentrated wort to cool water in the primary fermenter, mix well before taking a hydrometer sample.) Now is that magical time for pitching the yeast (throwing it in). But wait! Before you put those voracious little yeast beasties to work, take two minutes for a quick hydrometer reading.

First, clean and sanitize a brand-new turkey baster. Squeeze the ball of the turkey baster, lower it into the wort (a couple of inches below the surface), open your grip on the ball, withdraw the baster, and slowly squeeze the wort sample into the tall, narrow plastic sample tube. You may have to do this two or three times to fill the tube about ¾ full. Now slowly lower the hydrometer into the sample tube while giving the tip a quick twist with your thumb and index finger. This spinning action prevents clinging bubbles from altering the gravity reading by causing the hydrometer to float higher than it should.

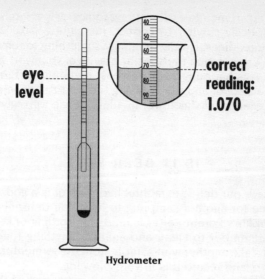

**Hydrometer**

Make sure the sample tube is resting on a level surface. The hydrometer should float centered in the tube. Because of surface tension, liquid surfaces tend to form a concave shape, called a meniscus; you have to sight the reading at the lowest point of the "bowl" or concavity to get an accurate reading (see Figure 1).

Read the scale marked "specific gravity" and write this number down in your brew log or recipe book. This number is called your *original gravity* (sometimes called *starting gravity*). You will probably see it abbreviated as OG. (It's also cool to say, "Hey, what's the OG on this batch?")

**Important point:** Do not return the sample to your wort because you might introduce infection. Either toss it out or drink it! (Ah, a sweet, cloying, syrupy taste of malt with just a hint of hops. I don't usually like the taste, but I drink it anyway.)

**Adjustment note no. 1:** Take a temperature reading of the sample. If your sample is 10 degrees F (5.5 degrees C) higher or lower than 60 degrees F (16 degrees C), you will need to adjust the gravity reading number up or down. Follow Table 1 for making gravity adjustments.

**Adjustment note no. 2:** This adjustment needs to be done only once. Place your hydrometer in the sample tube filled with water at exactly 60 degrees F (16 degrees C) or whatever temperature your particular instrument is rated for. You should get a reading of 1.000. If, for example, you get a reading of 1.002 (as with my hydrometer), remember all your readings will be off by 0.002 degrees specific gravity. To correct, just subtract 0.002 from all your future readings. No big deal. Just something to remember.

During fermentation the specific gravity of your wort will slowly decrease. This happens because yeast metabolizes fermentable sugars into alcohol and $CO_2$, causing the wort to become thinner. The hydrometer will sink lower in a less dense wort, giving you a lower reading.

## IS IT BEER YET?

This is where our delicate, faithful hydrometer is a godsend!

It is time for another sampling to help you determine if your Peter Pumpkin's Potent Porter is ready for bottling or kegging.

First, remember to clean and sanitize anything touching the wort. Next, take another wort sample from the fermenter, read the hydrometer, and record this in your brew log.

Refer to the recipe used to brew your creation. If the recipe called for an original gravity of 1.050 to 1.055 and a final gravity (FG), also called terminal gravity (TG), of 1.012 to 1.014, you can use this information to guide your next steps.

For example, if your original gravity was 1.054 and final gravity is 1.018, you can suspect your brew has not finished fermenting yet, so simply let it ferment for several more days, checking the gravity daily.

**Rule of thumb:** Once the final gravity has remained the same for two or three days, it is time to bottle.

## CALCULATING ALCOHOL LEVEL

To calculate the alcohol percent by weight, take the original gravity (for example, 1.045) and subtract the final gravity (let's say 1.010): 1.045 − 1.010 = 0.035. Then multiply by 105 to get 3.6 percent alcohol by weight. To calculate the alcohol percent by volume, multiply the alcohol by weight by 1.25. For example, 3.6 percent × 1.25 = 4.5 percent alcohol by volume.

## A SIMPLER SCALE

One neglected scale is the potential alcohol (percent by volume) scale that may also be on your hydrometer. My hydrometer measures from 0 to 16 percent alcohol by volume. Simply take a reading from this scale as you did before, prior to pitching the yeast. Let's say the number is 8 percent. When fermentation is complete take another reading. Maybe it is now 3 percent. These numbers represent the potential for alcohol in your batch at those two distinct times.

The difference between the two readings, in this case 5 percent, is the percent alcohol by volume in your finished brew.

One disadvantage when using this scale is that it can be hard to read the level when it falls between the numbers. (My hydrometer only shows whole-number gradation marks.)

## GLASS FERMENTER SAMPLING

If you ferment in glass carboys, you can modify your turkey baster for sampling purposes. Just insert one end of a long racking tube (or cane) about 1 inch into a 3- to 4-inch piece of plastic tubing (⅜-inch inside diameter). On the other end of the plastic tubing insert the turkey baster tip about 1 inch. Now you can reach down into the wort to sneak a sample.

**Note:** You may want to attach an adjustable metal tubing clamp over the joint where the plastic tubing overlaps the turkey baster tip. This way your baster does not lose drawing power due to leaking air. Leave the clamp on if possible, except for cleaning.

## IS IT OK NOT TO USE ONE?

Sure. Just listen to your beer à la Papazian. By listening to the sound and watching the air-lock activity, you can gauge what is happening without using a hydrometer.

At first the activity of the fermentation is slow. The calm before the storm. (This time frame varies depending on numerous factors including amount of yeast pitched, aeration, and fermentation temperature.) Twelve to 24 hours after pitching, the yeast fermentation really gets going. After 2 to 10 days (depending on yeast strain, fermentation temperature, and original gravity), things settle down. If you plan to lager, or don't think you will get around to bottling in the next week or two, transfer the beer to the secondary. (Note that during transfer to the secondary, a lot of $CO_2$ can come out of solution. It could take a day or two for the $CO_2$ to get resaturated in the almost-finished beer, and you may not see air-lock activity during this period.) When the $CO_2$ bubbling in the air lock has slowed to about one bubble every 2 minutes or so, you can now bottle. (Note also, if you are fermenting in cooler temperatures, you should probably wait until the airlock activity is down to one bubble every 4 or 5 minutes.)

Problems may arise, however, when you begin experimenting. You may try a different malt extract manufacturer. Maybe you want to try using liquid yeast. When experimenting with different brewing variables, you introduce the possibility of getting different results. For example, if you lower the fermentation temperature suddenly, you can shock the yeast out of suspension. This could result in a slow or stuck fermentation (wherein the fermentation has stopped before it is done). If you bottle this batch thinking it's done fermenting, you may end up with a very sweet, flat brew. Or you may get a wild gusher and never have a chance to taste it. (Overcarbonated beer tends to gush out, leaving not a drop in the bottle.)

At this point the benefits of using a hydrometer seem well worth the little time and effort needed to use it.

"*Hey, what's the OG on that batch?*"

*David A. Weisberg, homebrewer since 1989, is cofounder of the New Hampshire Biernuts homebrew club. During the day he is a promotions and research manager for two technical trade publications in Peterborough, New Hampshire. At night he's an intense homebrewer-author in search of that perfectly brewed batch. David is author of* 50 Great Homebrewing Tips *(Lampman Brewing Publications, 1994).*

## HYDROMETER CORRECTION TABLE

Noonan, Greg, from *Brewing Lager Beer*, Brewers Publications, 1986. For Temperatures Other than 60 degrees F

| If Temperature is: degrees F | Add to Hydrometer Reading: degrees S.G. |
|---|---|
| 32 | −0.0008 |
| 35 | −0.0009 |
| 40 | −0.0009 |
| 50 | −0.0007 |
| 70 | 0.001 |
| 80 | 0.002 |

| If Temperature is: degrees F | Add to Hydrometer Reading: degrees SG |
|---|---|
| 90 | 0.004 |
| 100 | 0.006 |
| 110 | 0.008 |
| 120 | 0.010 |
| 130 | 0.013 |
| 140 | 0.016 |
| 150 | 0.018 |
| 160 | 0.022 |
| 170 | 0.025 |
| 190 | 0.033 |
| 212 | 0.040 |

# WORT CHILLERS: THREE STYLES TO IMPROVE YOUR BREW

### BY MINDY AND ROSS GOERES

*This article originally appeared in* Zymurgy, Spring 1992 (vol. 15, no. 1)

At a recent meeting of the Dayton Regional Amateur Fermentation Technologists (DRAFT), one member demonstrated his whizbang high-performance wort chiller along with designs to produce an even more efficient model. During the demonstration the query arose from the novice brewers assembled: "So what's a wort chiller and why do I need one?"

This is a question that merits considerable attention. We need to step back and fill in the blanks, so to speak, or risk leaving newcomers and novice brewers in the dark and possibly discouraged by the apparent complexity of the craft. Remembering that it wasn't so long ago that we were novice brewers, we'll concentrate here on wort chillers.

A cursory literature review is in order. Let's start with the all-time great question: "Why?" John Alexander in *Brewing Lager* states: "Rapid cooling of the wort is deemed essential for the following reasons: to achieve the 'cold break,' to aerate worts, to rapidly lower the temperature so that yeast can be pitched and to reduce the chances of bacteria attacking the brew." Virtually any useful book on homebrewing says this in one way or another. Books suggesting the indiscriminate use of cane sugar, open fermenters, boiling yeast, or hydrometer gymnastics are not considered useful.

One aspect of the cold break is to force proteins to coagulate and drop out of solution to prevent chill haze from forming in the lighter-colored brews. The effects are largely aesthetic in nature and only count if you're grossed out by hazy beer or stand to lose a few appearance points in competition. Stouts and porters do not suffer from this effect because opaqueness hides the haze and the robust flavor subdues the more subtle off flavors.

The next question usually is, "Do I have to chill my wort quickly?" The best answer to this is: "It depends." Most folks start with malt extracts to facilitate the "incremental success approach" to brewing; i.e., start simple, see how you do, then get fancier if you like the beer and wish to fine-tune your recipe or methods. If not using a full-wort boil, putting the wort into the primary fermenter with a couple of gallons of cool water then into a sink of ice water probably will get the chilling job done in 15 to 30 minutes.

However, a full-wort boil, as well as light-bodied and light-colored styles, are more sensitive to off flavors. In his excellent book *Principles of Brewing Science* (Brewers Publications, 1989), George Fix explains how critically sensitive whatever may be suspended in the wort becomes to staling and oxidation. Dave Miller notes in *Continental Pilsener* (Brewers Publications, 1990): "Several pieces of equipment are particularly important in the brewing of Pilsener beer. One is a whirlpool or some other means of separating the trub (suspended proteins) from the boiled wort. Another is a wort chiller, which gives a rapid drop to pitching temperature. The clean taste of this beer style depends on a good cold break and trub removal." Thus the cold break and not shocking the yeast by adding it to hot wort are very important to lagers and the lighter styles.

Now you're ready to consider adding a wort chiller to your brewing process. Here's some information on what can be purchased and constructed.

Three wort chiller designs currently are popular. The two immersion types are seen more often than the counterflow type. They come in two varieties, depending on how you use them: (1) a copper coil with cold water running through it, to be immersed in the hot wort, and (2) a copper coil with the boiling wort flowing through it, to be immersed in ice water.

The third type is a counterflow wort chiller—cold water flowing through a jacket that surrounds the copper tube carrying the wort. The water and wort flow in opposite direction, hence the name *counterflow*, to provide the greatest temperature differential.

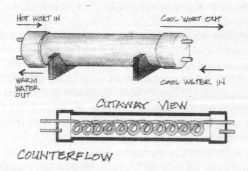

CUTAWAY VIEW

COUNTERFLOW

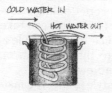

IMMERSION IN WORT

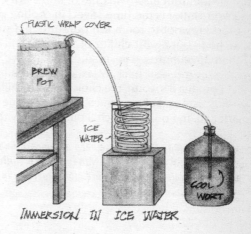

IMMERSION IN ICE WATER

Each type has unique advantages and drawbacks. The two immersion types are far easier to make, but can take longer to cool because of the formation of thermal layers. However, with immersion types constructed in a coil resembling a large spring, gentle compression of the spring reduces the layering effect and facilitates cooling.

The counterflow and immersion-in-ice-water types generally are more efficient, but need a good organic cleaner to prevent contamination of your beer. Because you can't see the inside of the tubing that will be in contact with your beer, cleaners such as trisodium phosphate (TSP) or clean-in-place (CIP, used by dairy farmers) are required to clean the inside of the copper tube.

By comparison, the immersion-in-wort type can be sterilized by placing it in the brewpot several minutes before the end of the boil. The inside doesn't have to be scrupulously clean because only cooling water flows through. (Note: Substitute stainless steel for copper if it is available and you know how to work with it. Don't even *think* about using aluminum!)

Aside from improving flavor and clarity, the advantage of using a wort chiller is the time saved in cooling your wort. Cooling a covered brewpot to room temperature in a sink of ice water can take an hour, but a wort chiller of any type will reduce the time to less than 15 minutes. The result: You'll have more time to brew or, even better—sit back, relax, and have a homebrew!

For what it's worth, the choice of wort chilling style is largely a matter of taste and of how much time and money you're willing to invest. After all, the point is to chill out and brew the kind of beer *you* like.

*Ross and Mindy Goeres have been a homebrewing team for many years. Ross is an electrical engineer and Mindy is a Recognized Judge in the Beer Judge Certification Program. They are often traveling in search of beers to be tasted.*

# ALL-GRAIN ON A SHOESTRING

## by Mark Moylan

*This article originally appeared in Zymurgy, Special 1995 (vol. 18, no. 4)*

The reasons I told folks I waited so long to begin all-grain brewing sounded honest enough. What if I made the investment and added

**Bucket with Holes**

the extra time to my brewing sessions and wasn't happy with the results? What if I jammed all that new gear into my storeroom and didn't make better beer? And what if all-grain brewing turned out to be a personal fiasco and each sip stung my already delicate ego?

So many what ifs. But I was fibbing. The real reason I held off was because I was broke. I had started a new career and at the same time my wife had decided not to go back to work after the birth of our baby boy, Brendan. Though money was tight, I was still spoiling for an all-grain brewing session. Lack of funds never stopped me from having fun in the homebrew arena before, and I sure wasn't going to let it stop me now. I was a homebrewer, dammit, and homebrewers are a resourceful bunch. If I made the decision to go all-grain, ready cash or no ready cash, it would happen.

I made the decision to make it happen and I'm glad I did. My all-grain setup is low cost, reliable, and beautifully Rube Goldbergesque. Every piece has been searched out, haggled over, cobbled, cursed, and cajoled. I just bottled my eleventh all-grain batch and the beer is very good. And I went all-grain for about a hundred bucks.

The biggest investment is a brewpot large enough to hold more than 5½ or 6 gallons (20.8 or 22.7 l.) of boiling wort. Shiny stainless steel pots are expensive, so I opted for a 33-quart (30.3-

l.) canning kettle I saw in a *Zymurgy* ad from a homebrew shop for $29. The metal could be heavier and it probably will wear out, but it fit my tight budget. Plus, it matches my old 4½-gallon brewpot that became my mash tun for stove-top step infusions, and I was always taught that economy should never be at the cost of fashion. Vanity, thy name is homebrewer.

A mill to crush grain is the next biggest expense. It can be avoided altogether if you order your grain precrushed through the mail or crush it when you buy it at your local homebrew supply store. I bought a Corona grain mill through a catalog from a homebrew shop in the Midwest to keep shipping costs low on such a heavy piece of equipment. I paid $35 less UPS charges. I have since seen this mill listed for as little as $26 and as much as $42.

I went with the Corona because I wanted to grind my grain fresh for each batch, and buying 50-pound bags is very cost-effective. You can figure about 1½ pounds of grain equals 1 pound of malt extract. U.S. two-row is about 50 to 60 cents a pound, so I'm paying about 90 cents per pound versus $3 for a basic ingredient. Homebrewers have been using these weighty iron Coronas for at least a decade, and they produce a pretty good grist if tended to properly. How well does it work for me? A lot better than "fair to middling," as the old miller's phrase goes.

Before I started buying stuff, I had put together a lauter tun for partial-mash recipes. The lauter tun lets me rinse the converted sugars from the mashed grains, and these runnings become the wort. The design is pure Papazian: two 6-gallon plastic buckets, a plastic spigot, and 3 feet of tubing. My neighbor makes wine and tossed out three plastic buckets his grape juice came in. One is my bottling bucket and the other two became a very workable lauter tun. (Most restaurants have 5-gallon plastic buckets for the asking or for a shared bottle of homebrew. Avoid pickle buckets because they will always smell of pickles.)

Inspired by the Zapap lauter tun system described by Charlie Papazian in *The New Complete Joy of Home Brewing* (Avon, 1991), I drilled a ton of holes (about ⅛ inch in diameter) in the bottom of one and attached a plastic spigot I bought at the local homebrew shop to the other. The two buckets fit together in a beautiful industrial-strength straining outfit that holds up to 15 pounds of grain. Luckily, this intricate piece of all-grain brewing equipment is the easiest one to put together. Elegant, it is not, but boy, it works.

I got the pot, ground the grain, mashed, sparged, and boiled; now, who's going to help me chill the wort? A major hardware chain to the rescue. Fifty feet of ⅜-inch copper tubing, hose clamps, and a threaded faucet connection ran me about $30. For the hoses I lucked out. My cousin Bill Moylan worked in Saudi Arabia for a few years and brought home all this good rubber tubing used when brewing beer to quench his thirst in the hot desert sun. Cost of the tubing was a few bottles of homebrew rather than about 50 cents a foot. Funny thing, his brother, Jim Moylan, got me started homebrewing. He sent me a bottle capper, hydrometer, tubing—what is it with my cousins and tubing?—Irish moss, and other sundries. Coming from a big family, I guess I'm just used to hand-me-downs.

Other bargains added to my setup include a dial thermometer for $5 from a surplus store in Chicago, tincture of iodine from the local drugstore to test starch conversion (less than $1), and a full-length white apron for which I traded a few homebrews from a friend who runs a restaurant.

The items I have not listed are the homebrewing books I have bought as I continue to brew. The information is invaluable. I am confident enough in my brewing skills to have two or three different books in the kitchen while I brew. Any time I get hit with self-doubt or excessive worry about how the beer is going to turn out, I flip open one of the books and it calms me down.

Each author is like a good friend who calls out through the print in either a church whisper or a megaphone shout, "Keep going, you're doing fine," while I'm puttering away over the mash, fretting about temperatures, or waiting for the wort to cool down and thinking I see uninvited bacteria going for my pot. Every homebrewing author has added something good to my homebrew.

Charlie Papazian is like a favorite coach coaxing the best out of any brewing effort. Dave Miller is an excellent instructor, though he describes extraction rates that only seem possible for homebrewers from the planet Krypton. Still, he's set a goal to shoot for and his recipes are delicious. Byron Burch always gives very sound advice. Greg Noonan makes an iodine test a study in color variation that rivals the nuances of the differences between D sharp and E flat on the musical scale. And my *Zymurgy* collection makes the brewing process fun because it's chock-full of helpful tidbits I use to tweak my own recipes. It's this information that helps me operate my inexpensive all-grain gear to limits I could never have imagined.

If you are thinking about going all-grain, but have been holding off because of the investment, hold off no more. Buy the best equipment you can afford, but if you're itching to brew all-grain beer and are short on funds, there is a way to get there. Search out the bargains. Improvise and adapt items to get you where you want to go. Put that Mensa membership to good use and invent a better way to make beer with what's on hand.

You'll get the feel for all-grain brewing the best way of all, by doing it. As you continue and decide you love all-grain brewing—yes, you will fall in love with the process and the beer—then you can replace your jury-rigged setup with fancier stuff. Or maybe you'll like your homemade gear so much you'll never part with it.

Myself, I'm waiting to replace it as I wear out my present equipment and will get new things piece by piece. Of course, it will have to be on sale. Force of habit, I suppose.

### TOP 10 WAYS TO KEEP COSTS DOWN

1. Don't be in a hurry. Good bargains, like good homebrew, cannot be rushed.
2. Use two pots to boil wort. You don't have to buy a big one. Just be sure to divide the hops correctly between the two pots, and watch carefully to avoid boilovers.
3. Borrow everything. This is good for a few batches, but don't forget to return stuff.
4. A nylon grain bag available from most homebrew shops turns one plastic pail—your bottling bucket?—into a lauter tun.
5. Two-liter pop bottles filled with water and frozen are a cheap wort chiller. Professor Surfeit liked this idea from a *Zymurgy* reader, but stressed the need to sanitize the pop bottles.
6. Use an aluminum pot. Aluminum pots are the least expensive, and aluminum's effect on beer flavor is still being debated in homebrew circles.
7. Watch the classified ads for used restaurant supplies. Big stainless steel pots are the first to go, so count on a little luck to get what you need.
8. Double up with a brewing buddy. This is half the cost and very little hassle if you get some basic rules on paper to avoid any misunderstandings.
9. Tell your spouse what you want for Christmas or an anniversary and leave the ad for the *exact* equipment on the kitchen counter. Trust me on this one.

**10.** Steal everything. (Plan on having a captive audience for your homebrew and try to make a beer the warden likes.)

---

*Mark Moylan is a freelance writer in Michigan who claims wrangling yeast is a spiritual experience.*

# A BOTTLER'S GUIDE TO KEGGING

## BY ED WESTEMEIER

*This article originally appeared in* Zymurgy, Summer 1995 *(vol. 18, no. 2)*

There's just something about fresh draft beer. It's not that the chore of bottling is such a big deal; after all, bottles are convenient to give to friends or send to competitions. It's great to be able to come home and grab a bottle of your own homebrew, but there's something even better about drawing a glass fresh from a keg.

You've probably admired those shiny cylinders many homebrewers use to "keg" their beers, but you may also have wondered how complicated it is to learn the techniques. In the words of a great homebrewer, relax. This article contains everything you need to know about dispensing your own beer from a keg. If you are a careful shopper, you should be able to put a complete draft homebrew system together for $150 or so. That may sound like a lot, but it's a lifetime investment that will give you many hours of pleasure and save many hours of bottling time.

## COLLECTING THE GEAR

### Kegs

The standard homebrewer keg is a 5-gallon (19-l.) container used by soft-drink bottlers for their pre- or postmix syrups. They're also made in 3- and 10-gallon (11- and 38-l.) sizes, but those can be hard to find on the used market. They are made by several manufacturers, but have several features in common: all stainless steel construction, inlet and outlet valves, a hatch cover for filling, and (usually) a safety valve to vent excess pressure. Some feature hard rubber protectors around both top and bottom.

Beyond those basics, they come in two styles: ball lock or pin lock, depending on whether the fittings on top of the keg use shallow grooves where the ball bearings in the quick-disconnects fit, or protruding pins to secure the disconnects. The two types are used by competing bottlers and are designed to be incompatible with each other to make it difficult for a restaurant to experiment with the rival brand. There are two types of lids, oval and racetrack (which is a flattened oval). They are not interchangeable.

It really doesn't matter which you use, but it is easier to stick with only one type. If you happen to obtain a keg of the other type, you could buy a second set of quick-disconnects, or trade with a friend who's in the same position.

A variety of new plastic kegs is beginning to come on the market and I've heard many good reports about them, but 90 percent of the kegs used by homebrewers are the soda syrup kind.

The good news is that many bottlers are switching to a newer dispensing system, so used stainless steel kegs are becoming available in large numbers. The bad news is that once they're gone, that's it. If you're even slightly interested in kegging homebrew, I'd advise buying some as soon as possible.

A few generalizations about kegs:

- Ball-lock kegs are usually easier to find.
- Ball-lock kegs are a little taller, so pin-lock kegs may fit more easily in your refrigerator.
- Ball-lock kegs are more likely to have a pressure-release valve in the lid.
- The standard ball-lock pressure-release valve is automatic, but can be used manually.
- The standard pin-lock pressure-release valve is only automatic (cannot be operated manually), and if it blows, it reseats automatically (doesn't require replacement). Your keg may or may not have a manual pressure-release valve, but models with this feature are recommended.
- There are many different keg designs on the market. Your gear may not look exactly like the kegs illustrated.

New kegs typically cost between $80 and $100, but used kegs are normally available from many sources for half that price. Try your local soft-drink bottler. Our club found thousands of used ball-lock kegs in a local Pepsi bottler's yard, and we were able to buy as many as we wanted for $10 each. All had some syrup still in them and many were dented and unattractive, but they held pressure and the valves were in good condition. Cleaning them

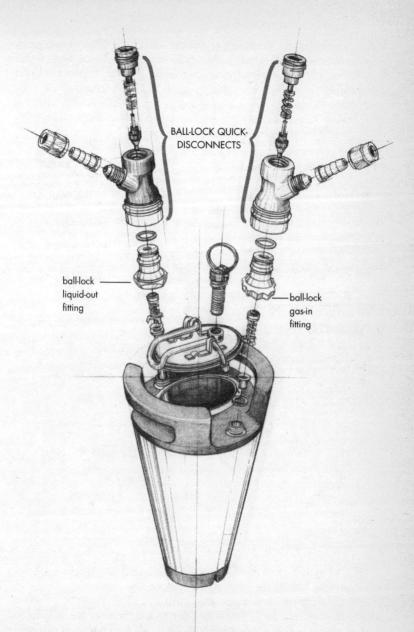

**Ball-lock Keg**
RANDY MOSHER

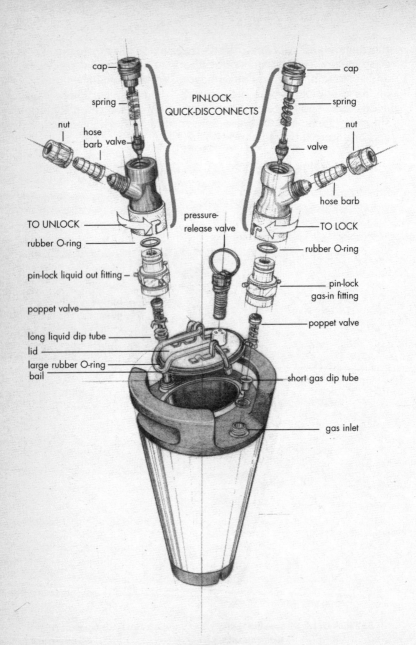

**Pin-lock Keg**
RANDY MOSHER

# A Bottler's Guide to Kegging

thoroughly and replacing the O-rings made them practically as good as new. Dealers in used restaurant equipment, scrap yards, and restaurant auctions often have used kegs available. If none of these sources works out, several homebrew suppliers carry used kegs. The price will be a little higher and often includes shipping costs, but their kegs have generally been inspected and are quite a bit cleaner.

## Connections

After acquiring your kegs (that's plural because you'll want a few), your next purchase probably will be some connectors, called quick-disconnects. There are two kinds for each style of keg: a gas-in connector and a liquid-out connector. If your kegs are the pin-lock style, it should come as no surprise that one fitting has two pins, the other

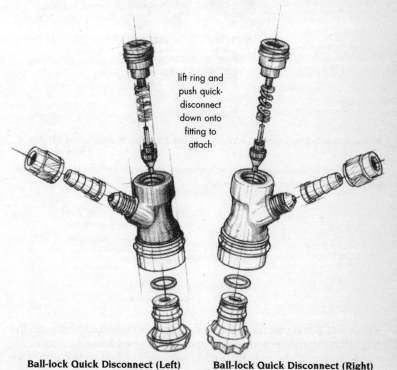

**Ball-lock Quick Disconnect (Left)**
RANDY MOSHER

**Ball-lock Quick Disconnect (Right)**
RANDY MOSHER

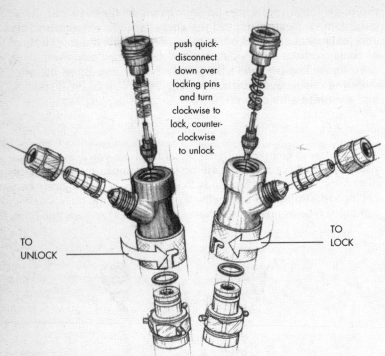

**Pin-lock Quick Disconnect (Left)**
RANDY MOSHER

**Pin-lock Quick Disconnect (Right)**
RANDY MOSHER

These are some examples of gas and liquid connections. The base of the gas fittings for ball-lock kegs are usually indented with small notches.

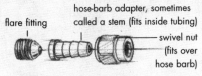

**Flare/Hose Barb/Swivel Nut**
RANDY MOSHER

has three. But if you have the more common ball-lock kegs, the fittings that the connectors attach to appear to be identical. If you look carefully, you'll see that their dimensions are slightly different. Trying to force a gas quick-disconnect onto a liquid keg fitting is a frustrating experience. (That's the voice of experience speaking!)

Quick-disconnects are sometimes available in different colors to help you distinguish gas from liquid, which is a very good idea. Plastic quick-disconnects are fairly inexpensive at $4 to $7 each, but are prone to crack and break from frequent use. Stainless steel versions generally are twice as expensive but should last a lifetime.

Quick-disconnects are available with either a hose barb to which you can clamp the end of your plastic tubing, or a threaded male flare fitting. Or by using a male flare fitting, you can attach a swivel nut and hose-barb adapter to the hose. A swivel nut that is part of a hose-barb adapter screws onto the flare fitting. Either method works well, but I prefer the male flare/swivel nut assembly. That way I can simply screw a keg connector onto the swivel nut and I'm ready to use my keg. This way it is also easier to clean and sanitize the hoses and connectors separately.

## $CO_2$ Tank

The next item on your shopping list is a carbon dioxide ($CO_2$) cylinder. The most common size for homebrewers seems to be the 5-pound tank. It is the shape of a typical fire extinguisher and is made of either steel or aluminum. My $CO_2$ tank is aluminum, about 5 inches in diameter and 18 inches high. This is a handy size for taking with you when you travel with a keg, and one full tank this size should be enough to carbonate and dispense up to a year's worth of typical homebrewing output. The 5-pound size sells for $50 to $100 new, half that amount if used. Check bar and restaurant suppliers, welding shops, and fire extinguisher companies.

Filling a 5-pound tank costs about $9 to $12 in my area, depending on where I go. Industrial gas suppliers, welding supply dealers, fire extinguisher companies, and soft-drink bottlers are all good sources, so check your phone directory and call around. Not all are willing to bother with such small orders and some refuse to fill aluminum tanks.

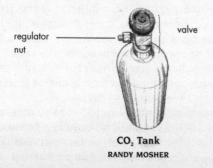

**$CO_2$ Tank**
RANDY MOSHER

When you buy a $CO_2$ tank, be sure it has a current certification. It is required by law to have a pressure test every 5 years, and to be stamped with the certification date. For example, mine has "5-91" stamped on it, so I'll have to get it recertified in May 1996 by taking it in to the place where I get it filled. If it fails the certification, which rarely happens, they'll drill a hole in the side, making it necessary to buy a new tank.

In many areas you can avoid the hassle of recertifying a tank by simply leasing one instead of buying it. With a leased tank, you merely exchange it for a full one when yours is empty. The supplier is responsible for having the tanks certified. The exchange costs a little more than a fill, but it may be worth it to you.

The empty (tare) weight of the cylinder also is stamped on it. For example, "TW 7.50" means it weighs 7.5 pounds when empty. After filling with 5 pounds of $CO_2$, it should weigh 12.5 pounds. Weigh your tank at home right after you get it filled so you'll be able to tell how much $CO_2$ is left in it by weighing it again.

If you don't plan to take your kegs to parties, or if your production of homebrew is starting to increase, you should consider a 20-pound $CO_2$ tank. My 20-pound tank is about 8 inches in diameter, 27 inches high, and weighs 50 pounds when full, including the regulator. The advantage of a 20-pound tank is that it only costs a few dollars more to fill it than a 5-pounder, but it holds four times as much. If you shop around, the 20-pound size can often be found for less than $100 new or $50 to $60 used, but in any case shouldn't cost more than $125.

Always keep $CO_2$ cylinders secured in the upright position to avoid accidents and injury. Keeping the keg upright will keep liquid $CO_2$ out of the regulator, which could damage it.

### Regulator

A full $CO_2$ tank holds a pressure of 800 pounds per square inch (psi) at room temperature. That's a bit more than the 10 to 30 psi I need to carbonate or dispense my beer, so a regulator is a must. The regulator screws onto the tank valve and reduces the pressure to safe levels. A set screw lets you adjust the regulator's output pressure with an ordinary screwdriver (or even a dime), and a gauge shows the working pressure coming out of it. (You rarely need more than 25 to 30 pounds.)

The $CO_2$ in the tank starts out as liquid, and the pressure of the gas in the head space of the tank will be between 700 and 800 psi, depending only on the temperature of the cylinder. The high-pressure gauge on the regulator will only begin to fall when all the

# A Bottler's Guide to Kegging

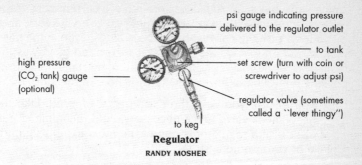

**Regulator**
RANDY MOSHER

liquid is gone. Depending on your tank size, this could be enough $CO_2$ to dispense only part of a keg. The accurate way to estimate how much $CO_2$ is left in a tank is by weight, not by pressure.

The regulator may have a hose barb adapter to which you can clamp a piece of tubing. Otherwise, it has a flare fitting to which you can screw a hose barb attached to your tubing. Regulators come in many styles and cost anywhere from $25 to $75 new. You can generally find them at any place that sells $CO_2$ tanks.

## Miscellaneous Parts

The faucet is the familiar plastic gizmo with a lever that you press to start the flow of beer. You'll find them at most homebrew suppliers, beer distributors, or bar-supply dealers. It's inexpensive at $4 to $7 and works well provided you have the correct pressure behind it. It may not be obvious, but the typical faucet easily disassembles into three pieces for cleaning.

Flexible plastic tubing is used between the $CO_2$ tank and the keg's gas-in connection, and between the keg's liquid-out connection and the faucet. Your best bet is 3/16-inch ID (inside diameter) food-grade vinyl (not polyethylene) tubing. You can probably find this at a local hardware store for less than 50 cents per foot. Buy at least 10 feet to start. Cut it in half and use 5 feet for each line.

A gauge cage is a nice investment for $10 to $15. This is a steel wire frame that attaches to the regulator and protects the gauges from breakage in case the tank falls over.

Some regulators come with a one-way check valve that makes it impossible for beer to accidentally flow back into the regulator and ruin it. Check valves come in many shapes and sizes and can be difficult to spot. Check with the manufacturer to see if your equipment includes this feature. You can add it later to any reg-

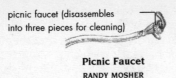

**Picnic Faucet**
RANDY MOSHER

ulator if you decide you want the extra peace of mind. Like the gauge cage, a check valve is not necessary, but I recommend getting both because a regulator is one of the most expensive parts to replace of your whole system.

## DISASSEMBLY

When you get your used kegs home, the first order of business is a thorough cleaning. First, vent any pressure by lifting the tab or ring attached to the pressure-release valve. If there's no safety valve, press down on the poppet in the center part of the gas fitting (labeled "in") with a key or small screwdriver. If you vent the liquid or "out" fitting first, you will get an unpleasant shower. Kegs are almost always shipped with some pressure in them: be certain to release all of it for safety reasons.

With the pressure vented, you'll be able to open the hatch. Lift up on the bail. There are many different types, but you'll see a steel wire frame that obviously is meant to be lifted up. Using this bail as a handle, push the hatch cover down into the keg an inch or so (don't drop it). If you can't budge it, there is still pressure in the keg, so go back to the previous step and release the pressure. Caution: When new, these kegs are rated to hold up to 130 psi of pressure, so you could injure yourself if you try to force the lid open with pressure inside. After pushing the cover down a little, rotate it a quarter turn and you'll be able to lift it right out.

Notice the large rubber O-ring on the lip of the cover. After years of contact with soft-drink syrup, the rubber is thoroughly impregnated with the stuff, and I've never found an effective way to remove the aroma. You don't want that flavor in your beer, so it's best to replace the O-ring. Many homebrew suppliers carry them, and some even offer kits to replace all the O-rings in a keg. If you're lucky enough to have a well-stocked hardware store nearby, you may be able to find them, but take an old one along to be sure you get the same size and one of food-grade quality. A complete set of new O-rings for a keg shouldn't cost more than $5 to $7 in most areas.

# A Bottler's Guide to Kegging

Using an open-end (your only option for pin-lock fittings) or deep-socket wrench, loosen both the gas and liquid fittings on top of the keg. There are several sizes of fittings, but the most common are 7/8-inch diameter for ball-lock kegs and 13/16-inch for pin locks. Loosening the fittings might take some strength, but once you have them loosened, they should be easy to unscrew with your fingers.

Remove both fittings as well as the tubes beneath them. The gas fitting has a short tube and the liquid fitting has a long tube that goes all the way to the bottom of the keg. Each tank fitting has a small O-ring around it on the outside, and each tube has an even smaller O-ring around it. All four of these O-rings should be replaced for the same reason as the large one. You may be able to find food-grade-quality replacements at the local hardware store, but a homebrew supplier will be your best bet. Keep in mind that pin- and ball-lock kegs use slightly different size O-rings. They are sometimes interchangeable.

Inside the fittings you'll find the actual valve, called a poppet, which may need to be replaced if you find out it leaks.

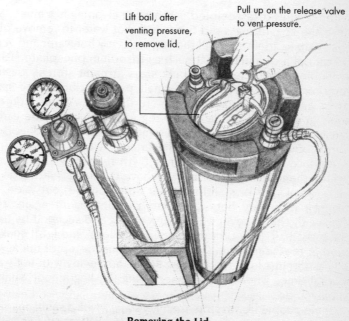

**Removing the Lid**
RANDY MOSHER

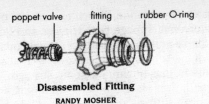

**Disassembled Fitting**
RANDY MOSHER

If either dip tube is plastic, try to replace it with a stainless steel version. This can't always be done because some nylon dip tubes are a different size from their stainless counterparts, and are not interchangeable. Plastic will have absorbed soda smells—something you don't want. While you have the dip tubes disassembled for cleaning, trim the gas tube with a tubing cutter down to about ½ to ¾ inch if it is longer. This will prevent beer from pushing into the gas tubing.

# CLEANING

With the kegs disassembled, a thorough cleaning is a must. First, rinse the keg, fittings, and tubes with hot water to remove obvious syrup residue. Then fill the keg with very hot water and ¼ cup (59 ml.) of a cleaning agent such as trisodium phosphate (TSP is available as wall cleaner in bulk sizes at paint stores), washing soda, or B-Brite. Drop in the small parts, including the hatch cover, and let it all soak for no more than a few hours. You may need to scrub the inside of the keg with a nylon bristle carboy brush or nylon scrubbing pad to remove stubborn residue. Don't worry about scratching the inside surface, just get it clean.

Replace the tubes and fittings (with their new O-rings) and tighten them securely. Be sure you have put the gas fitting on the "in" side of your keg, and the beer fitting on the "out" side. Fill the keg with very hot water and cleaning agent again, then replace the hatch cover with its new O-ring and secure it by tightening the bail. Now turn the keg upside down and let it soak for another few hours. This step cleans the inside top of the keg as well as the inside of the fittings. Rinse thoroughly with hot water several times when you're done, giving the keg a good shaking when you do.

Some people like to store their kegs filled with a sanitizing solution such as ½ ounce (15 ml.) of iodophor in 5 gallons (19 l.) of water. Don't use hot water because it will reduce the effectiveness of the iodophor. Don't use chlorine bleach for this purpose because prolonged exposure can damage the stainless steel—even short exposure can pit the kegs at the bleach-water/air interface.

## USING THE GEAR

### Filling the Keg

Among the advantages of kegging your homebrew is no longer having to wait a couple of weeks until your beer is carbonated. Another is being able to completely forget about the sediment on the bottom of the bottle. The solution to both problems is forced carbonation with $CO_2$ from your tank.

As soon as your fermentation has completely finished and your beer is clear, you're ready to keg. Here's the procedure I use: To sanitize the keg thoroughly, I fill it with 5 gallons (19 l.) of water and ½ ounce (15 ml.) of iodophor, seal the hatch, and let it sit for at least 10 to 20 minutes. Then I turn it upside down and let it sit for another 10 to 20 minutes. Finally I turn it right side up, open the hatch, and empty it. I leave it upside down in the sink while I prepare everything else. That allows it to air-dry, so rinsing isn't necessary. If you use a different sanitizer than iodophor, a final rinse with preboiled water would be advisable.

To connect the $CO_2$ tank to the keg, attach one end of a length of tubing to the regulator and the other end, with the gas-in quick-disconnect, to the gas fitting on the keg. You can leave the gas tubing attached to the regulator for future use.

To avoid oxidation, purge the air from the keg by turning on the $CO_2$ for 10 seconds or so at 5 psi. $CO_2$ will enter the keg, sink to the bottom, and push the air out the open top. Do this just before racking the beer into the keg because the $CO_2$ and air will mix together after a while.

Turn the regulator off and rack the beer from the fermenter to the bottom of the keg to avoid splashing. As the beer fills the keg, the $CO_2$ is gradually pushed out, leaving a blanket of carbon dioxide to protect the surface of the beer from the air.

Replace the hatch cover. The gas-in line is still connected to the keg, so set the regulator to about 5 psi and fill the head space of the keg with $CO_2$. (Listen for the gas to stop flowing.) Turn off the $CO_2$. Open the safety valve to let almost all the pressure out, then fill the keg with $CO_2$ again. Do this three times to purge any remaining air from the head space in the keg. Now you're ready to carbonate the beer.

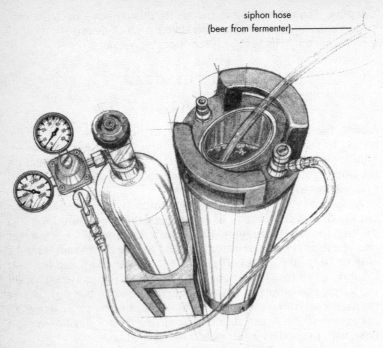

**Filling the Keg from Fermenter**
RANDY MOSHER

## Carbonation

The first step is to determine how much carbonation you want. The dial gauge on your regulator is calibrated in pounds per square inch, but carbonation is generally measured in volumes of $CO_2$ (for the quantity of gas that is actually dissolved in the beer). For English styles like bitter, about 1.5 to 1.8 volumes of $CO_2$ are about right. For effervescent styles like German Weizens, 2.8 to 3.0 volumes can be used. For most other beers, something in the 2.4 to 2.6 range seems to work best.

The colder the beer, the more easily $CO_2$ can be dissolved in it, so it's important to know the temperature of the beer in your keg before you begin. Let's assume the beer you just kegged is 42 degrees F (6 degrees C). With most beers, you want carbonation to be in the range of 2.2 to 2.6 volumes of $CO_2$. By using Table 3-1, you learn you'll need about 10 psi at 42 degrees F (6 degrees C) for 2.2 volumes of $CO_2$. For this style at this temperature, you

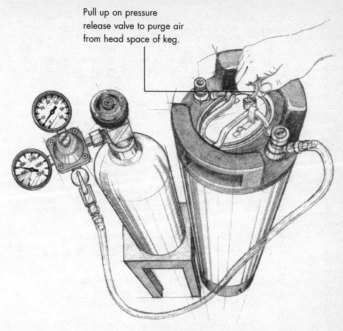

Pull up on pressure release valve to purge air from head space of keg.

**"Burping" the Keg**
RANDY MOSHER

want to apply 10 psi of $CO_2$ pressure until as much $CO_2$ as possible is dissolved in the beer. The colder your beer is, the less pressure it will take to carbonate it.

So you are ready to carbonate. Your $CO_2$ tank is hooked up to the gas-in connection, hatch cover is on, and air is purged from the keg's head space. Now turn on the $CO_2$. Turn the screw on the regulator to set it for (in this example) 10 pounds. Now listen to the $CO_2$ flow. As the pressure reaches equilibrium, the flow will slow down and eventually stop. (This won't take more than a few minutes.) The reason it stopped so soon is that when the keg is upright there is only a small surface area where the gas can dissolve into the beer.

Shaking the keg will agitate the surface and start the flow again. To make this process more efficient, I like to roll the keg on its side back and forth on the floor with my foot while listening to the gas flow. Don't try this unless you have a check valve in the system, though, otherwise there is a chance that beer will back up the gas line. If beer gets into the regulator, the regulator could be ruined.

## Table 3-1. Pressure Required for Desired Carbonation

Directions: Look down the left column to find your keg temperature and read across to the number in the column corresponding to the desired carbonation level. That number is the $CO_2$ pressure to apply to the beer, in psi.

| Temp (°F) | Volumes of $CO_2$ desired | | | | | | | | | | | | | |
|---|---|---|---|---|---|---|---|---|---|---|---|---|---|---|
| 32 | 0.6 | 1.6 | 2.5 | 3.5 | 4.4 | 5.4 | 6.3 | 7.3 | 8.2 | 9.2 | 10.1 | 11.0 | 12.0 | 12.9 |
| 34 | 1.3 | 2.3 | 3.3 | 4.3 | 5.3 | 6.3 | 7.3 | 8.2 | 9.2 | 10.2 | 11.2 | 12.1 | 13.1 | 14.1 |
| 36 | 2.1 | 3.1 | 4.1 | 5.1 | 6.2 | 7.2 | 8.2 | 9.2 | 10.2 | 11.2 | 12.3 | 13.3 | 14.3 | 15.3 |
| 38 | 2.8 | 3.9 | 4.9 | 6.0 | 7.0 | 8.1 | 9.1 | 10.2 | 11.2 | 12.3 | 13.3 | 14.4 | 15.4 | 16.5 |
| 40 | 3.6 | 4.7 | 5.7 | 6.8 | 7.9 | 9.0 | 10.1 | 11.2 | 12.3 | 13.4 | 14.4 | 15.5 | 16.6 | 17.7 |
| 42 | 4.3 | 5.5 | 6.6 | 7.7 | 8.8 | 10.0 | 11.1 | 12.2 | 13.3 | 14.4 | 15.5 | 16.7 | 17.8 | 18.9 |
| 44 | 5.1 | 6.3 | 7.4 | 8.6 | 9.7 | 10.9 | 12.1 | 13.2 | 14.4 | 15.5 | 16.7 | 17.8 | 19.0 | 20.1 |
| 46 | 5.9 | 7.1 | 8.3 | 9.5 | 10.7 | 11.8 | 13.0 | 14.2 | 15.4 | 16.6 | 17.8 | 19.0 | 20.2 | 21.3 |
| 48 | 6.7 | 7.9 | 9.1 | 10.4 | 11.6 | 12.8 | 14.0 | 15.3 | 16.5 | 17.7 | 18.9 | 20.1 | 21.4 | 22.6 |
| 50 | 7.5 | 8.7 | 10.0 | 11.3 | 12.5 | 13.8 | 15.0 | 16.3 | 17.6 | 18.8 | 20.1 | 21.3 | 22.6 | 23.8 |
| 52 | 8.3 | 9.6 | 10.9 | 12.2 | 13.5 | 14.8 | 16.1 | 17.3 | 18.6 | 19.9 | 21.2 | 22.5 | 23.8 | 25.1 |
| 54 | 8.9 | 10.4 | 11.8 | 13.1 | 14.4 | 15.7 | 17.1 | 18.4 | 19.7 | 21.1 | 22.4 | 23.7 | 25.0 | 26.3 |
| 56 | 9.9 | 11.3 | 12.6 | 14.0 | 15.4 | 16.7 | 18.1 | 19.5 | 20.8 | 22.2 | 23.6 | 24.9 | 26.3 | 27.6 |
| 58 | 10.7 | 12.1 | 13.6 | 15.0 | 16.4 | 17.8 | 19.2 | 20.6 | 21.9 | 23.3 | 24.7 | 26.1 | 27.5 | 28.9 |
| 60 | 11.6 | 13.0 | 14.5 | 15.9 | 17.3 | 18.8 | 20.2 | 21.6 | 23.1 | 24.5 | 25.9 | 27.4 | 28.8 | 30.2 |

**Temp (°F)** | | | | | | **Volumes of $CO_2$ desired** | | | | | | |
---|---|---|---|---|---|---|---|---|---|---|---|---|
62 | 12.4 | 13.9 | 15.4 | 16.9 | 18.3 | 19.8 | 21.3 | 22.7 | 24.2 | 25.7 | 27.1 | 28.6 | 30.0 | 31.5
64 | 13.3 | 14.8 | 16.3 | 17.8 | 19.3 | 20.8 | 22.3 | 23.8 | 25.3 | 26.8 | 28.3 | 29.8 | 31.3 | 32.8
66 | 14.2 | 15.7 | 17.3 | 18.8 | 20.3 | 21.9 | 23.4 | 25.0 | 26.5 | 28.0 | 29.6 | 31.1 | 32.6 | 34.1
68 | 15.1 | 16.6 | 18.2 | 19.8 | 21.4 | 22.9 | 24.5 | 26.1 | 27.6 | 29.2 | 30.8 | 32.4 | 33.9 | 35.5
70 | 15.9 | 17.6 | 19.2 | 20.8 | 22.4 | 24.0 | 25.6 | 27.2 | 28.8 | 30.4 | 32.0 | 33.6 | 35.2 | 36.8
72 | 16.8 | 18.5 | 20.1 | 21.8 | 23.4 | 25.1 | 26.7 | 28.4 | 30.0 | 31.6 | 33.3 | 34.9 | 36.5 | 38.2
74 | 17.8 | 19.4 | 21.1 | 22.8 | 24.5 | 26.2 | 27.8 | 29.5 | 31.2 | 32.9 | 34.5 | 36.2 | 37.9 | 39.5
76 | 18.7 | 20.4 | 22.1 | 23.8 | 25.5 | 27.2 | 29.0 | 30.7 | 32.4 | 34.1 | 35.8 | 37.5 | 39.2 | 40.9
78 | 19.6 | 21.4 | 23.1 | 24.9 | 26.6 | 28.4 | 30.1 | 31.8 | 33.6 | 35.3 | 37.1 | 38.8 | 40.5 | 42.3
80 | 20.5 | 22.3 | 24.1 | 25.9 | 27.7 | 29.5 | 31.2 | 33.0 | 34.8 | 36.6 | 38.3 | 40.1 | 41.9 | 43.7

(Table developed by Alan Edwards. Used here with permission.)

The more vigorously I agitate the keg, the more gas flows. Eventually it stops flowing no matter how much I shake the keg. That means my beer is fully carbonated. This process can take up to 15 minutes at 42 degrees F (6 degrees C), with faster results at lower temperatures.

Even though carbonation is complete, I know the beer will be foamy because of all the agitation, so I set the keg back upright and disconnect it. After a few hours the beer settles and is ready to serve.

An alternate method if you don't have a check valve or you don't have time to shake the keg is to simply leave the keg standing upright with the $CO_2$ connected and the pressure set at 10 psi. (If you can't keep the keg at 42 degrees F, check Table 3-1 for the psi you should apply.) The potential drawback with this method is that unless all the connections in the system are very tight, you may lose some of your $CO_2$ to leakage. It can take a few days to carbonate the beer if the keg is standing upright at 42 degrees F, but you can reduce that time if you give the keg a bit of a shake every time you pass by.

If you plan to cool the keg to an even lower temperature, put a few more pounds of pressure on it first. The gas in the head space of the keg will dissolve quickly when the beer cools down, and it's possible that this will let the O-ring leak on an old keg. If the seal isn't tight (from the internal pressure of the keg), all the gas can come back out of solution and escape through the leaky O-ring, leaving you with flat beer.

The third way to carbonate beer in a keg is to simply treat it as a giant bottle, using ½ cup (118 ml.) of corn sugar to prime. This method works as well as any other, but with two possible drawbacks. First, there's the problem of sediment on the bottom of the keg. That can be avoided to some extent by cutting off the bottom half inch of the liquid dip tube. Second, the large O-ring in some older kegs may not form a perfectly tight seal, and as carbonation is developed it could leak out through the seal, leaving you with flat beer.

Regardless of which method you use, you'll notice that the carbonation level improves with time. After a few days, the bubbles will seem finer and the head on your beer will probably be longer-lasting. After a week or two, the carbonation will be so perfect you may never go back to bottling. Science still hasn't completely explained this effect, but it's clear that something wonderful is happening to the carbonation quality as the beer matures.

# A Bottler's Guide to Kegging

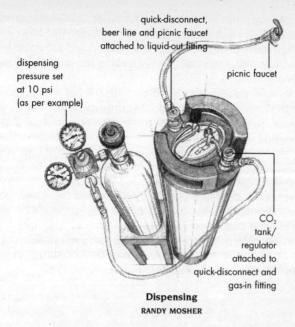

**Dispensing**
RANDY MOSHER

### Table 3-2. Pressure Drop in Pounds per Foot

| ID                | 3/16"  | 3/16"  | 1/4"  | 1/4"  |
|-------------------|--------|--------|-------|-------|
| Material          | vinyl  | poly   | vinyl | poly  |
| Restriction Factor| 3      | 2.2    | 0.85  | 0.5   |

## DISPENSING BASICS

So you are ready to sample a glass of your freshly kegged and carbonated homebrew—what do you do now? The simplest way involves a little trial and error, but don't worry.

Attach the beer-out quick-disconnect to the flexible dispensing hose and beer faucet, then attach this assembly to the beer-out fitting of the keg. The pressure on your regulator should be at 10 psi (the carbonating pressure from our example). Depress the lever on the faucet completely and fill your beer glass. If your glass fills with foam, turn the pressure on the regulator down 1 or 2 psi and try again. If no beer comes out, double-check to make sure the gas is turned on. If the beer just trickles out or seems undercarbonated or flat, increase the pressure 1 or 2 psi and try again. Believe it or not, this is the easiest way to dispense for the first time.

In reality, proper dispensing depends on several variables, including length and diameter of the dispensing line and the material it is made of.

As a general rule you want to wind up with 0.5 psi at the picnic faucet. To do this, multiply the length of your dispensing line by its restriction factor from Table 3-2 and set your regulator 1 or 2 psi above the resulting number. The pressure will drop as the beer travels up the keg's dip tube, through the length of line, and into your glass. For fine-tuning, you can raise and lower the level of the faucet and glass to make minor pressure changes. By raising the glass, you lower the dispensing pressure slightly; by lowering the glass, you raise the pressure slightly. Proper dispensing also depends on the temperature of your beer and dispensing line. You may need to adjust your pressure to perfect your pour. For more information see Dave Miller's chapter, "Setting Up Your Home Draft System," in *Just Brew It! Beer and Brewing*, vol. 12 (Brewers Publications, 1992).

## STORAGE

When you are done serving you can turn off the $CO_2$ tank and remove the gas quick-disconnect and beer line. Now store your beer in a cool location until next time. In the winter it may be enough to simply put a plastic bag over the top of the keg to protect it from dust and store it outdoors or in the garage. If that isn't possible, keep it at room temperature and dispense it into a frozen mug. Once you start kegging you may want to pick up a used refrigerator. Many keggers store their kegged beer in a spare refrigerator to keep it at serving temperature. Others like to keep the refrigerator even colder for lagering. Some even keep the beer they're currently serving standing next to the fridge, with the beer line running through one wall of the fridge, into a cold plate stored inside with the lagering tanks, and then out the other side to a beer tap mounted on the opposite wall of the fridge. (Caution: If you do this, be absolutely certain there are no cooling coils inside the wall of the fridge where you make the holes, and seal the holes with silicone caulk so the insulation does not get wet from condensation, ruining the efficiency of the refrigerator.) Even if you don't have the luxury of a brewing refrigerator (although the freezer compartment makes a great place to store your hops!), you can still enjoy your own draft beer. A cold plate or jockeybox (coiled copper tubing packed with ice in a picnic cooler) will do the job nicely. See *Zymurgy*, Fall 1991 (vol. 14, no. 3), for a great article by Teri Fahrendorf on building and using a jockeybox.

There's a lot of information here and you'll probably discover your own tricks and techniques for working with kegs. I could discuss many more things, like counterpressure bottle filling and transporting kegs for parties, but there's enough here to get you started. In addition to what you've learned here, find a local club of homebrewers and ask them for ideas. If you have any lingering doubts, they'll put them to rest immediately. Homebrewers who begin kegging their beer rarely look back. There's just something about fresh draft beer.

## FURTHER READING

Other articles that discuss draft systems:

"Closed System Pressurized Fermentation," by Teri Fahrendorf, *Zymurgy*, Special Issue 1992 (vol. 15, no. 4).

"Counterpressure Bottling," by Dan Fink, *Zymurgy*, Special Issue 1992 (vol. 15, no. 4).

"A Great System for Draft Beers," by Byron Burch, *Beer and Brewing*, vol. 10, pp. 177–89, Brewers Publications, 1990.

"Kegging Basics and a Buyer's Guide to Kegging Equipment," by Dan Fink, *Zymurgy*, Special Issue 1992 (vol. 15, no. 4).

"Kraeusening and Cold-Hopping Soda Kegs," by Cy Martin, *Zymurgy*, Winter 1989 (vol. 12, no. 5).

"Racking from Carboys to Soda Kegs," by Cy Martin, *Zymurgy*, Winter 1989 (vol. 12, no. 5).

"Setting Up Your Home Draft System," by Dave Miller, *Just Brew It! Beer and Brewing*, vol. 12, pp. 201–14, Brewers Publications, 1992.

"A Simple Keg System," by Malt Disney, *Zymurgy*, Winter 1986 (vol. 9, no. 5).

"Soda Keg Draft Systems—Better than Bottles?" by Jackie Rager, *Zymurgy*, Winter 1989 (vol. 12, no. 5).

*Ed Westemeier (hopfen@iac.net) is a National BJCP Judge and a member of the Bloatarian Brewing League of Cincinnati. He writes about beer, computers, and other things for a number of publications and has been a homebrewer since 1987.*

# THE COUNTERPRESSURE CONNECTION—WHERE BOTTLES AND KEGS UNITE

BY DAVID RUGGIERO, JONATHAN SPILLANE, AND DOUG SNYDER

*This article originally appeared in Zymurgy, Fall 1995 (vol. 18, no. 3)*

Bottling beer is a necessary part of beer-making. It requires about 1 or 2 hours to clean and sanitize fifty-two bottles, transfer beer to a bottling bucket, prime, then fill and cap the bottles. In comparison, kegging beer is simpler and requires only 15 minutes to clean and fill a keg. Priming isn't necessary, there are no multiple transfers and no capping. Let's face it, bottling takes time and effort while kegging doesn't, and the more time you spend bottling, the less time you have for brewing and tasting.

With that said, why would a homebrewer want to put kegged beer into bottles, the very bottles that kegging is supposed to do away with? It's simple: homebrewers want to share their beer. They want to give it to friends and enter it in competitions. They want to do these things without lugging around a heavy keg or carbon dioxide ($CO_2$) tank, and they want to do it with ease.

A counterpressure bottle filler (CPBF) is the piece of equipment a homebrewer needs to transfer beer from the pressurized $CO_2$ environment of a keg into a bottle. In an effort to explain the CPBF process and show homebrewers what kind of equipment is available, we used seven commercial bottle fillers and one homemade filler. All were evaluated during one bottling session with the same beer. Strict adherence to the manufacturer's directions was observed. In total, twenty-four bottled samples were collected, three from each filler. We noted the range of each filler's application, home or commercial depending on both cost and ease of operation, and whether it could simply be attached to a $CO_2$ tank and used, or if assembly and additional parts were required. The fillers were then rated and described with regard to the following: additional equipment required, ease of setup, ease of operation, effectiveness of operation, quality of materials, and quality of construction. The coded samples were then forwarded to Dr. George Fix for laboratory analysis of $CO_2$ and oxygen amounts. The findings follow, but first, an introduction and explanation of counterpressure bottle filling.

## COUNTERPRESSURE BOTTLE FILLER PRINCIPLE

The term *counterpressure bottle filling* may seem ominous, but it really is not. When you use a CPBF you are filling a bottle from a keg of carbonated beer. If you were to bottle by conventional means, siphoning the carbonated beer into an open bottle, the $CO_2$ in the beer would come out of solution and the beer would foam up and out. The same release of $CO_2$ occurs when you pour yourself a glass of beer. But when you apply $CO_2$ top pressure to the bottle, or counterpressure, the release of $CO_2$ from the beer can be decreased, if not eliminated. To achieve this balance the beer must be kept cold and at a constant pressure. The standard method used by a majority of U.S. bottle fillers is to fill bottles with beer that is between 32 and 37 degrees F (0 and 3 degrees C).

The need for counterpressure filling is based on the following: When carbonated beer experiences a pressure drop, a temperature increase, or some turbulence, it cannot hold as much $CO_2$ in solution, and therefore $CO_2$ bubbles out of solution. Simply pouring from a keg faucet into a room-temperature bottle causes all three, so there will be foaming, carbonation loss, and oxidation

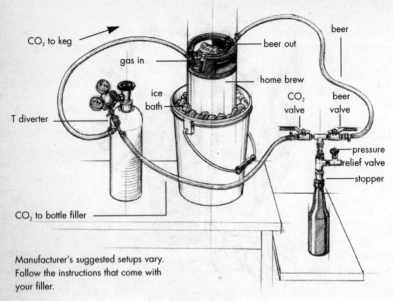

Manufacturer's suggested setups vary. Follow the instructions that come with your filler.

**Sample CPRF Setup**
RANDY MOSHER

(because of the air in the bottle). CPBFs allow the user to transfer beer without a pressure drop (the beer is moved from pressurized keg to pressurized bottle) and, in some models, to purge air from bottles, thus minimizing oxidation. Although turbulence during transfer cannot be eliminated, long fill tubes minimize this and the $CO_2$ under pressure in the bottle increases the tendency of the $CO_2$ in the beer to stay in solution.

All eight fillers we used required the beer to be processed cold. Seven incorporate the principle of counterpressure filling. Of those seven, six instruct the operator to evacuate air from the bottle prior to filling.

## EVALUATION CRITERIA

With more than fifteen years of homebrewing experience between the authors, an evaluation of counterpressure bottle fillers seemed an interesting task. We relied on the individual with the least kegging/CPBF experience to review all of the instructions and coordinate the bottling activities for two reasons: first, we assumed that most CPBFs are purchased by brewers unfamiliar with CPBF operations, and second, Jonathan wouldn't let us play with all the neat toys we were sent.

Each filler was assessed on a scale of one to ten (with ten being the best) for the first five criteria:

### Instructions

Practical and theoretical information was presented with each filler. How easy to follow were the step-by-step instructions? How did they stack up against the physical principles of CPBFing?

### Construction and Materials

What was the filler made of? How durable did the parts appear? What degree of detail and level of craftsmanship were evident?

### Ease of Setup

Was the filler ready to use right out of the box? What extra equipment was needed and at what cost? How compatible to typical homebrewing keg equipment was the bottle filler?

### Ease of Operation

How comfortable was the filler to hold and operate? Was it heavy, awkward, or unruly? Does it require one, two, or three hands to operate?

### Efficiency

How easily did the filler purge the bottle, pressurize it, bleed the pressure, and fill the bottle according to the manufacturer's instructions?

### Effectiveness

Dr. Fix used a Zahm and Nagel $CO_2$/Air Tester to measure the effectiveness of each CPBF. The results indicate how well $CO_2$ was retained and the total volume of air in milliliters introduced during the filling process. Effectiveness percentages were determined by taking the average volume of $CO_2$ per bottle divided by the original volume of $CO_2$. The higher the percentage, the greater the effectiveness. (Special note: Not all bottles tested were filled per manufacturers' instructions. If the fillers were uncooperative or the instructions were poorly documented, liberties were taken to fill the bottles.)

How much air was brought into solution during the CPBF process? Air, of course, contains oxygen, and high levels of oxygen result in a higher rate of oxidation-related spoilage and subsequently a shorter shelf life. According to the Master Brewers Association of the Americas, the standard of 1 milliliter of air is often sighted as the commercial brewing industry's acceptable high-end level. Homebrewers use higher levels, about 2 milliliters of air, and this analysis uses the 2-milliliter level as its standard. The levels of air are listed in percentage form above or below 2 milliliters. A level of air at or below 2 milliliters is optimal for homebrewers provided the homebrew did not suffer thermal abuse, according to *Brewing Science* (Academic Press, 1981).

**Zahm & Nagel Co.**
**74 Jewett Ave.**
**Buffalo, NY 14214**
**(716) 833-1532**

|  | Zahm & Nagel |  |
|---|---|---|
| Retail Price: | $370 (w/¼" adapters) | |
| Application: | Home √ | Commercial √ |
| Ready to use: | Yes | No √ |

*Instructions* (10)

Zahm and Nagel covered all the bases with a first-class, single-page, step-by-step hose hookup and bottle-filling fact sheet. The company also sent a copy of their catalog, which proved to be an invaluable reference and tutorial for preparing this article. A supplier to the commercial brewing industry, Zahm and Nagel manufactures high-quality $CO_2$ volume meters and bottle fillers, among other items. This hand filler is part of the series 9000R Zahm Pilot Plant filter, carbonator, and filler product.

### Construction and Materials (10)

(All stainless steel construction.) This unit is impeccably made and easy to handle. The polished stainless steel fittings and Delrin plastic handles not only look good but feel good and are easy to use. From setup and filling through cleanup, the filler's design and materials make it first-rate.

### Ease of Setup (2)

The filler we received came with ⅜-inch tubing stems, which is not typical for homebrew kegging systems. Retrofitting quick-disconnects and obtaining the proper-sized pressure-rated tubing may be difficult and could cost around $10. When ordering, specify the interior diameter you require for your setup.

The unit was shipped with filler tubes of three different lengths. While the brewer has the option of choosing the correct length for the bottle size being used, the effort needed to install the tube was dramatic. Also, the rubber stopper that seals the bottle is almost too big. Replacement stoppers are only available from Zahm and Nagel.

### Ease of Operation (10)

Perfect scores are tough to get, but this filler has three. The most outstanding feature is its valves. Both the $CO_2$ and liquid-feed valves are stainless steel one-quarter-turn plug valves with Teflon coating. They operated perfectly. The $CO_2$ bleeder needle valve was easy to use and gave us no trouble.

### Efficiency (7)

Two points worth mentioning are the stopper size and the volume of $CO_2$ required to operate this unit. The rubber stopper provided by the supplier was slightly oversized for the bottles we were using. This seemed odd because we were using standard 12-ounce bar bottles, the same kind that nearly every commercial brewery uses. Though a tight seal between bottle and stopper was achieved, there was little room for error. Regarding the volume of $CO_2$ required during bottling, the instructions call for the $CO_2$ supply valve to be open during the filling process, unlike some of the other fillers we used.

### Effectiveness

$CO_2$ retained: 91.25 percent of original volumes.
Air: 25 percent above recommended levels.

**The Beverage People**
840 Piner Road #14
Santa Rosa, CA 95403
(707) 544-2520

| | The Beverage People |  |
|---|---|---|
| Retail Price: | $59.95 | |
| Application: | Home√ | Commercial |
| Ready to use: | Yes | No√ |

### Instructions (1)

The instructions were clearly written and concise. A helpful list of additional parts is included. The instructions seem flawed in that a release or even a recycling of $CO_2$ was impossible to achieve. Without the release of $CO_2$ from the bottle, beer will not flow.

[Editor's Note: Curiosity got the best of us, so we tried using the filler ourselves. After a few unsuccessful attempts and a call to the manufacturer, the AHA staff was able to get the filler to perform as intended. Users should note the operating differences between two- and three-valve fillers.]

### Construction and Materials (8)

(Two-way trigger valves and stainless steel beverage path.) A nice design in a clean, simple package. The trigger valves are the best feature of this unit.

### Ease of Setup (5)

As with most of the other fillers, additional $CO_2$ and beer lines are required. An estimated cost of $5 for extra parts is all you need to make this unit operational.

### Ease of Operation (8)

The unit was by far the easiest to handle. The ergonomics are fantastic; you need only two hands to operate it.

### Efficiency (5)

When following the instructions provided, each time the beer line was opened, no beer would flow. Repeated attempts were made to fill bottles, and changes were made to the system each time. Different pressures, $CO_2$ paths, bottle heights, and keg heights were used, but all proved ineffective.

Despite the problems with the instructions, we were able to fill bottles using an improvised method. By simply lifting the rubber stopper away from the bottle lip and creating a very small opening for the $CO_2$ pressure to escape, our test bottles were filled very easily. Fortunately, the design of this filler lent itself to this improvisation. On further research we found that several professional brewers used this method when filling bottles with the Zahm and Nagel filler.

### Effectiveness

$CO_2$ retained: 93.65 percent of original volumes.
Air: 8 percent below recommended levels.

**Benjamin Machine Products**
1121 Doker Dr.
Modesto, CA 95351
(209) 523-8874

# Equipment and Gadgets

**Benjamin Machine 3BF**
Retail Price: $60
Application: Home√  Commercial
Ready to use: Yes  No√

## Instructions (6)

Brief but effective, these instructions include interesting and useful facts about sanitation, which we applied to all of the fillers used.

## Construction and Materials (8)

(Plastic body, stainless steel beverage path.) Perhaps the most unique of all of the fillers we reviewed. Two-way trigger valves are used instead of the ever-present ball and check valves that most fillers use. These valves give greater flow control of liquid and gas, which is a requirement when filling. The other unique feature about this unit is its one-piece plastic body. Advertised as leak-proof and chill-proof, the plastic body in fact did not leak, but none of the others did either. As for it being chill-proof, which would eliminate the "frozen hand syndrome," only three bottles were filled and our hands felt no different. In fact, none of the fillers froze our hands, so an assessment cannot be made regarding this claim.

## Ease of Setup (6)

This unit is ready to receive standard-size ¼-inch tubing. An estimated cost of $5 for tubing clamps and T-fittings is all that is required to get you started.

### Ease of Operation (7)

Though the flow valves on this filler are superior, the pressure-relief needle valve was tougher to use in comparison to the trigger valves. Also, the fill stem was much too short for a standard 12-ounce bottle. The manufacturer says the fill stem length will accommodate any size bottle, "and if proper pressures are used, foaming will not occur." We found this to be true. However, filling the bottles took almost twice as long, about 2 minutes each, because a slower pressure-relief flow rate was required.

### Efficiency (6)

This unit is quite effective and easy to use. The only drawback was the slow fill rate.

### Effectiveness

$CO_2$ retained: 88.91 percent of original volumes.
Air: 27 percent above recommended levels.

**Benjamin Machine Products**
1121 Doker Dr.
Modesto, CA 95351
(209) 523-8874

| Benjamin Machine 3S | | |
|---|---|---|
| Retail Price: | $125 | |
| Application: | Home√ | Commercial√ |
| Ready to use: | Yes | No√ |

### Instructions (6)

Same as the BFD 3B model.

### Construction and Materials (8)

Same as the BFD 3B model with the following exceptions: the BFD 3S comes with a pressure gauge mounted on the pressure-relief valve, allowing for faster filling by monitoring the bottle pressure, and the pressure-relief needle valve is superior and easier to use.

### Ease of Setup (6)

Same as the BFD 3B model.

### Ease of Operation (7)

Same as the BFD 3B model. The only exception is the improved pressure-relief valve.

### Efficiency (6)

Same as the BFD 3B model. However, once mastered, the 3S model's pressure-relief gauge and valve may be used to speed the process.

### Effectiveness

$CO_2$ retained: 90.52 percent of original volumes.
Air: 3 percent below recommended levels.

**Foxx Equipment Company**
**421 Southwest Blvd.**
**Kansas City, MO 64108**
**(816) 421-3600**

### Instructions (8)

The step-by-step instructions were clear, concise, and didn't cause any confusion.

### Construction and Materials (4)

(Brass valves, stainless steel and brass beverage path.) Three needle valves, stainless steel tubing, and a rubber stopper were all tied together in an economical package.

# The Counterpressure Connection—Where Bottles and Kegs Unite

|  |  |  |
|---|---|---|
| Retail Price: | Foxx |  |
|  | $40 |  |
| Application: | Home√ | Commercial |
| Ready to use: | Yes | No√ |

### Ease of Setup (6)

This unit is ready to receive standard-size tubing onto stems that fit standard homebrew keg systems. An estimated cost of $5 for tubing clamps and T-fittings is all you need.

Note: The unit we received was greasy and required extra time to clean, and one of the brass fittings was not sufficiently tightened. Two of the authors own Foxx Fillers and never encountered this situation before, so we assume the filler sent for testing slipped by the quality assurance people.

### Ease of Operation (2)

This filler uses needle valves that have a very rough finish and are quite difficult to turn. The person operating the filler complained of cuts on his thumbs. The sharp edges and numerous turns required to open the liquid and gas lines were cited as the cause of the cuts and a major drawback of the unit.

### Efficiency (7)

Overall, this unit does what it advertises. It does what a CPBF should do and does it well. Unfortunately, its rough valves make this effective tool uncomfortable to operate.

## Effectiveness

$CO_2$ retained: 89.53 percent of original volumes.
Air: 12 percent above recommended levels.

**Vinotheque**
2142 Trans Canada Highway
Dorval, Quebec H9P 2N4 Canada
(514) 684-1331; (800) 363-1506

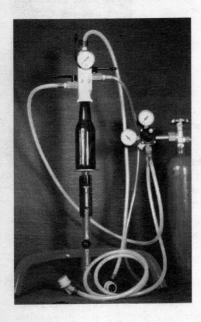

|  | Melvico |  |
|---|---|---|
| Retail Price: | $350 (Canada) | |
| Application: | Home√ | Commercial√ |
| Ready to use: | Yes√ | No |

## Instructions (6)

The step-by-step instructions given in French and English were complete and all of the basics were covered.

## Construction and Materials (9)

(Plastic and stainless steel beverage path, plastic ball valves and filling head.) Perhaps this unit is overkill for most people, but we found it to be the best and most enjoyable to use. It has a great design with only one small drawback: the adjustable bottle support is not easy to operate.

### The Counterpressure Connection—Where Bottles and Kegs Unite 123

### Ease of Setup (10)

The only things this unit didn't provide were the keg and the $CO_2$ tank. Complete with a regulator, it is 100 percent homebrew-compatible with ¼-inch ID hoses. It requires no extra parts. Just hook up the beer lines and the regulator and fill.

### Ease of Operation (9)

The unit is great. You look and feel like a professional when you bottle with it. The two drawbacks are its unstable frame construction (bolt it down and this shouldn't be a problem) and the touchy pressure-relief valve. (With practice it becomes easier to use.)

### Efficiency (10)

From hookup to cleanup, there were no problems. Because the bottle is locked into place, both hands are free to work the valves, which work very well. Purging and filling are done quickly. This unit caused less foaming than the others.

### Effectiveness

$CO_2$ retained: 95.94 percent of original volumes.
Air: 8 percent above recommended levels.

**Braukunst**
**Homebrewer's Systems**
**55 Lakeview Dr.**
**Carlton, MN 55718-9220**
**(218) 384-9844**

### Instructions (10)

Complete and easy to read. Though not in a step-by-step format, all of the information, parts lists, and helpful hints are included in a well-written two-page document.

### Construction and Materials (8)

(Stainless steel and brass beverage path, brass $CO_2$ and beer ball valves, brass pressure-relief needle valve.) A complete package with all the parts, from $CO_2$ to beer hookups. Compatible to homebrew keg systems. One great additional feature is

**Braukunst**
Retail Price: $44.95
Application: Home√ Commercial
Ready to use: Yes√ No

the rubber grips on the ball valves. They provide a solid grip without any slipping.

### Ease of Setup (10)

This was the one unit that required no additional equipment to set up. Just clean the unit before use. Braukunst even manufactures a tool for this task. (The K-63 hose cleaner retails for $36.95.) Attach their beer line quick-disconnect to your keg and their $CO_2$ fitting to your regulator, then bottle.

### Ease of Operation (6)

The configuration and the use of ball valves makes operation easy and provides a straight path for the beer, which helps prevent foaming. The placement of the ball valves was the only problem we had with this unit. Arranged in a V shape, the ball valves are close together and are awkward to operate.

### Efficiency (10)

One of the most enjoyable units to use. From the ease of setting up to the effective valve operation, no problems were encountered.

## Effectiveness

$CO_2$ retained: 91.98 percent of original volumes.
Air: 8 percent below recommended levels.

## Homemade Bottle Filler

**Homemade Bottle Filler**
Retail Price:   20 cents
Application:    Home√   Commercial
Ready to use:   Yes√    No

### Instructions (10)

Simply slip a piece of ¼-inch inside-diameter tubing into the mouth of your keg's squeeze faucet. Insert the tube into the bottle, then squeeze the trigger and fill the bottle.

### Construction and Materials (6)

(One-quarter-inch inside-diameter plastic tubing.) As simple a design as you can get. No frills, no added expense, and no complicated procedures.

### Ease of Setup (10)

Beyond putting the plastic tubing into the squeeze faucet, none is required. The same procedures for chilling beer and bottles as with the other fillers must be observed.

### Ease of Operation (10)

If you can serve beer from your keg's squeeze faucet, you can use this method to fill bottles.

### Efficiency (4)

While this filler is easy to use and affordable, it fails to purge the bottle of oxygen and it does not fill via the counterpressure principles. It was included in this review because of its low cost and ability to put beer into bottles quickly. If you fill with low enough temperatures and pressures, minimal $CO_2$ will be lost. Regarding the unit's inability to purge the bottle of air, few homebrewers filling bottles via conventional means purge the bottles. It is assumed that the same rate of oxidation will occur. Minimal in a worst-case basis.

### Effectiveness

$CO_2$ retained: 76.72 percent of original volumes.
Air: 19 percent below recommended levels.

## SUMMARY

The process of putting beer into a keg and then into bottles may not be for everyone. Most brewers would rather take beer from kegs and put it directly into pint glasses. If you are in the market for a CPBF, the information presented here and summarized in Table 3-3 should help you make an informed purchase. Considering the findings of this project, we believe cost and unit effectiveness should be the most closely scrutinized. You should purchase a unit that will give you the results you need for the money you have to spend. Remember, regardless of the filler you buy, your $CO_2$ and oxygen effectiveness will improve with practice.

### COUNTERPRESSURE BOTTLE FILLER PARTS LIST

While differences among kegging systems and CPBF exist, the basic design and function of the equipment will be the same. A kegging system must have a pressurized container to hold the beer and a $CO_2$ source to dispense it. A CPBF allows the user to transfer beer from the pressurized keg to a pressurized bottle. Tying these two pieces of equipment together is not difficult, but it may be a uniqe fit depending on the equipment you are using. Listed below are the minimum requirements that homebrewers will need to use a CPBF.

At the heart of this enterprise would be a homebrew kegging systyem. Generally these systems cost beween $175 and $250. They should include the following:

1. soda keg
2. $CO_2$ cylinder
3. $CO_2$ regulator
4. $CO_2$ quick-disconnect
5. liquid quick-disconnect
6. pressure-rated tubing and hose clamps
7. faucet

Once your kegging system is assembled and operational, kegging, not bottling, may be your preferred method of packaging. The savings of time, both in filling bottles and waiting for the beer to carbonate, can be worth the expense of the kegging system. However, requests for an occasional bottle and the headache of lugging around all your equipment will eventually create an interest in counterpressure bottle filling. Along with a CPBF you will need:

1. extra pressure tubing and clamps
2. a hose-barb T (plastic or stainless steel) or a $CO_2$ regulator manifold
3. additional liquid tap quick-disconnect

**COUNTERPRESSURE BOTTLE FILLING STEP BY STEP**

1. Assemble you CPBF (if required).
2. Connect a line to both the CPBF and to the keg of beer by using a plastic or stainless steel T.
3. Connect a beer line from the keg to the CPBF.
4. Insert the CPBF filling tube into the neck of a sanitized and chilled bottle, checking to make sure that the rubber stopper forms a tight seal.

## Table 3-3. Counterpressure Bottle Filler Evaluation Summary

| Model | Instructions | Construction/ Material | Ease of Setup | Ease of Operation | Efficiency | Effectiveness Volumes $CO_2$* | Air** | Manufacturer's Suggested Retail Price | Home Use | Commercial Use | Ready to Use |
|---|---|---|---|---|---|---|---|---|---|---|---|
| Zahm & Nagel | 10 | 10 | 2 | 10 | 7 | 91.25% | 25%+ | $370.00 | | √ | No |
| Beverage People | 1 | 8 | 5 | 8 | 5 | 93.65% | 8%– | $59.95 | √ | | No |
| Benjamin Machine BFD 3B | 6 | 8 | 6 | 7 | 6 | 88.91% | 27%+ | $60.00 | √ | | No |
| Benjamin Machine BFD 3S | 6 | 8 | 6 | 7 | 6 | 90.52% | 3%– | $125.00 | √ | √ | No |
| Foxx | 8 | 4 | 6 | 2 | 7 | 89.53% | 12%+ | $40.00 | √ | | No |
| Melvico | 6 | 9 | 10 | 9 | 10 | 95.94% | 8%+ | $350.00 (Can) | √ | √ | YES |

| | | | | | | | |
|---|---|---|---|---|---|---|---|
| Braukunst | 10 | 8 | 10 | 6 | 10 | 91.98% | 8%— | $44.95 | √ | Yes |
| 20¢ Homemade Filler | 10 | 6 | 10 | 10 | 4 | 76.72% | 19%— | $0.20 | √ | Yes |

*The first five criteria were rated on a scale of one to 10 by the authors, 10 being the highest score.*

*The effectiveness of each filler was tested by Dr. Fix with a Zahm and Nagel $CO_2$/Air Tester.*

*\*Volumes of $CO_2$ are expressed as a percentage of $CO_2$ retained during transfer from keg to bottle.*

*\*\*The amount of air picked up during transfer is expressed as a percentage above (+) or below (−) 2 milliliters, the accepted standard for homebrewers.*

5. Purge air from the bottle following manufacturer's instructions. Some fillers may not include this step.

6. Fill the bottle with $CO_2$ and position the $CO_2$ valve according to manufacturer's directions.

7. Open the beer-line valve. If all has gone well and the keg and bottle are at equal pressure, no beer will flow when the beer-line valve is opened.

8. Slowly open the $CO_2$ pressure-relief valve and start the flow of beer into the bottle. Do not allow the beer to foam.

9. Fill the bottle to the manufacturer's recommended level, then turn off pressure-relief, beer, and $CO_2$ valves.

10. Remove the filler and cap the bottle.

**TOP 10 TIPS FOR TROUBLE-FREE COUNTERPRESSURE BOTTLE FILLING**

1. Read instructions carefully and follow them completely.

2. Chill your beer and bottles to between 34 and 37 degrees F (0 and 3 degrees C).

3. Invest in a $CO_2$ regulator manifold. Multiple $CO_2$ lines can then be run simultaneously from one regulator to the keg and CPBF, maintaining proper pressure (unless the manufacturer's design prohibits this setup).

4. Work with another person; one will bottle and the other will cap.

5. Wear old clothing and safety glasses.

6. Try your system with water first. No sense wasting good beer.

7. Purge air from the bottle. Oxidized beer can be avoided.

8. If the beer foams too much, start again.

9. Remove the fill tube slowly. Excessive movements will cause the beer to gush.

10. Relax, don't worry, have a homebrew.

## REFERENCES

*Beer Packaging*, Master Brewers Association of the Americas, 1982, pp. 150–51, 580.

*Brewing Science*, vol. 2, edited by J. R. A. Pollock, Academic Press, Inc., 1981.

David Ruggiero has been a homebrewer for over twelve years. He owns and operates Barleymalt and Vine, a homebrew supply shop in Boston, Massachusetts. In April 1995 he opened a brew-on-premise at his shop.

Doug Snyder is a videographer/editor and founding member of Echo Bridge Productions. He has been involved in homebrewing since 1980, but didn't brew a batch in his home until 1993.

Jonathan Spillane has been a self-employed carpenter for twelve years and an avid homebrewer since 1993. In 1995 Jonathan installed a 10-gallon grain brewery in his basement, and was involved in the design and building of the Barleymalt and Vine U-Brew.


# HOW TO OXYGENATE YOUR WORT— CHEAPLY

### BY RUSS KLISCH

*This article originally appeared in* Zymurgy, *Special 1992 (vol. 15, no. 4)*

Put oxygen in my wort? This will eventually give my beer a stale flavor, won't it? How can it be that sometimes you want to put oxygen in your wort and other times you do everything possible to prevent it? Oxygen in wort is one of brewing's paradoxes. Sometimes it's good and sometimes it's bad. Knowing the difference is critical to making good beer.

Oxygen is needed right after the wort has been cooled to pitching temperature and has just entered the fermenter. At this time the yeast is in an aerobic stage. Early in life yeast needs oxygen, an essential ingredient in the cell membrane and for its sex life. Without oxygen, yeast will not reproduce as quickly, and there will be a higher percentage of dead yeast cells, which have a marked effect on flavor. Knowing this, it's hard to deny yeast the proper amount of oxygen needed to carry on a decent sex life. Rapid yeast growth is

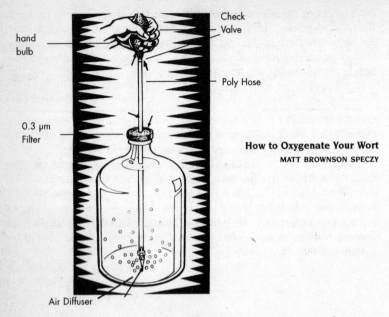

**How to Oxygenate Your Wort**
MATT BROWNSON SPECZY

one of the best methods of fighting off contamination. Another way of thinking about this is that yeast is doing its beer-obics at this stage and needs oxygen.

After the first few hours of yeast growth, oxygen should never be introduced into the wort.

But how do you get oxygen into the wort? For those who don't have any equipment, take a few gallons of cold wort and slowly pour it from one container into another. This is normally done with the yeast in the wort.

There are two things to remember when doing this. First, the cooler the wort, the more air it will absorb. Second, the more small bubbles you create, the greater the surface area that will absorb oxygen. When pouring the wort, the object is to create a slow, steady stream that trickles into the other container and causes a lot of splashing and small bubbles. Wort in the two containers should be poured back and forth at least ten times. This pitching of the wort and yeast is how the term "pitching yeast" came about. The great fear of doing this is that airborne bacteria will get into the wort. This should be a concern, but until a year ago, this is how the wort was aerated at my brewery, Lakefront, with no contamination problems.

Anyone who has tried this method knows it can be very tiring and messy. An easier method is to use a wort aeration device that is nothing more than a squeeze bulb with two check valves on both sides, food-grade tubing, a bacterial air vent, and a cindered air muffler with hose barb at the end. This will allow you to pump small bubbles of sanitized air into your wort with great ease. The parts you need for the aerator are listed below.

| Part/Example | Source | Estimated Cost |
|---|---|---|
| Air muffler/Parker EM25 | Pneumatic fittings outlet | $7.00 |
| Female hose barb/ Parker 126 HBL-6-4 | Hardware store | $1.50 |
| Bulb with check valves/S73125 | Sargent-Welch | $6.50 |
| Food-grade tubing/ 5 feet of ⅜ inch | Homebrew supply store | $3.00 |
| Bacterial filter/ L-29701-00 Vacu-Guard | Cole-Parmer | $5.70* |
| *Total Cost* | | $23.70 |

You can probably substitute some equivalent items you already have. Some suppliers may only sell wholesale, so find out what items are available at retail.

The aerator can be sanitized with a chlorine solution and washed by rinsing hot water through the muffler. Chlorine is a very corrosive chemical, and the cindered muffler should not sit in the solution for more than a few minutes or the brass will start to corrode. The cindered air muffler will create the smallest bubbles when a slow, steady stream of air is passed through it. Do not pump high air pressure through the aerator because it may damage the filter membrane. Try to pass air through the wort for at least 5 minutes. Make sure the carboy is not too full because the wort will foam a lot. If the filter is not abused, it should last for at least ten batches before needing replacement.

All of the "wizards" out there probably know that because of Henry's Law, wort saturated with air cannot contain any more than the 21 percent oxygen that air contains (20 percent is optimum). To increase the amount of saturated oxygen in your wort, you

---

*The filters are only available in packages of ten. Cole-Parmer sells filters that will filter 0.3 micron or less but are more expensive.

must use pure oxygen. You can do this by replacing the squeeze bulb with an oxygen cylinder and regulator. This will take the strain off your hand and free it up for a homebrew.

Unfortunately this also becomes more expensive for most homebrewers. The smallest cylinder you can buy is an R size, which, with regulator, will cost about $185 at a welding supply store. The good thing about buying oxygen is that a cylinder will last a long time and it is relatively inexpensive—a refill costs about $9.

One important thing to know when buying oxygen is that there are different grades. The top grade is medical, and to get it legally you must have a prescription. The next grade is industrial, and anyone can purchase it at a welding supply store. Technically there is no difference between the medical and industrial oxygen, and both tanks are filled from the same line. The reason for the different grades is that medical oxygen can only be put into a cylinder that has been cleaned and has never been used industrially. If you buy a new tank and only use it for aeration, using industrial gas would be fine. I would not recommend using an old industrial cylinder because acetylene or some other gas could have backflowed into the tank and will contaminate the beer. If you use pure oxygen, beware, because excessive fermentation will result if the wort is fully saturated. Before we were able to determine the correct dosage at Lakefront, the blob that came out of the fermenter looked like something out of a sci-fi movie.

The methods mentioned here are only a few ways a homebrewer can aerate wort. Another method is to use a clean aquarium pump with a bacteriological filter, such as cotton. Other types of filters, air pumps, or bellows are within the reach of homebrewers and can be put to proper use with a little ingenuity.

*Russ Klisch started his homebrew career shortly after graduating in chemistry from the University of Wisconson, Eau Claire, in 1981. Russ turned professional in 1987, as one of the founders of Lakefront Brewery, Milwaukee, Wisconsin.*

# THE JOCKEYBOX

## BY TERI FAHRENDORF

*This article originally appeared in* Zymurgy, Fall 1991 *(vol. 14, no. 3)*

Draft beer is wonderful to have at a picnic or barbecue—but how do you keep the kegs cold? A giant bucket of ice will work, but with several kegs, huge amounts of ice are required to cover the kegs and replace what melts. What's a poor homebrewer to do?

The answer is to chill the beer, not the keg! A device called a jockeybox is widely used by professional servers at sporting events and the like. It consists of a picnic cooler with stainless steel coils inside. The cooler is filled with ice, and beer from the warm kegs is chilled as it flows through the coils and out the taps on the side of the cooler. Usually two or three inlet connections and faucets are mounted on the box, though miniature one-tap jockeyboxes also are used.

## USING A THREE-KEG JOCKEYBOX

- jockeybox
- $CO_2$ cylinder
- $CO_2$ regulator with one-way check valve
- $CO_2$ lines that T into three lines, or two more $CO_2$ cylinders (optional—three miniregulators, one for each T line)
- ice
- water bucket
- crescent wrench and tap wrench
- extra rubber washers (see equipment needed to build your own, below)

## SETTING UP

Set the jockeybox on a sturdy surface. A table is best. Fill to the top of the coils with ice. Never put food in the jockeybox. It may look like a picnic cooler, but it is not. Only ice and water should go inside. Food may cause bacterial contamination later.

Fill with water to the top of the coils. Make sure the drain plug is closed. Set the three Cornelius kegs out of your way. Under the table is best. Attach the beer hoses to the inlet ports on the side of the jockeybox. Be sure there is a small rubber washer for each connection. This prevents beer from leaking under pressure. You will probably need a crescent wrench to connect the hoses.

Close the jockeybox taps. Attach the beer lines to the Cornelius kegs. You should now be able to pull a little beer from each of the taps.

Attach $CO_2$ hoses to the $CO_2$ cylinder. Close the one-way check valve and set the regulator to 25 or more psi. Open the check valve. If you have three miniregulators in-line, set them the same way for about 15 psi each. (If you have double coils, set the main regulator to about 40 psi, and the miniregulators to about 22 psi.)

Open any toggle switches or check valves in the $CO_2$ lines so that $CO_2$ is rushing out of the lines. (You can use a pen to push in the pin on each of the Cornelius fittings to make sure the $CO_2$ is on and flowing.) Connect the $CO_2$ lines to the Cornelius kegs. You are now pushing beer through the jockeybox.

## DISCONNECTING

Close the jockeybox taps. Disconnect the $CO_2$ connections from the Cornelius kegs, while the $CO_2$ is still flowing. Then turn off the large valve on the top of the $CO_2$ cylinder. Disconnect the beer hoses from the Cornelius kegs first, then disconnect the beer hoses from the jockeybox. Don't lose the washers. Disconnect the $CO_2$ hoses from the $CO_2$ cylinder.

## CLEANING THE JOCKEYBOX

As you might imagine, it is important to keep the jockeybox clean so that bacteria do not grow inside the lines and coils and change the flavor of the beer as it is served. The jockeybox can be cleaned with a little help from a Cornelius keg and $CO_2$ cylinder.

Fill a Cornelius keg half full with trisodium phosphate (TSP) or another cleaning solution. Hook up the $CO_2$ cylinder and connect it to one of the jockeybox inlet connections. Pull the cleaning solution through to the tap until all the beer is out and let it run a few seconds. Reconnect the $CO_2$ cylinder to the next inlet connection and pull the solution through to the next tap. Repeat for the third inlet and tap. You have just cleaned all the coils.

Now rinse the Cornelius keg and fill it half full with a mild bleach solution (½ ounce bleach in 2½ gallons of water). Attach $CO_2$ and pull the bleach solution through to each tap as described above.

Now rinse the Cornelius keg again and run cold water through the jockeybox to each tap as described above. To purge the final rinse water, let all water flow out of the last tap until $CO_2$ is spitting through. Hook the empty but pressurized Cornelius kegs up to the other two taps and let the $CO_2$ push the rinse water through them too.

The coils are now clean and clear. Just wipe the inside and outside of the jockeybox with a damp cloth to remove beer stains. Store the jockeybox with the drain plug open, the taps open, and the lid slightly ajar (until it's dry inside).

Before the next use, just run the bleach water and then rinse water through the jockeybox. Remember, *always run rinse water after bleach water. Never leave bleach water in the coils overnight.* They are thin and expensive, and the bleach will eat tiny holes in them. The next morning you could have a beer sprinkler instead of a beer server.

If you don't let the beer sit in the coils overnight, you can skip the TSP or cleaning step; just do the bleach water and rinse water steps. Every now and then the jockeybox coils should be thoroughly cleaned with TSP or another cleaning solution.

Don't forget to clean your beer hoses!

## BUILDING A THREE-KEG JOCKEYBOX

- One large plastic picnic cooler.
- Drill, crescent wrench, and screwdriver or pincher (depending on the type of hose clamps you use).
- Three stainless steel coils. Single coils are 50 feet by $5/16$-inch inside diameter (ID) stainless tubing. Double coils use both ⅜-inch ID and ¼-inch ID to increase pressure flow. Double coils cool more efficiently but cost about $100 each. (Note: Copper coils have been used successfully by homebrewers, but are not recommended. Copper can react with beer and cause oxidation if it sits in the coils.)
- Three beer faucet assemblies with tap handles (beer outlet outside box).
- Three beer shank assemblies (beer outlet through box).
- Three beer cooler couplings (beer inlet through box).
- Three faucet coupling hex or wing nuts and hose tailpieces to attach to your three beer hoses (beer inlet outside box).

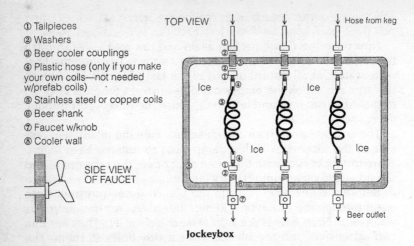

① Tailpieces
② Washers
③ Beer cooler couplings
④ Plastic hose (only if you make your own coils—not needed w/prefab coils)
⑤ Stainless steel or copper coils
⑥ Beer shank
⑦ Faucet w/knob
⑧ Cooler wall

**Jockeybox**

- Optional: six more faucet coupling hex or wing nuts and hose tail pieces if you make your own coils (beer inlet and outlet inside box). See Note.
- Neoprene coupling washers for hex or wing nut connections.
- One tap or faucet wrench.
- Hose clamps for hose tailpieces and beer hoses.
- Beer hose.

Note: If you buy the coils, they come with hex nut fittings to attach to the cooler couplings (beer inlet) and beer shank assemblies (beer outlet) inside the cooler box. If you bend your own coils, you will have to connect short pieces of hose to both ends of the coils and then to a hose tail piece and faucet coupling hex or wing nut. This will allow you to connect the coils to the beer inlets and outlets inside the box.

Jockeyboxes, parts, and catalogs are available from many homebrew retailers, soda distributors, and beer distributors.

*Teri Fahrendorf is Brewmaster at the Steelhead Brewery in Eugene, Oregon. She is a member of the Master Brewer's Association of the Americas and received training at the Siebel Institute of Brewing Technology in Chicago. She has been a homebrewer since 1985, a home wine-maker since 1980, and a home bread baker before that. (She likes yeast a lot.)*

# ON THE SURPLUS TRAIL

## by Randy Mosher

*This article originally appeared in* Zymurgy, Special 1992 *(vol. 15, no. 4)*

As corporate America upgrades, shuts down, rebuilds, and replaces, it tosses off vast quantities of used but otherwise perfectly fine parts and materials. Let's face it, folks, stainless steel is very durable stuff, it's not going to simply melt away.

Every city of any size will doubtless have one or more of these types of surplus equipment outlets available to you. The Yellow Pages are your best bet. Generally the type of surplus available matches the industry in that area. If you're looking for junk near an aircraft plant, you're likely to find aerospace junk.

Most of this material is useful to the more advanced brewer who is scaling up to a 10-gallon or larger system, but there are plenty of items useful to all serious brewers. Knowing what to look for is important, but the general rule is, if it's cool, stainless steel, and cheap, buy first and ask questions later.

Do be careful to avoid leaded metals and make sure whatever you scrounge is free of oil, grease, and gasoline or can be thoroughly cleaned. It is a good idea to have any technical equipment safety-checked.

The industrial surplus and salvage category is the first place to look. You'll always find things like motors, relays, electrical parts, nuts, bolts, casters, gears, and other ubiquitous items. If you get lucky, you might stumble on stainless steel valves, pumps, filter housings, and other invaluable components.

Electrical surplus may offer much more than just electrical parts. There is one in Southern California that has a mountain of used Cornelius tanks, for example. And of course, what's a modern brewery without switches, motors, pumps, and remote temperature measurement—all electrical parts? Look for them in both electronics and surplus categories of the Yellow Pages.

When searching scrapyards the important thing is to find one that has high-class stuff. Generally they list their specialties in Yellow Pages ads. What you want is one that specializes in non-ferrous metals, and if you can find one that specializes in exotic metals, so much the better. If you're looking for castoffs from one particular factory, you can even call that plant and ask who buys their scrap.

Expect to find chunks, strips, bars, rods, sheet, and tubing of various metals, at least. Sometimes you can find tanks and other vessels as well. Perforated sheet metal and screen wire can sometimes be found. I find it useful to take a set of tools with me—wrenches, pliers, screwdrivers, etc.—to remove parts and fittings from big hunks of wreckage. Be sure to wear old clothes and sturdy shoes, and carry a rag with you, as there is no dirt quite as filthy as junkyard dirt.

Check the university surplus outlets. Scientists like to have the very newest and best laboratory equipment, so there are a lot of perfectly workable, but not state-of-the-art, instruments that need to be disposed of. You can find pH measurement equipment, spectrophotometers, microscopes, incubators, and a host of other lab equipment at these outlets. Check with the university for location and hours.

Corporate surplus outlets are the same idea as university outlets, but specialize according to the industry. Drug companies, for example, cast off very expensive pilot fermentation plants rather than do the extensive work required to recertify them to FDA standards.

You never know what will show up at flea and antique markets. I once turned down a whole box of specific ion probes for an ion meter at a giveaway price, thinking "Where on earth will I ever find an ion meter?" Later, of course, I found the meter for 20 bucks, but by then the probes were long gone. All kinds of useful lab stuff shows up, and beverage-handling equipment is quite common. Old beer taps can be found at flea and antique markets, as can old grocery-store scales that are useful for weighing grains.

Used material shops are a step up from scrapyards. They have done the picking for you and generally offer neat pieces of sheet, tube, bar, plate, rod, perforated sheet metal, and screen, plus other goodies like brass and stainless steel nuts, bolts, and more. They are sometimes listed under building material, used.

What do drug dealers use that homebrewers need? Scales! Police and sheriff's auctions of seized contraband usually are chockablock with expensive laboratory balances, perfect for weighing hops. You may have to buy a few at a time, so it's a good idea to get several people together before you go.

More than a few microbreweries have been started with tanks and other equipment from used process equipment dealers. Most of the equipment is a little large for homebrewers, but we can dream, can't we?

Stoves, sinks, pots, tanks, kettles, and other useful items can be had for prices considerably lower than full retail from used restaurant equipment dealers.

Much of the preceding stuff has passed through an auction before you get to it. Check your local newspaper for current auction listings. Restaurants, dairies, laboratories, food-processing plants, and other kinds of places are always biting the dirt and ending up in liquidation.

Many laboratory and science supply houses are for industrial customers only, but most cities have one that will sell over the counter. You'll pay full retail, which will be a shock if you've been shopping the surplus joints. Sometimes you just need that one little part to put something together and then it's worth it. Laboratory supply houses are a good source of chemical flasks, chemicals, filter paper, pipettes, and other yeast-culturing supplies, all of which are reasonably priced in any event. Be sure whatever you buy is free of radioactivity, viruses, or any other dangerous substances.

So there you have it. Combine this with the ability to put it all together and you can have a world-class brewery for $19.95, or slightly higher in some states, batteries not included.

*Randy Mosher began homebrewing in 1983. He is the creator of Doctor Bob Technical's Amazing Wheel of Beer and author of The Brewer's Companion (Alephenalia Publications). Randy enjoys tinkering with homebrew systems. Professionally he runs his own graphic design and product development company.*

# four

# Brewing Theory

## AN EASY GUIDE TO RECIPE FORMULATION
### A For the Beginner column
BY MONICA FAVRE AND TRACY LOYSEN

*This article originally appeared in Zymurgy, Winter 1989 (vol. 12, no. 5)*

Authors' note: *We're writing for those brewers who want to do more than follow the directions on a can, but who rely on basic brewing techniques such as boiling on the stove, straining out certain ingredients, transferring wort to a carboy, and fermenting at room temperature. Accordingly, we will direct our discussions to people who brew ales with malt extract (or dried malt) and additional ingredients, such as adjunct malts and hops, that can be thrown into the brewkettle without any special treatment.*

It was time to brew again. Winter was approaching and we wanted to make a stout for the holiday season. But where to begin? First we had to clarify what we liked in a stout. After some thought, we decided to create a brew that was not too bitter, but still dark and rich in flavor.

As beginners we had little idea of what ingredients and procedures would result in these characteristics. What makes a beer dark?

What gives it body? What gives it bitterness? We looked up the style definition for some clues. It told us about the average percent alcohol, color, and general flavor characteristics, all of which were interesting but not helpful. Of course, we knew we could simply follow someone else's recipe, or ask our local brewing professor to help us write one, but we wanted to understand how to create a specific beer style to match our own unique specifications.

Next we tried looking at other people's recipes. This was a useful process. Though the recipes varied in many ways, we saw that certain ingredients showed up with regularity: dark malt extract, roasted barley, and black patent malt. We formulated our preliminary recipe with these common ingredients, and decided on quantities using intuition and by averaging the amounts that other brewers used in their recipes.

Armed with our painstakingly created recipe, we went to our local homebrew shop. The owner, however, didn't have all our ingredients. "Don't worry," he said, "use these instead." He poured them into plastic bags without measuring the specific amounts we had carefully determined. We left his store feeling that we had the makings for a good brew, but one that was not our own. We were back where we started, with someone else's recipe. We realized that without understanding why we needed certain ingredients and what they did, we had no idea how last-minute modifications would influence our end product. We wanted to be better informed.

In our search for information we found a variety of good sources, including two one-page freebies from the AHA called "How to Use Specialty Malts" and "The Zymurgist's Guide to Hops"; the 1986 malt extract special issue of *Zymurgy* (vol. 9, no. 4); charts in *The Complete Joy of Home Brewing* (Avon, 1984); winning recipes; and other, more experienced homebrewers. We recommend exploring any and all of these avenues. (We keep copies of stuff like this, as well as blow-by-blow accounts of our brewing endeavors, in our handy brewing notebook.)

Through our reading and conversations, we discovered information about the basic ingredients of beer that will be helpful to you in your recipe formulation.

Color, body, and maltiness are derived from grains. Darker malts will produce darker beers. In small quantities grains will affect color and add little or no flavor. Your brew's color will darken, flavors will become evident, and the contribution to body will increase as you use larger quantities. Body and maltiness are created from malts with sugars. Most of the sugars in roasted barley and black patent malt have been burnt out, so these grains do not contribute significantly to body. The AHA's guide to malts (mentioned earlier) outlines what color and flavor will result from specific amounts of the specialty grains.

Maltiness and body describe different things—one refers to taste and the other to mouth feel—but we have combined them into one category for ease of understanding and use.

The hop amounts suggested are based on hops of average bitterness. We recommend that beginners stay away from the very bitter varieties because it is easy to add too much. The use of medium-bitterness hops will give you more room for error. Five commonly available varieties in this range are Willamette, Tettnanger, Cascade, Styrian Goldings, and Hallertauer. For further information on differences between hop varieties, consult the "Zymurgist's Guide to Hops."

As a beginner, how do you assimilate all this information into something you can use? We believe that you can write your own recipe by using the following chart, as well as the steps we've outlined in the next section..

Before you formulate your recipe, a good first step is to list the general characteristics of the style you wish to brew. Then refine the list by considering which characteristics you like and want to emphasize, which you want to tone down, and any other special features you want to add. We have found it useful to think in terms of three dimensions along which any beer can vary: color, body/maltiness, and bitterness. Remember, you're brewing for you! Unless you are brewing specifically for a competition, there is no need to adhere strictly to what someone else says your beer should be like.

With your list and the chart in hand, you are now ready to write your recipe. We will take you through the steps of formulating the stout we described earlier, as an example of how to use the chart. We can describe the beer we want as dark-colored, medium- to full-bodied, with medium to high maltiness and low to medium bitterness.

- Color. Knowing that we want a dark color, and that color comes from malt, we'll choose the high end of the base malt range. In addition, we'll choose enough black malt (½ pound) to add to the color without adding a charcoal bitterness.
- Body/Maltiness. We want medium to high for both body and maltiness. The adjuncts we can pick from, according to the chart, are chocolate, roasted barley, black patent, and crystal malts. Of these, only crystal malt contributes significantly to body. So we want a relatively high amount of crys-

tal malt. However, we cannot determine the specific quantity at this stage.
- Bitterness. We want low to medium bitterness. Bitterness is created by both hops and some dark malts. Since we don't want a lot of bitterness, we will only use a small quantity of roasted barley (⅓ pound). We'll use 2 ounces of hops, the lower end of the range, to keep the bitterness down.

Having determined the amounts of roasted and black malts, we can figure the amount of crystal malt to use. The total amount of dark grain adjuncts should be around 1⅓ pounds, leaving us with ½ pound for crystal malt.

The completed list of ingredients for a 5-gallon batch looks something like this:

6½ lbs. dark malt extract

½ lb. black patent malt

½ lb. crystal malt

⅓ lb. roasted barley

2 oz. hops (choose type by personal taste) for boiling

Of course, there are additional ingredients you can add that aren't specific to any beer style, such as finishing hops and spices. You can experiment with these as the mood strikes you.

The three dimensions we present in Table 4-1 are sufficient for tailoring a beer to your specifications, with two exceptions. High alcohol content is a major distinguishing characteristic of barley wine. It results from brewing with a lot of malt; the more malt you use, the higher your alcohol percentage or strength will be. A special feature of some pale ales is "dryness," which results from brewing with certain types of water and/or from brewing with a highly attenuating yeast. If you like this character in beer, you can modify your water by adding 2 to 4 teaspoons of gypsum or choose an appropriate yeast strain. However, these steps are not necessary to create a great pale ale.

Part of the fun in homebrewing is experimenting. It's even more fun (and more relaxing) when you have a general idea of what you're experimenting with.

*Monica Favre and Tracy Loysen (Travre Brewing Company) are former AHA staff members. They don't want to discuss any personal information as it is all rather vague, contradictory, and incriminating.*

Table 4-1. Guidelines for Beginning Recipe Formulation (for 5-gallon batch)

| STYLE | CHARACTERISTICS | | | | INGREDIENTS | | oz. boiling hops* |
|---|---|---|---|---|---|---|---|
| | identifying characteristic | color | body | malti-ness | | lbs. malts | |
| Barley Wine | strong, high alcohol (a lot of malt) | copper to dark brown | full | high | base:<br>adjuncts: | 8-10 light<br>black, chocolate or combination—<$1/3$ for color<br>crystal—no amount limit | 4-6 |
| Brown Ale | brown color, sweet | dark brown | med. to full | low | base:<br>adjuncts: | 5-6 light or amber<br>chocolate—>$1/3$<br>black, roasted or combo—<$1/4$ for color, crystal—$1/4$-$3/4$ for color, body, sweetness | 1-1.5 |
| Pale Ale | hop bitterness | light to copper | low to med. | high | base:<br>adjuncts: | 6-7 light or amber<br>crystal—$1/2$-1 optional for color, body, sweetness<br>choc., roasted or black—<$1/8$ optional for color | 2-3 |

| Style | Color | Bitterness | Malt | Base & Adjuncts | OG (approx) |
|---|---|---|---|---|---|
| **Porter** | dark color, no burnt flavor | dark almost black | med. | high | base: 7-8 lbs. dark<br>adjuncts: black—<1/2 for color, 1/2-2/3 for taste<br>roasted—<1/4<br>chocolate—<1<br>crystal—1/4-1 for color, body, sweetness | 1.5-2 |
| **Stout** | roasted barley character | dark brown to black | med. to full | high | base: 6-6.5 dark<br>adjuncts: black—<1/2 for color, 1/2-2/3 for taste<br>roasted—1/3-1<br>chocolate—optional, <2/3<br>crystal—1/3-1 for color, body, sweetness<br>(Use up to a total of 1 1/3 pounds dark grains in total adjuncts) | 2-2.5 |

*The amounts listed refer to fresh, whole hops of average bitterness (4 to 6% alpha acid)

**Adjunct Malt Characteristics**
black: dry, burnt, charcoal bitterness
crystal: sweet, amber color, body
roasted barley: coffee-like bitterness, roasted flavor
chocolate: dark color, toasted flavor

**Malt Extract Conversions**
5 lbs. Very Dark Extract = 5 lbs. light or amber malt extract + 2 cups roasted, black or chocolate
5 lbs. Amber Extract = 5 lbs. light malt extract + 2 cups crystal = 5 lbs. light malt extract + 1/2 cup black, roasted or chocolate
5 lbs. Medium Brown Extract = 5 lbs. light or amber malt extract + 1 cup black, roasted or chocolate malts + (optional) 1 or 2 cups crystal

# BREW BEER YEAR-ROUND

## BY BYRON BURCH

*This article originally appeared in Zymurgy, Winter 1985 (vol. 8, no. 5)*

It's always a bit sad when brewers feel they should stop beer-making activities during the summer months, because in many cases it is not necessary. True, it may be more fun to stand at the old stove stirring a boiling wort when there's snow on the ground outside than when you can see heat waves rising off the sidewalk. However, it is possible, with a little extra care, to brew very good beer when the weather is warm.

For those fortunate enough to have cellars or spare refrigerators, of course, it's relatively easy. A good cellar will stay somewhat close to 60 degrees F except during heat waves, and that's just about perfect for most fermentations. Where I live, near Santa Rosa, California, I've done some nice lagers in August by fermenting in the cellar.

I've also gone the refrigerator route, and that works superbly if properly set up. The main problem with using a refrigerator for fermentation is that you have to make an adjustment to keep the wort warm enough, because most refrigerator thermostats max out at 45 degrees F. You want the temperature, in most cases, to be as close to 55 degrees F as possible.

The easiest way to accomplish this is to hook the refrigerator up to an automatic timer so it only works part of the day. Get one of those timers people buy to turn their house lights on and off when they're away from home. The liquid temperature in the fermenter changes much more slowly than the air temperature around it, so it won't normally fluctuate very much.

A little experimentation should show you the best way to set up your system. Just remember that if your unit has the freezer at the top, you should use something like cheap vodka in the fermentation lock so it won't freeze solid and build up pressure in the carboy. Too much pressure buildup, of course, would lead to weeping, wailing, and work, and an ounce of prevention beats crying over spilled beer.

One final thing to remember is not to put your wort in the refrigerator until the yeast has actually started to work. Keep the wort temperature between 70 and 80 degrees F until that happens.

Many brewers, of course, have neither a spare refrigerator nor a cellar, but all is far from lost. The solution is really quite simple—

a water bath. Get a large plastic trash container to put your fermenter in, and keep water between the two while fermentation is going on. Not only does the water provide insulation, but also the small amount of evaporation that occurs provides a bit of additional cooling as well. If things get really tough because of a severe heat wave, you can even add a small amount of ice, though this will not be necessary as often as you might think.

I first suggested this approach to a couple who were desperate to begin making beer even though their west-facing apartment hit 90 degrees F every afternoon without fail (which is probably why they were so desperate). As it turned out, they made an excellent California Common beer.

When it was done, they not only brought me one, but showed me the meticulous, twice-a-day temperature records they had kept. The beer generally fermented in the high 60s, a few times getting as high as 70 degrees F. They were pleased and so was I.

In any case, it works. You can also take wet towels, leave one end in water and drape them over your secondary fermenter so that they continue to pick up moisture. The evaporation will provide additional cooling.

I generally try to do at least two brews each summer. In July or August I might be putting together an Imperial Stout, getting ready for drinking next winter and spring. In June I might be working on a quick, light beer for later in the summer.

If you're in the mood to brew something like this next June, you should. Don't let the weather stop you. Here's a simple recipe to get you started. My business partner, Jay Conner, used a simple recipe much like this and almost won "Best of Show" in the 1984 Home Wine and Beer Trade Association Judging.

## American Pilsener

*Ingredients for 5 gallons*

- 3 lbs. British hopped light dry malt
- 1 lb. white rice syrup
- Water to make 5 gal.
- 1 tsp. yeast nutrient
- 1 tsp. gypsum
- ½ oz. Hallertauer hop pellets (aromatic)
- 14 g. Red Star lager yeast
- ¾ c. corn sugar (for priming)

Dry malt, rice syrup, gypsum, and nutrient should be added to the water and boiled 30 minutes, with the hop pellets added at the end. You can enjoy this one as soon as it is carbonated, but it may take six weeks or so in the bottle to clarify fully.

*Byron Burch is the author of* Brewing Quality Beers: The Home Brewer's Essential Guidebook, *Joby Books, 1993. He owns* The Beverage People, *a supply outlet for homebrewers and wine-makers in Santa Rosa, California. His articles have appeared in Zymurgy, Celebrator Beer News, All About Beer, Brew Your Own, The Moderation Reader, Practical Winery & Vineyard, and others.*

# BREW BY THE NUMBERS—ADD UP WHAT'S IN YOUR BEER

## BY MICHAEL L. HALL, PH.D.

This article originally appeared in Zymurgy, Summer 1995 (vol. 18, no. 2)

In the following article I will develop equations to describe many different facets of homebrewing. In each section I first give an equation that is as accurate as possible, including the effects of all of the important parameters. This equation is sometimes rather complicated, but the idea is to program it into a computer spreadsheet and never look at it again. Then, where possible, I simplify the equations using appropriate assumptions so that calculations can be made on the fly in the brewery or kitchen.

One thing I should explain at the outset is curve fitting. There are many occasions when a functional relationship for a set of data is needed but not available. Curve fitting (also called regression or least-squares fitting) is nothing more than a mathematical way to draw the best curve through a set of experimental data points. In this article I use quadratic ($a + bx + cx^2$) and cubic ($a + bx + cx^2 + dx^3$) functions to draw the curves.

## SPECIFIC GRAVITY

Specific gravity is defined as the density of a substance relative to the density of water. This is not quite sufficient as a definition because the density of water varies with temperature. Above 39 degrees F, water expands as it heats up, making it necessary to specify the temperature when indicating a specific gravity.

The most common way for a homebrewer to measure specific gravity is to use a hydrometer. A hydrometer measures the density of a fluid in units of the density of water at some reference temperature (the temperature at which the hydrometer is calibrated). In other words, the measurement you get from your hydrometer is really

$$\text{Measured SG} = \frac{\text{density of wort at temperature T}}{\text{density of water at reference temperature}} \quad (1)$$

For most of the hydrometers available to homebrewers, the reference temperature is 60 degrees F, which is the value that will be used in the rest of this section.

The measured specific gravity value must be temperature-corrected before it is meaningful. The ideal way to do this would be to convert the density of the wort to the reference temperature, like this:

$$\text{Ideal SG} = \frac{\text{density of wort at 60 °F}}{\text{density of water at 60 °F}} \quad (2)$$

Modifying the wort density to account for temperature is not that easy, however, because we don't have tabulated data for the variation of wort density with temperature for every possible wort composition. Fortunately, the temperature coefficients of expansion for wort and water are practically the same, making the density ratio roughly invariant to temperature. Making this assumption leads us to this equation:

$$\text{Corrected SG} = \frac{\text{density of wort at T}}{\text{density of water at T}}$$

$$= \left(\frac{\text{density of wort at T}}{\text{density of water at 60 °F}}\right) \left(\frac{\text{density of water at 60 °F}}{\text{density of water at T}}\right) \quad (3)$$

$$= (\text{Measured SG at T})\,(\text{SG correction factor})$$

This correction factor is solely a function of the density of water at various temperatures, which is well known. I made a curve fit to some data from a couple of sources (De Clerck, Weast) that yielded the SG correction factor as a function of measuring temperature (T in Fahrenheit):

SG correction factor = $1.00130346 - 1.34722124 \times 10^{-4} T$
$$+ 2.04052596 \times 10^{-6} T^2 - 2.32820948 \times 10^{-9} T^3 \quad (4)$$

This equation is accurate over the entire region from 32 degrees F to 212 degrees F.

If you're putting this equation in a spreadsheet, the most accurate way to do it would be to multiply your measured specific gravity by the correction factor evaluated at the temperature at which the SG measurement was made, as shown in Equation 4. However, for the times when it is easier to just add or subtract a point or two, we can make the following approximation. Both the measured SG and the correction factor are numerically very close to one. If we represent them by one plus a small number ($e_{SG}$ or $e_{CF}$), then

$$\begin{aligned} \text{Corrected SG} &= \text{(Measured SG at T) (SG correction factor)} \\ &= (1 + e_{SG})(1 + e_{CF}) \\ &= 1 + e_{SG} + e_{CF} + e_{SG} E_{CF} \end{aligned} \quad (5)$$

Because both e terms represent small quantities, multiplying them together makes a number that is very small with respect to the other terms and can be neglected. This yields

$$\begin{aligned} \text{Corrected SG} &\cong 1 + e_{SG} + e_{CF} \\ &= \text{(Measured SG at T)} + \text{(SG correction factor} -1) \end{aligned} \quad (6)$$

So instead of multiplying by the correction factor, you can just add the correction factor minus one. Table 4-2 gives the additive correction factors (in points; see SG in Nomenclature/Glossary for explanation) as a function of the measuring temperature. Using an additive correction factor is not as accurate as the multiplicative correction factor, but it is easier and is adequate for most circumstances.

## EXTRACT

Since the early days of brewing, brewers (and drinkers) have been concerned about the strength of their concoctions. One measure of wort strength is how much material has been extracted from the malted grains into solution by the mashing and lautering processes. The term *extract* refers to the weight percent of dissolved materials in the wort. Weight percent is percent by weight. For example, if you have 5 pounds of sugar in a solution that weighs 100 pounds total, the solution is 5 weight percent sugar. (Incidentally, this is also 5 °Plato.)

## Table 4-2. Specific Gravity Temperature Correction

| Measuring Temperature (°F) | Addition to Specific Gravity in Points |
|---|---|
| 32 | −0.83 |
| 40 | −0.97 |
| 45 | −0.84 |
| 50 | −0.69 |
| 55 | −0.38 |
| 60 | 0.00 |
| 65 | 0.53 |
| 70 | 1.05 |
| 75 | 1.69 |
| 80 | 2.39 |
| 85 | 3.17 |
| 90 | 4.01 |
| 95 | 5.01 |
| 100 | 5.91 |
| 105 | 6.96 |
| 110 | 8.08 |
| 115 | 9.26 |
| 120 | 10.50 |
| 125 | 11.80 |
| 130 | 13.16 |
| 140 | 16.07 |
| 150 | 19.15 |
| 160 | 22.45 |
| 170 | 25.93 |
| 180 | 29.59 |
| 190 | 33.40 |
| 200 | 37.35 |
| 212 | 42.42 |

Note: *Subjecting a room temperature hydrometer to 212 °F temperatures can break your hydrometer.*

The only difficulty with this is that there is not a good way to measure the amount of dissolved materials, short of evaporating your wort until you have dry malt extract, measuring the weight of the powder, and then dividing by the weight of the original solution. One thing that you can measure is the specific gravity, which is the density of the solution relative to the density of water. But there is still a problem: the dissolved materials are made up of fermentable sugars, nonfermentable sugars, proteins, and other goodies, with proportions that vary from wort to wort. How can the relationship between weight percent of solids and specific gravity be determined if the identity of the solids isn't even known?

In 1843 Carl Joseph Napoleon Balling determined a way around this problem. He noticed that the specific gravity of a wort increased with the weight percent of dissolved materials in almost the same manner as if the dissolved materials were entirely sucrose. He could do all of his experiments with pure sucrose and water, and they would be a good approximation to beer worts. After making up sugar solutions, Balling developed a table relating the density of the solution (at 17.5 degrees C) to the weight percent of sugar. He measured the weight percent in degrees Balling (°B), defined as the number of grams of sugar per 100 grams of wort. Several years later, around 1900, Dr. Fritz Plato corrected some slight mistakes and developed his own set of tables, calling the corrected unit a degree Plato (°P). To convert back and forth between extract (E) in degrees Plato and specific gravity (SG), you can use these equations:

$$E = -668.962 + 1262.45\, SG - 776.43\, SG^2 + 182.94\, SG^3 \qquad (7)$$

$$SG = 1.00001 + 0.0038661\, E + 1.3488 \times 10^{-5}\, E^2 + 4.3074 \times 10^{-8}\, E^3 \qquad (8)$$

These equations are accurate cubic fits that I did to Plato's data (Timmermans) over the range of 0 to 33 °P, which covers specific gravities between 1.000 and 1.144. A simpler formula that is good for most applications is:

$$E = 1000\, (SG - 1) / 4 \qquad (9)$$

In other words, just take the number of specific gravity points and divide by four to get the extract value in degrees Plato. For example, if you measured the specific gravity of your wort to be 1.065, then the simple formula (Equation 9) gives an extract of 16.25 °P, while the cubic formula (Equation 7) gives an answer of 15.88 °P.

But there's still a little more to extract than this. If you measure the specific gravity before fermentation starts and convert that number to degrees Plato, that's called the *original extract* (OE). If you take a measurement after fermentation is finished, that's called the *appar-*

ent extract (AE). The reason it isn't the real extract is that your beer is no longer just a solution of solids and water. Now you have alcohol in there too, and alcohol is less dense than water (specific gravity of 0.794 at 15 degrees C), so it changes all the neat equations that we've just come up with. Just as before, the hard way to determine the real extract (RE) of your beer would be to boil it until all of the alcohol is boiled off, replace the volume with water, and then measure the specific gravity. Fortunately, Balling once again comes to the rescue with an empirical relationship between the real extract, the apparent extract, and the original extract (De Clerck):

$$q = 0.22 + 0.001 \; OE$$
$$RE = (q \; OE + AE) / (1 + q) \quad (10)$$

The variable $q$ is called the *attenuation coefficient*, but you can just think of it as an intermediate value. Many sources quote this equation in the form $RE = 0.8192 \; AE + 0.1808 \; OE$, but this form assumes that $q$ is calculated at an OE of zero, which is not very accurate. For our simplified version, we'll calculate $q$ assuming a reasonable OE of 12.5 °P, which gives us

$$RE = 0.8114 \; AE + 0.1886 \; OE \quad (11)$$

Let's say our beer has finished fermenting and we have measured the specific gravity to be 1.014. The cubic formula for extract (Equation 7) gives an apparent extract of 3.57 °P, but the simple formula does very well and gives a value of 3.50 °P. Calculating the real extract using the most accurate RE formula (Equation 10) and the cubic extract formula gives 5.92 °P. Making the same calculation with the simple version of both formulas (Equations 9 and 11) gives 5.90 °P. The difference in real extract calculated by these two methods is small, but the two methods show greater differences in subsequent calculations. If you're not interested in extreme accuracy, then the simple formulas are probably adequate. This is especially true of beers with original gravities less than 1.070, because the two formulas only diverge significantly for high specific gravities. You might want to go through all the complicated calculations for a barley wine or a mead (or if you're putting all of this in a spreadsheet), but for regular beers the simple divide-by-four rule is all you need.

The simple formulas will be used in the rest of the examples, with the results from the more accurate formulas in parentheses for comparison.

## ATTENUATION

A closely related term that is often used to describe a yeast strain or the dryness of a beer is *attenuation*. Attenuation is simply the percentage of sugar that has been converted to alcohol. Attenuation comes in two forms, just like the final extract value. Apparent attenuation (AA) is calculated using the apparent extract value:

$$AA = \frac{(OE - AE)}{OE} \times 100\% \qquad (12)$$

Real attenuation (RA) is calculated using the real extract value:

$$RA = \frac{(OE - RE)}{OE} \times 100\% \qquad (13)$$

For our sample beer, the apparent attenuation is 78.5 percent (77.5 percent) and the real attenuation is 63.7 percent (62.7 percent). Most beers will have a real attenuation that falls in the 60 to 80 percent range.

## ALCOHOL CONTENT

One might think that the alcohol percentage of the final beer would be directly proportional to the difference in the original and final (real) extract values. After all, the chemical equation for the conversion of a monosaccharide to ethanol is

$$C_6H_{12}O_6 \rightarrow 2C_2H_5OH + 2CO_2 \qquad (14)$$

so the weight of the sugar molecule (180 amu) should be converted into the weight of two ethanol molecules (92 amu) and two carbon dioxide molecules (88 amu). This would give us an equation for the alcohol percent by weight (A%w) of:

$$A\%w = (OE - RE)\frac{92}{180} = \frac{(OE - RE)}{1.9565} \qquad (15)$$

The fly in the ointment here is that fermentation is not that simple. It's a biological process with all kinds of intermediate products and side reactions that don't lead to our desired result.

## Table 4-3. Alcohol Percent by Weight (Using Most Accurate Equations)

### Final Specific Gravity

| Original Specific Gravity | 0.990 | 0.995 | 1.000 | 1.005 | 1.010 | 1.015 | 1.020 | 1.025 | 1.030 | 1.035 | 1.040 | 1.045 | 1.050 |
|---|---|---|---|---|---|---|---|---|---|---|---|---|---|
| 1.030 | 4.17 | 3.63 | 3.10 | 2.57 | 2.05 | 1.53 | 1.01 | 0.51 | 0.00 | — | — | — | — |
| 1.035 | 4.69 | 4.15 | 3.62 | 3.09 | 2.56 | 2.04 | 1.52 | 1.01 | 0.50 | 0.00 | — | — | — |
| 1.040 | 5.22 | 4.68 | 4.14 | 3.61 | 3.08 | 2.55 | 2.03 | 1.52 | 1.01 | 0.50 | 0.00 | — | — |
| 1.045 | 5.75 | 5.21 | 4.67 | 4.13 | 3.60 | 3.07 | 2.55 | 2.03 | 1.51 | 1.01 | 0.50 | 0.00 | — |
| 1.050 | 6.28 | 5.73 | 5.19 | 4.65 | 4.12 | 3.59 | 3.06 | 2.54 | 2.02 | 1.51 | 1.00 | 0.50 | 0.00 |
| 1.055 | 6.82 | 6.26 | 5.72 | 5.17 | 4.64 | 4.10 | 3.58 | 3.05 | 2.53 | 2.02 | 1.51 | 1.00 | 0.50 |
| 1.060 | 7.35 | 6.80 | 6.25 | 5.70 | 5.16 | 4.62 | 4.09 | 3.56 | 3.04 | 2.52 | 2.01 | 1.50 | 1.00 |
| 1.065 | 7.89 | 7.33 | 6.78 | 6.23 | 5.68 | 5.14 | 4.61 | 4.08 | 3.55 | 3.03 | 2.52 | 2.01 | 1.50 |
| 1.070 | 8.42 | 7.86 | 7.31 | 6.75 | 6.21 | 5.67 | 5.13 | 4.60 | 4.07 | 3.54 | 3.02 | 2.51 | 2.00 |
| 1.075 | 8.96 | 8.40 | 7.84 | 7.28 | 6.73 | 6.19 | 5.65 | 5.11 | 4.58 | 4.06 | 3.53 | 3.02 | 2.50 |
| 1.080 | 9.50 | 8.94 | 8.37 | 7.82 | 7.26 | 6.72 | 6.17 | 5.63 | 5.10 | 4.57 | 4.04 | 3.52 | 3.01 |
| 1.085 | 10.05 | 9.48 | 8.91 | 8.35 | 7.79 | 7.24 | 6.70 | 6.15 | 5.62 | 5.08 | 4.56 | 4.03 | 3.51 |
| 1.090 | 10.59 | 10.02 | 9.45 | 8.88 | 8.33 | 7.77 | 7.22 | 6.68 | 6.14 | 5.60 | 5.07 | 4.54 | 4.02 |
| 1.095 | 11.14 | 10.56 | 9.99 | 9.42 | 8.86 | 8.30 | 7.75 | 7.20 | 6.66 | 6.12 | 5.59 | 5.06 | 4.53 |
| 1.100 | 11.69 | 11.11 | 10.53 | 9.96 | 9.39 | 8.83 | 8.28 | 7.73 | 7.18 | 6.64 | 6.10 | 5.57 | 5.04 |
| 1.105 | 12.24 | 11.65 | 11.07 | 10.50 | 9.93 | 9.37 | 8.81 | 8.26 | 7.71 | 7.16 | 6.62 | 6.09 | 5.56 |
| 1.110 | 12.79 | 12.20 | 11.62 | 11.04 | 10.47 | 9.90 | 9.34 | 8.78 | 8.23 | 7.69 | 7.14 | 6.60 | 6.07 |

Balling comes through for us again with an empirical formula for the alcohol content (De Clerck):

$$A\%w = \frac{OE - RE}{2.0665 - 0.010665\, OE} \qquad (16)$$

If we insert the simple extract equation (Equation 9) and the simplified version of Balling's equation for real extract (Equation 11), we can derive a relationship for the alcohol content as a function of the original and final specific gravities:

$$A\%w = \frac{76.08\,(OG - FG)}{1.775 - OG} \qquad (17)$$

Either of these equations can be converted to alcohol percent by volume (A%v) by the following formula:

$$A\%v = A\%w\,(FG / 0.794) \qquad (18)$$

where 0.794 is the specific gravity of ethanol. For our sample brew, the alcohol percentages are 5.46 percent (5.25 percent) by weight and 6.98 percent (6.71 percent) by volume. Table 4-3 shows the alcohol percent by weight as a function of original and final specific gravities calculated with the most accurate of the preceding equations (Equations 7, 10, and 16).

## CALORIE CONTENT

Before all that alcohol goes to your head, let's calculate how your homebrew adds to your beer belly. The following equations all give calories (C) per one 12-ounce bottle of beer. First, there's a contribution due to the residual sugar (extract) in the beer:

$$C_{ext} = 3.55\, FG\,(3.8)\, RE \qquad (19)$$

The 3.8 factor is the number of calories per gram of sugar, the final specific gravity converts from grams of solution to grams of water, and the factor of 3.55 is the number of grams of water in a 12-ounce bottle, divided by 100 to cancel with the implicit percent of the real extract. There's also a contribution because of the alcohol present:

$$C_{alc} = 3.55 \text{ FG } (7.1) \text{ A\%w} \qquad (20)$$

The 7.1 factor here is indicative of alcohol's higher number of calories per gram. Finally, there's a small contribution because of the protein in the beer:

$$C_{pro} = 3.55 \text{ FG } (4.0) (0.07) \text{ RE} \qquad (21)$$

The 4.0 factor represents the calories per gram of protein, and the percentage of protein has been estimated at 7 percent of the percentage of sugar. This estimate is a median value for protein estimates I have seen in the literature, which range from 5 to 10 percent of the real extract value. The calorie-per-gram values are from De Clerck. The total number of calories in your homebrew is then:

$$\begin{aligned} C &= C_{ext} + C_{alc} + C_{pro} \\ &= 3.55 \text{ FG } [3.8 \text{ RE} + 7.1 \text{ A\%w} + (4.0)(0.07) \text{ RE}] \\ &= 3.55 \text{ FG } (4.08 \text{ RE} + 7.1 \text{ A\%w}) \end{aligned} \qquad (22)$$

If we convert the calorie equation to a function of specific gravities using both of Balling's approximations and the simple equation for extract (Equations 9, 11, and 17), we get:

$$C = 3621 \text{ FG} \left[ (0.8114 \text{ FG} + 0.1886 \text{ OG} - 1) + 0.53 \frac{\text{OG} - \text{FG}}{1.775 - \text{OG}} \right] \qquad (23)$$

For our example beer this gives us a calorie count of 226.5 (221.2) per 12-ounce bottle. Table 4-4 shows the calorie content as a function of original and final specific gravities calculated with the most accurate of the preceding equations (Equations 7, 10, 16, and 22).

Okay, now, hang on tight for some fast and furious approximations. First, note that a gram of sugar gives 3.8 calories, but when it converts to alcohol it gives roughly 7.1 (92/180) = 3.63 calories. This means that the sugar doesn't lose a lot of calories by converting to alcohol, and therefore the calorie count is primarily a function of the original specific gravity. We can take advantage of this fact by calculating the calories of the unfermented wort (setting the FG = OG), and realizing that our estimate will be a little high. Better yet, we can make an educated guess about the final specific gravity, setting it equal to one fourth of the point value of the original specific gravity:

$$\text{FG} = \frac{\text{OG} - 1}{4} + 1 \qquad (24)$$

Table 4-4. Calories per 12-ounce Bottle of Beer (Using Most Accurate Equations)

Final Specific Gravity

| Original Specific Gravity | 0.990 | 0.995 | 1.000 | 1.005 | 1.010 | 1.015 | 1.020 | 1.025 | 1.030 | 1.035 | 1.040 | 1.045 | 1.050 |
|---|---|---|---|---|---|---|---|---|---|---|---|---|---|
| 1.030 | 93.6  | 96.0  | 98.3  | 100.7 | 103.1 | 105.5 | 107.9 | 110.2 | 112.6 | —     | —     | —     | —     |
| 1.035 | 110.1 | 112.5 | 114.8 | 117.2 | 119.6 | 122.0 | 124.3 | 126.7 | 129.1 | 131.5 | —     | —     | —     |
| 1.040 | 126.7 | 129.0 | 131.4 | 133.8 | 136.1 | 138.5 | 140.9 | 143.2 | 145.6 | 148.0 | 150.4 | —     | —     |
| 1.045 | 143.3 | 145.6 | 148.0 | 150.3 | 152.7 | 155.0 | 157.4 | 159.8 | 162.1 | 164.5 | 166.9 | 169.3 | —     |
| 1.050 | 159.9 | 162.2 | 164.6 | 166.9 | 169.3 | 171.6 | 174.0 | 176.4 | 178.7 | 181.1 | 183.4 | 185.8 | 188.2 |
| 1.055 | 176.6 | 178.9 | 181.2 | 183.6 | 185.9 | 188.3 | 190.6 | 193.0 | 195.3 | 197.7 | 200.0 | 202.4 | 204.7 |
| 1.060 | 193.3 | 195.6 | 197.9 | 200.3 | 202.6 | 204.9 | 207.3 | 209.6 | 212.0 | 214.3 | 216.6 | 219.0 | 221.3 |
| 1.065 | 210.0 | 212.3 | 214.7 | 217.0 | 219.3 | 221.6 | 224.0 | 226.3 | 228.6 | 231.0 | 233.3 | 235.6 | 238.0 |
| 1.070 | 226.8 | 229.1 | 231.4 | 233.7 | 236.1 | 238.4 | 240.7 | 243.0 | 245.3 | 247.7 | 250.0 | 252.3 | 254.7 |
| 1.075 | 243.6 | 245.9 | 248.2 | 250.5 | 252.8 | 255.2 | 257.5 | 259.8 | 262.1 | 264.4 | 266.7 | 269.1 | 271.4 |
| 1.080 | 260.5 | 262.8 | 265.1 | 267.4 | 269.7 | 272.0 | 274.3 | 276.6 | 278.9 | 281.2 | 283.5 | 285.8 | 288.1 |
| 1.085 | 277.3 | 279.6 | 281.9 | 284.2 | 286.5 | 288.8 | 291.1 | 293.4 | 295.7 | 298.0 | 300.3 | 302.6 | 304.9 |
| 1.090 | 294.3 | 296.6 | 298.8 | 301.1 | 303.4 | 305.7 | 308.0 | 310.3 | 312.6 | 314.9 | 317.2 | 319.5 | 321.7 |
| 1.095 | 311.2 | 313.5 | 315.8 | 318.1 | 320.4 | 322.6 | 324.9 | 327.2 | 329.5 | 331.8 | 334.0 | 336.3 | 338.6 |
| 1.100 | 328.3 | 330.5 | 332.8 | 335.1 | 337.3 | 339.6 | 341.9 | 344.2 | 346.4 | 348.7 | 351.0 | 353.3 | 355.5 |
| 1.105 | 345.3 | 347.6 | 349.8 | 352.1 | 354.4 | 356.6 | 358.9 | 361.1 | 363.4 | 365.7 | 367.9 | 370.2 | 372.5 |
| 1.110 | 362.4 | 364.7 | 366.9 | 369.2 | 371.4 | 373.7 | 375.9 | 378.2 | 380.4 | 382.7 | 385.0 | 387.2 | 389.5 |

An *Easy Guide to Recipe Formulation*　　　　161

### Table 4-5. Carbon Dioxide Equilibrium Concentration

| Temperature (°F) | Volumes of $CO_2$ in Solution |
|---|---|
| 32 | 1.71 |
| 35 | 1.61 |
| 40 | 1.46 |
| 45 | 1.32 |
| 50 | 1.20 |
| 55 | 1.09 |
| 60 | 0.99 |
| 65 | 0.91 |
| 70 | 0.83 |
| 75 | 0.78 |
| 80 | 0.73 |

Second, that (1.775 - OG) term in the denominator of Equation 23 is going to give us trouble, so let's set that particular OG to a midrange value of 1.050. Making those approximations and fiddling around with the numbers a bit yields:

$$C = 851 \, (OG - 1)(OG + 3) \qquad (25)$$

For our example beer this gives a calorie count of 224.8 (221.2) per 12-ounce bottle. For most beers, this equation will be a reasonable approximation of the number of calories.

## CARBONATION LEVEL

If you are a homebrew bottler instead of a homebrew kegger, you're probably a little bit jealous of all the control that a kegger has over carbonation levels. If you're a kegger, it's relatively easy: you just put your beer (whether or not it has finished fermenting) into a keg and adjust the temperature to the desired serving temperature and adjust your pressure so that the carbonation level is what you want. You determine the desired carbonation level by reading the "volumes of $CO_2$," off a chart as a function of the temperature and pressure. You can always readjust things if they are not to your liking. (Okay, it's not really that simple, but you get my point.)

A bottler has a more difficult life when it comes to carbonation.

Once the cap is on, everything is fixed. If you overcarbonate, you have to chill all your bottles down and drink them as fast as you can. If you undercarbonate, your beer suffers from a lack of aroma and tingly mouth feel, and overall aesthetics suffer. Clearly, this situation could benefit from a little more control.

First, what exactly is a "volume of $CO_2$"? The number of volumes of $CO_2$ is a measure of the amount of dissolved carbon dioxide. It is equal to the volume occupied by the carbon dioxide if it were taken out of solution and put at standard temperature and pressure (STP = 32°F or 0 °C and 1 atmosphere) divided by the volume of the beer. In other words, if you took all the carbon dioxide out of your 5-gallon batch of beer, changed it to STP, and got 10 gallons of gas, the $CO_2$ level in the original beer was 2 volumes. The amount of carbon dioxide that will dissolve is a fixed quantity that depends on temperature and pressure (and also the other things in solution, but that effect is negligible for our purposes). A desirable carbonation level for most beers is 2.2 to 2.6 volumes of $CO_2$. Miller recommends 1.8 to 2.2 volumes for British ales, 2.5 volumes for lagers and German ales, 2.6 to 2.8 volumes for American beers, and 3.0 volumes for wheat beers and fruit ales.

Now that our target carbonation level is set, we need to determine our starting point. If your beer has been bubbling away in a carboy since the last time you racked it, it has an overpressure of carbon dioxide at one atmosphere. This means that a considerable amount of carbon dioxide is already in solution in your beer. The amount that is in solution ($CD_{init}$ in volumes) is a function of temperature (T, in Fahrenheit). I fit a function to some empirical data (Linke) to give this relationship:

$$CD_{init} = 3.0378 - 5.0062 \times 10^{-2}\, T + 2.6555 \times 10^{-4}\, T^2 \qquad (26)$$

Table 4-5 shows that this function varies from 1.71 to 0.73 volumes over the range of possible bottling temperatures, which indicates that determining the amount of dissolved $CO_2$ at the start of bottling is crucial.

The next step is calculating how much sugar is necessary to get from the initial carbon dioxide level to the target level. I want to point out here that measuring your priming sugar by volume (the old ¾-cup-per-5-gallon-batch rule) is not very accurate. We're going to need a certain mass of sugar to get the desired mass of $CO_2$, so if you measure by volume, you'll need to know the density to determine the mass. I've seen density estimates for corn sugar that varied between 133 and 193 grams per cup. I measured it myself to be 151 grams per cup. In the calculations that follow, it is assumed that the sugar is measured by weight instead of volume, and I strongly recommend that you measure it that way too.

Given a weight of priming sugar (PS) in grams and the volume of beer (VB) in gallons, we can estimate the amount of carbon dioxide generated through fermentation ($CD_{gen}$):

$$CD_{gen} = PS \left(\frac{88g\ CO_2}{180g\ C_6H_{12}O_6}\right) \left(\frac{1}{7.4287g\ CO_2\ /\ gallon\ at\ STP}\right) \left(\frac{1}{VB}\right)$$

$$= 6.5811 \times 10^{-2}\ \frac{PS}{VB}\ \text{volumes} \tag{27}$$

This formula makes use of the simplistic chemical equation of fermentation (Equation 14) and assumes that 100 percent of the sugar is fermented. This assumption is valid for corn sugar and table sugar, but if you're priming with something else, like honey or dry malt extract, you should include a factor for the fraction of the priming substance that is fermentable, and increase your priming rate accordingly.

The total amount of carbon dioxide in our primed and conditioned beer is then:

$$CD = CD_{gen} + CD_{init}$$
$$= 6.5811 \times 10^{-2} \left(\frac{PS}{VB}\right) + 3.0378 - 5.0062 \times 10^{-2}\ T + 2.6555 \times 10^{-4}\ T^2 \tag{28}$$

Inverting this formula, we can get an equation for the weight of priming sugar:

$$PS = 15.195\ VB\ (CD - 3.0378 + 5.0062 \times 10^{-2}\ T - 2.6555 \times 10^{-4}\ T^2)$$

So there's the complicated formula. Table 4-6 gives values from this formula for various combinations of carbonation level and temperature. What would an "average" case look like? Let's assume that you have 5 gallons of fully fermented beer at 65 degrees F and you want to have 2.5 volumes of $CO_2$ in the final product. Running the numbers gives a priming sugar weight of 121 grams. If you insist on having a volume of priming sugar to work with, this comes out to 0.8 cups (using the density for corn sugar that I measured). Lest you think that it always comes out close to ¾ of a cup, here's another example: This time we'll assume that you've been making a lager and your beer is waiting to be primed at 40 degrees F, with everything else the same.

Table 4-6. Priming Sugar Weight (in grams) for a Five-Gallon Batch of Beer

| Temp. of beer °F | Desired Carbonation Level (Volumes of $CO_2$) | | | | | | | | | | |
|---|---|---|---|---|---|---|---|---|---|---|---|
| | 1.8 | 2.0 | 2.1 | 2.2 | 2.3 | 2.4 | 2.5 | 2.6 | 2.7 | 2.8 |
| 32 | 7.0 | 22.2 | 29.8 | 37.4 | 45.0 | 52.6 | 60.2 | 67.8 | 75.4 | 83.0 |
| 35 | 14.4 | 29.6 | 37.2 | 44.8 | 52.4 | 60.0 | 67.6 | 75.1 | 82.7 | 90.3 |
| 40 | 25.8 | 41.0 | 48.6 | 56.2 | 63.8 | 71.4 | 79.0 | 86.6 | 94.2 | 101.8 |
| 45 | 36.3 | 51.5 | 59.1 | 66.7 | 74.3 | 81.9 | 89.4 | 97.0 | 104.6 | 112.2 |
| 50 | 45.7 | 60.9 | 68.5 | 76.1 | 83.7 | 91.3 | 98.9 | 106.5 | 114.1 | 121.7 |
| 55 | 54.1 | 69.3 | 76.9 | 84.5 | 92.1 | 99.7 | 107.3 | 114.9 | 122.5 | 130.1 |
| 60 | 61.5 | 76.7 | 84.3 | 91.9 | 99.5 | 107.1 | 114.7 | 122.3 | 129.9 | 137.5 |
| 65 | 68.0 | 83.2 | 90.8 | 98.3 | 105.9 | 113.5 | 121.1 | 128.7 | 136.3 | 143.9 |
| 70 | 73.4 | 88.6 | 96.2 | 103.8 | 111.3 | 118.9 | 126.5 | 134.1 | 141.7 | 149.3 |
| 75 | 77.8 | 92.9 | 100.5 | 108.1 | 115.7 | 123.3 | 130.9 | 138.5 | 146.1 | 153.7 |
| 80 | 81.1 | 96.3 | 103.9 | 111.5 | 119.1 | 126.7 | 134.3 | 141.9 | 149.5 | 157.1 |

# An Easy Guide to Recipe Formulation

The result for this case is 79 grams of priming sugar, considerably less than before.

I should mention here the caveat that if your beer hasn't finished fermenting, none of this analysis is accurate and you may have significant amounts of sugar left that will create large amounts of carbon dioxide. The amount of priming sugar that is normally used will only raise the specific gravity by 0.002, so you can see that it is imperative that there is not an extra bit of fermentable sugar hanging around. In practice, termination of fermentation is easy to discern by watching your fermentation lock or taking specific gravity readings.

## CONCLUSION

I hope that these explanations have made the confusing world of real extracts and apparent attenuations a little bit clearer. Using the equations in this article (and keeping good records) should enable you to duplicate your beer successes and to avoid reinventing your beer duds. And now that you've been armed with information about how your homebrew affects your head and body, you will be able to make a better decision about whether or not to have that next beer.

## NOMENCLATURE / GLOSSARY

A%w—Alcohol percent by weight.
A%v—Alcohol percent by volume.
AA—Apparent attenuation (%), apparent percentage of sugar that converted to alcohol.
AE—Apparent extract (degrees Plato), the apparent weight percent of dissolved solids in the beer, before correcting for the lower density of the alcohol.
AMU—Atomic mass unit.
C—Calories in a single 12-ounce beer.
$C_{alc}$—Calories in a single 12-ounce beer attributed to alcohol.
$C_{ext}$—Calories in a single 12-ounce beer attributed to extract (residual sugar).
$C_{pro}$—Calories in a single 12-ounce beer attributed to protein.

CD—Total carbon dioxide concentration in the conditioned beer (volumes).
$CD_{init}$—Initial carbon dioxide concentration in the beer before priming (volumes).
$CD_{gen}$—Incremental carbon dioxide concentration caused by the fermentation of the priming sugar (in volumes).
$C_2H_5OH$—Chemical formula for ethanol, the primary alcohol in beer.
$C_6H_{12}O_6$—Chemical formula for a monosaccharide sugar (glucose).
$CO_2$—Chemical formula for carbon dioxide.
E—Extract (degrees Plato), the weight percent of dissolved materials in the wort.
FG—Final specific gravity.
OE—Original extract (degrees Plato).
OG—Original specific gravity.
PS—Weight of the priming sugar (grams).
RA—Real attenuation (%), real percentage of sugar that is converted to alcohol.
RE—Real extract (degrees Plato), the real weight percent of dissolved solids in the beer, after correcting for the lower density of the alcohol.
SG—Specific gravity (density relative to water). Specific gravity in points is equal to $1,000 (SG - 1)$.
STP—Standard temperature (0 degrees C or 32 degrees F) and pressure (1 atmosphere).
T—Temperature of the beer (Fahrenheit).
VB—Volume of the beer (gallons).

# REFERENCES

De Clerck, Jean, A *Textbook of Brewing*, Chapman & Hall Ltd., 1958.

Miller, Dave, *Brewing the World's Great Beers*, Storey Communications Inc., 1992.

Linke, William F., *Solubilities of Inorganic and Metal Organic Compounds*, American Chemical Society, 1958.

Timmermans, Jean, *The Physico-Chemical Constants of Binary Systems in Concentrated Solutions*, vol. 4: *Systems with Inorganic and Organic or Inorganic Compounds*, Interscience Publishers, 1960.

Weast, Robert C., *CRC Handbook of Chemistry and Physics*, CRC Press, Inc., 1980.

*Michael L. Hall, Ph.D., is a computational physicist at Los Alamos National*

Laboratory in New Mexico. Mike has been brewing for five years and is a Certified Judge in the BJCP. He was one of the founding members of the Los Alamos Atom Mashers and has worn many hats in the club (newsletter editor, treasurer, librarian, organizer).

# five

# Malt

## THE ENCHANTING WORLD OF MALT EXTRACT—MAKE THE MOST OF IT

BY NORMAN FARRELL

*This article originally appeared in Zymurgy, Winter 1994 (vol. 17, no. 5)*

Homebrewers everywhere are striving to improve their beers. We have an unprecedented variety of ingredients, suppliers, publications, and seminars. Homebrewers are devouring information as fast as they get it. But there seems to be a black hole of sorts when it comes to information on some of our raw materials. Study the label next time you pick up a container of malt extract. What can you tell about the contents? Chances are you can't tell a great deal.

Almost every brewer I know started out using malt extracts, and all-grain brewers use extracts to make starters, adjust original gravities, or turn a bock into a doppelbock, so there are a lot of us who want to know what's in that can. Malt extracts are expensive because ingredients, processing, packaging, and transportation all cost money. The materials used and the concentration techniques are but two causes of variation between products. Yet from outside the can two malt extracts may look quite alike.

In an effort to peel away the labels and shed a little light, the members of the Just Brew It homebrew club in Bartlesville, Okla-

# The Enchanting World of Malt Extract—Make the Most of It

homa, worked with the American Homebrewers Association to prepare and sample twenty commercial malt extracts under homebrewing conditions. The wort samples were sent to Siebel Institute of Technology in Chicago for analysis. The point of this experiment was to find out what to expect in general from modern malt extracts and kits in a typical homebrew setting.

## SETTING THE SCENE

Twenty different malt extracts were prepared 1 gallon of wort at a time. The leftover extract was put to good use, of course. The malt extracts and kits were processed on three successive Sundays (seven on the first day, seven on the second, and six on the last) in order to make the work manageable and allow as many club members to help as possible. After the samples were collected, unused extracts were combined to make several composite beers: a fitting reward. After all, who could let all that nascent beer go down the drain?

For access to experimental know-how and equipment, you just can't beat having club members who experiment for a living. Many of our club members are technicians, scientists, or engineers who represent decades of research and development experience. Their expertise, in addition to guidance from Siebel, was very helpful in selecting equipment and designing procedures.

## PROCEDURE

Each extract was given a code number keyed to a master list. All work after sample preparation was done using the code numbers.

Each container (can, kit, or bag) was opened and 1 pound (454 grams plus or minus 1 gram) of extract was collected. All weights were determined with a triple-beam balance. One gallon of preheated distilled water was added to a restaurant-grade, 2-gallon cast-aluminum pot. The preweighed extract was added to the pot. The cup used to hold the extract was repeatedly dipped into the heating water in order to get every drop of extract. The mixture was brought to a boil (at approximately 300 feet above sea level) over medium heat and kept at a gentle rolling boil for 50 minutes. Temperature readings were taken at 15, 30, and 45 minutes using a digital thermometer with a thermocouple probe. Readings varied from 208 to 209 degrees F (98 degrees C), but never above or below.

The pot was removed from the heat at the end of the boiling period, the wort transferred to a 1-gallon heat-proof glass jar, and the pot thoroughly rinsed between batches. The wort was immediately siphoned through a freshly sanitized counterflow chiller into an acrylic pitcher with a 1-gallon reference mark. The chiller was drained completely and cleaned between each extract batch.

The collected wort was topped up with distilled water (the same brand each time) to the 1-gallon mark and gently stirred. One sample was withdrawn to measure specific gravity and a second 250-milliliter sample was withdrawn to an HDPE (high-density polyethylene) plastic bottle sanitized with iodophor. As per Siebel's instructions, 1 milliliter of chloroform was added with a syringe to each jar as a preservative/antimicrobial. The sample was capped, lightly agitated, and kept refrigerated until shipment the following day.

Two identical pots were used to streamline the process. Two wort boils were going at a time, staggered because we wanted to use the same chiller. Specific gravity was measured with an ordinary brewing hydrometer. (Calibration with distilled water showed 1.000 corrected to 60 degrees F or 15.5 degrees C.) Samples were shipped each Monday, the day after brewing, via next-day air. Siebel froze the samples and held them until all twenty were received before doing any analyses.

## THE RESULTS

The test results for all of the malt extracts are shown in Table 2. Siebel analyzed for extract, pH, acidity, color, iron, reducing sugar, nitrogen, and free amino nitrogen (FAN). The extracts are grouped by type: liquid, dry, or kit. Some of the test results give two values. The first is the actual test result; the numbers in the columns titled "compensated" have been compensated for differences in wort specific gravity. Let's look at each test and how the results can affect your beer.

## COMPENSATING THE RESULTS

Because most recipes target a specific starting gravity and because each malt extract had a different extract potential, we thought it would be useful to look at making the same starting gravity beer with each one. What if we had made an 11 °Plato (1.044) wort with each extract instead of using the flat rate of 1 pound per gallon? How would the test results look? That would put all of the malt extracts on equal footing. Fortunately we do not have to redo the experiment to answer these questions.

The properties Siebel tested are concentration-dependent. The test results for a given extract are proportional to the amount of malt extract used to make the wort. Twice as much malt per gallon would give twice the results for FAN or nitrogen. A simple ratio of 11.0 to the extract percent will give the conversion factor for each malt extract. The conversion for extract number 1008 would be 11 divided by 8.89, which equals 1.2373. The factor for extract 1011 would be 0.9532 (11 divided by 11.54). The numbers in the columns labeled "compensated" were created by multiplying the raw test results by the conversion factors. Thus the FAN of 140.0 for extract number 1008 is compensated to 173.23 (140 times 1.2373).

We chose 11.0 degrees Plato (1.044) because it is a typical starting gravity and gives an idea of what to expect in a real homebrewing situation. We did not compensate the results for pH because the effect would be negligible, or the results for color because there is no way to do so with accuracy. Adjusting the results in this way helps make up for the concentration advantage of the dry extracts and for the various dilution instructions in the kits.

# EXTRACT AND SPECIFIC GRAVITY

Extract is the basic measurement of fermentable material in the wort expressed as weight percent dissolved sucrose or degrees Plato (or Brix). Sucrose is the standard reference because it gives a larger increase in specific gravity than other sugars, ensuring that results are always below 100 weight percent to avoid confusion. Specific gravity of the samples ranged from 1.028 to 1.046. Keep in mind that each of the extracts will contain a different amount of water depending on how much water the manufacturer wanted to remove before packaging. The liquid extracts will contain about 20 percent water by weight. The dry extracts will be closer to 1 percent water by weight, thus pound-for-pound the dry extracts will contain more fermentable material than liquid extracts.

Because the worts were all prepared at the rate of 1 pound extract per gallon of water, they can be compared directly with the general values for grains and extracts in *The New Complete Joy of Home Brewing*, by Charlie Papazian (Avon Books, 1991). See Table 5-1.

Dry extract number 1013 could be used at the rate of 1 pound per gallon of water to make an amber ale with a starting gravity of 1.044. You get 44 specific gravity points per pound per gallon with this extract. The number to the right of the decimal point is the specific gravity yield. Expressed in degrees Plato, this quantity is called *extract potential*. Extract 1013 has 11.03 degrees of extract potential at the rate of 1 pound malt extract per gallon of water.

Knowing the specific gravity yield, it's easy to estimate starting gravity with the following formula.

$$SG = \frac{\text{points per pound . per gallon} \times \text{pounds}}{1{,}000 \times \text{gallons}} + 1.0$$

This is an important mathematical relationship for brewers to know. Thus 6.75 pounds of extract number 1001 would give a starting gravity of 1.050 for a 5-gallon batch of wheat beer $37 \times 6.75 \div 5{,}000 + 1.0$. Later we'll see how to calculate the amount of extract needed to adjust starting gravity.

If you have a hydrometer, you can check the specific gravity of your wort. Specific gravity can be converted to degrees Plato by dividing the last two digits to the right of the decimal point by four. Take the specific gravity measurement at 60 degrees F (16 degrees C) or adjust the reading if taken at higher temperatures. See pages 80–81 for a hydrometer correction table.

## Table 5-1. Extract and Specific Gravity

One pound of the following ingredients and enough water to make 1 gallon will yield (approximately) the specific gravity indicated.

| Ingredient | Specific Gravity | (Balling) |
|---|---|---|
| Corn sugar | 1.035 – 1.038 | (9 – 9.5) |
| Malt extract (syrup) | 1.030 – 1.038 | (8 – 9.5) |
| Malt extract (dry) | 1.038 – 1.045 | (9.5 – 10.5) |
| Malted barley* | 1.025 – 1.030 | (6 – 7.5) |
| Munich malt* | 1.020 – 1.025 | (5 – 6) |
| Dextrine malt | 1.015 – 1.020 | (4 – 5) |
| Crystal malt | 1.015 – 1.020 | (4 – 5) |
| Grain adjuncts* | 1.020 – 1.035 | (5 – 9) |

*These grains must be mashed to obtain the indicated extract.

## pH

The acidity or alkalinity of a solution can be expressed as pH on a scale from one (highly acidic) to fourteen (highly alkaline). pH is the natural logarithm of the hydrogen ion concentration in the wort. The pH of an all-barley malt wort will naturally fall in the 5.0 to 5.5 range with no need for adjustment. Yeast activity will quickly drop the pH to the mid-4s, which helps protect the fermenting beer from some bacterial infections.

Several of the worts had a pH below 5.0. This is not a problem and does not require corrective action. The low pH may be perfectly natural. The extract producer may have acidified the mash during production or may have used dark specialty grains (chocolate malt, black malt), either of which can lower wort pH.

Note that hop utilization (how much bitterness is achieved in the beer) decreases as pH decreases. A pH below 5.0 may require additional bittering hops to get your usual result. Lower pH worts may yield a rounder, more well-defined malt character, but this is highly subjective. Experiment with wort pH and decide for yourself. If you like the results, food-grade lactic or phosphoric acid may be used to acidify wort. Both of these acids are naturally occurring in beers. Exercise caution when handling phosphoric acid.

You can test the pH of your own wort at home. pH test strips are available for the typical brewing range of 4.0 to 6.0 from laboratory suppliers or beer and wine supply stores.

## ACIDITY

Acidity can be expressed in units other than pH. In this test an alkaline solution (such as dilute sodium hydroxide) is added to counteract the acidity of the wort. This is called *titration*. The amount of acid can be calculated from the amount of alkaline solution added. Acidity in the wort may be naturally occurring from dark malts or the result of bacterial infection. A stout kit may have a higher acidity, or lower pH, than a pale lager kit. Acidity above 0.3 weight percent as lactic acid can indicate bacterial infection.

Titration kits are available at beer and wine supply stores so you can determine wort acidity at home.

## IODINE REACTION

Remember this test from chemistry class? Drop iodine from the medicine cabinet on a potato slice and look for dark blackish purple, signaling the presence of starch. Yeast will not process starch. The starch will remain in the finished beer, making for a hazy final product with poor flavor stability. Although it was surprising to see four out of the twenty extracts tested give a slight starch reaction, Dr. Joe Power of the Siebel Institute did not believe this would be a problem in homebrewing use. A trace reaction indicates that the presence of starch was just barely discernible.

Get some iodine from the drugstore and try this test yourself. Siebel fills a test tube with wort and carefully adds a layer of iodine on the surface of the wort. Hold it up to the light and note any blackish purple color change. Pour the test tube's contents down the sink, not back into the boil pot.

## COLOR

In degrees SRM (Standard Research Method), color tells you whether you can expect to create a pale ale (8 to 10 SRM) or a stout (35 and up). Lower numbers in degrees SRM indicate a lighter color. Amber or dark extracts will get higher values. The lightest-colored extracts were none too light for a Pilsener (2 to 5). Our standardized dilution of 1 pound extract per gallon is fine for making a comparison between extracts but might not be the actual dilution recommended for a kit.

I do not know of any rules of thumb for estimating color based on blending extracts or using different amounts of the same extract.

The best option is to check the color of the wort before the boil using the techniques described in the chapter by George Fix in *Evaluating Beer* (Brewers Publications, 1994), pages 133 through 141.

## IRON

An all-malt wort will typically contain some iron. About 0.05 parts per million (ppm) is typical. This is a really small amount. Thinking of percent as parts per hundred may help understand ppm. One percent is 10,000 ppm. Iron can be introduced from a rusty can or rusty brewing gear or even from your tap water. Iron in wort is mostly a negative. High iron content can cause oxidation haze in the finished beer, and can also promote the oxidative production of diacetyl from acetolactic acid (more on diacetyl later). Iron can usually be tasted in amounts greater than 0.15 ppm, according to Ted Konis in *Evaluating Beer*.

Yeast will absorb some iron from the wort. Joe Power reports that yeast can cope with iron levels as high as 0.2 ppm. Greg Noonan in *Brew Free or Die* (Brewers Publications, 1991) states that iron levels in brewing water greater than 0.3 ppm will damage yeast. Iron levels greater than 0.3 ppm in the wort are undesirable.

The high iron levels in some of the extracts are a cause for concern, especially since the brewer cannot correct for this defect. A review of our procedure, equipment, water, and schedule showed no common element for the high-iron extracts. Furthermore, all of the original containers appeared to be in good condition with no damage or rust. Possible sources of iron from the extract manufacturing process are equipment and brewing water.

## REDUCING SUGAR

This quantity indicates how much fermentable material is in the wort and hints at its character. The analytical method used is adequate for comparison purposes but does not give an absolute number you can hang your hat on. A 2 percent difference between extracts is considered significant. All sugars will register to some extent, and the test is twice as sensitive to glucose as maltose and dextrins. Thus an adulterated extract, an extract to which sugar has been added, would give a much higher number than an all-malt extract. A highly adulterated extract (20 percent glucose) would give a reducing sugar result of 10 percent in an 11 °Plato wort.

Malt extract is the largest single cost in a batch of beer, and it's natural to want as much malt for your money as you can get. Concern over adulteration of malt extract is supported by a University of Saskatchewan report by Paik, Low, and Ingledew. The detailed study found adulteration even in extracts labeled "all malt."

While you cannot test directly for reducing sugar at home, you can test for the presence of other sugars, such as glucose. Glucose test strips (designed for use by diabetics) are available at most pharmacies and can be used to check your wort at home. Be sure the strips you buy are selective for glucose.

## NITROGEN AND FAN

The nitrogen test results give the total nitrogen in the wort. The nitrogen exists in the form of proteins and amino acids. Free amino nitrogen (FAN) is a measure of amino acid nitrogen. Only the nitrogen contained in the amino acids is usable by the yeast. Amino acids, the building block of proteins, are organic acids containing an $NH_2$ group. A typical brewing wort will have a FAN in the 240 to 275 ppm range. A FAN of 150 ppm is the minimum acceptable.

Ingredients have an important effect on the amount of nitrogen and FAN in an extract wort. Malting procedure, mashing temperature program, and the variety of barley malt all play a part. The amount of malt may be the overriding factor. Use more malted barley, get more FAN. Wheat has a higher protein content than barley, so using wheat will significantly increase the nitrogen and FAN, as you can see by looking at the numbers for the two wheat extracts.

FAN for most of the extracts tested lower than 200 ppm and well below the normal range for an all-malt wort. There are a number of possible reasons for the low FAN numbers. The effects of malting and mashing techniques and variety of barley have already been mentioned. Heavy use of adjuncts, like rice, corn, or sugar, can result in a lower FAN because those adjuncts contribute poorly if at all to the FAN pool. Lastly, protein and FAN are lost in the process of making the malt extract. Wort is normally whirlpooled and filtered after the boil and before concentration to remove trub, which harbors spoilage organisms. While this step increases stability and lowers the risk of bacterial contamination, it results in the inevitable loss of some material we would rather have in the wort.

Brewers care about FAN because low free amino nitrogen is a cause of off flavors. Not all amino acids are equal in the eyes of your yeast. At low FAN levels, yeast is forced to use the less desirable amino acids, resulting in higher levels of off flavors in the finished beer. Low FAN also contributes to slow fermentation starts and incomplete fermentation (Paik et. al.; Lodahl).

Diacetyl may be the most noticeable off flavor to result from low FAN, according to George Fix. Diacetyl is a ketone, one of many by-products of fermentation. Its buttery or butterscotch flavor and aroma can be okay at low levels but becomes unpleasant at high levels and as beer ages. Diacetyl is inappropriate in many beer styles, especially many lagers. Unfortunately, lager yeasts are more sensitive to low FAN than are ale yeasts. Avoid the use of continental lager yeasts with low FAN worts.

There are effective measures you can take to increase FAN in your extract brewing. If a kit calls for additional sugar, add dry malt extract instead. Sugar contributes no FAN. Liquid malt extracts can be used, but dry extracts are easier to handle. The more adventurous can do a partial mash with a pale malt. About 2 pounds of barley malt in addition to the kit for a 5-gallon batch will increase FAN by 60 ppm.

Whether you brew from kits or from your favorite recipe, you can do what some big brewers do: ferment a strong wort and then dilute to the desired final strength. A given amount of malt extract contains a given amount of FAN. The less you dilute the extract with water, the greater the FAN concentration will be. To do this at home, brew your regular extract recipe adding only enough liquid to bring the fermentation volume up to 3 gallons. Ferment as usual and then dilute the batch to 5 gallons with boiled/chilled water at bottling time. Boiling 3 gallons instead of 5 will increase FAN levels in the fermenter while sacrificing some hop utilization because of the lower-volume/higher-gravity boil. The higher-gravity boil is likely to be no greater than the gravity equivalent of a barley wine. If you employ a minimash to boost FAN, include a 30-minute protein rest at 120 to 135 degrees F (49 to 57 degrees C) before raising the temperature to 150 degrees F (66 degrees C) for conversion.

I could not find a practical nitrogen or FAN test for home use.

## USING THE MALT EXTRACT TABLE

Some examples will illustrate how you might use Table 5-2 on pages 180–181. What if you want to make a Pilsener? Only two of the extracts in the table, 1007 and 1015, are light enough to make a credible Pilsener. Both of these extracts have borderline FAN results, but you really want a light Pilsener and you want it bad. Solve the low FAN problem by brewing a strong wort and diluting later, as described previously.

What if you're a kit brewer? One of the easiest ways to improve a kit is to add malt extract when the kit calls for sugar. Extract number 1012 is a stout kit weighing 3.75 pounds that calls for the addition of 2.2 pounds of sugar. Mix the kit contents with 5 gallons of water and check the specific gravity. Using the formula above, the specific gravity should be about 1.026. Assuming you wanted an original gravity of 1.040 and were going to use dry extract number 1019 (SG yield equals 44 from Table 2), you would need to add 1.6 pounds of dry extract. The following formula makes the calculation easy:

$$\text{Pounds of extract} = \frac{(\text{change in SG} \times 1{,}000 \times \text{gallons})}{(\text{SG yield})}$$

Example:

$$1.6 \text{ pounds} = \frac{1.040 - 1.026 \times 1{,}000 \times 5 \text{ gallons}}{44 \text{ points/pounds/gallons}}$$

The same trick will work for liquid extracts, but they are messier to handle than dry extracts.

Sometimes you should look at a combination of properties. The reducing-sugar test is not accurate enough on its own to indicate adulteration unless the results are up around 10 percent. An adulterated extract would also be expected to have a low FAN because of lower malt content. The combination of low FAN and high reducing sugar together are a cause for concern. But don't worry, get out the glucose test strips if you can't relax with a particular extract.

## THE NEXT STEP

We only sampled twenty malt extracts while there are several times that many available worldwide. There is no way we could test them all. And we certainly don't have access to the Siebel lab every time there's a question about extracts. The good news is that most of the wort evaluation tests can be done at home, except nitrogen and FAN. If you run the tests (gravity, pH, color, iron, iodine reaction, and glucose) with every new extract or kit you buy, you'll have an impressive database. This could certainly be done as a club activity.

Although several of the extracts listed ingredients, only two gave typical analysis results (numbers 1018 and 1020). The printed specifications agreed with the Siebel test results. The test results on the cans appeared to be for the malt extract "as is," so we had to "dilute" the numbers to the 1-pound-per-gallon level in order to make a fair comparison. Manufacturers will not know how much brewers will dilute their extracts, so listing bulk properties is a realistic compromise.

I tried tracking down manufacturers' specifications through homebrew shops but did not have any luck. When I called one extract maker, I was told that more data are generally available for their food-grade extracts but not for the brewer-grade extracts. If you want to see more information on the label, let manufacturers know and encourage them to make it available. This is especially true for tests that we can't do at home, such as FAN.

## FINAL REMARKS

We weren't looking to find bad extracts. We did this experiment to learn something. Even though most of the samples looked pretty good, the test results showed a wide variation in some properties, close agreement in others, and a few surprises, especially the iron numbers.

Just Brew It hopes that you find this information useful. We had fun with the experiment and learned a few things too. Many people outside the club helped us along the way. Special thanks to Fisher Scientific for donating equipment and the American Homebrewers Association for initiating the project and paying for the tests and materials. Thanks to Ilse Shelton and Joe Power at Siebel, Martin Lodahl, and George Fix for their help with technical details.

The Just Brew It club members who did the experimental work are Tim Harper (president), Lee Kobylinski, Adrienne Kobylinski, Layne Hair, Oscar Sarlandt, Mike Watkins, Laurie Watkins, Rob Dahlgren, Tod Taylor, Joe Bergmeister, and Fred Wenzel. Without them, there would have been no data to write about.

Having good data is important, but data is not beer. The point, after all, is to make enjoyable beer and to enjoy the process. While you are studying all of the numbers, remember that the most important test is the taste test.

## REFERENCES

Fix, George J., "Diacetyl: Formation, Reduction and Control," *BrewingTechniques*, vol. 1, no. 2, July/August 1993.

## Table 5-2. Malt Extract Analysis Results

| Sample Number | Extract Name | Extract % | Specific Gravity | pH | Acidity as % Lactic Acid | Acidity, Compensated | Iodine Reaction | Color, degrees SRM | Iron, ppm | Iron, Compensated | Reducing Sugar % | Reducing Sugar, Compensated | Nitrogen, % | Nitrogen, compensated | FAN, ppm | FAN, Compensated |
|---|---|---|---|---|---|---|---|---|---|---|---|---|---|---|---|---|
| **Liquid Malt Extract** | | | | | | | | | | | | | | | | |
| 1007 | Alexander's Hopped Extract | 7.09 | 1.028 | 4.8 | 0.06 | 0.09 | negative | 4.5 | 0.02 | 0.03 | 4.16 | 6.45 | 0.05 | 0.08 | 92.00 | 142.74 |
| 1015 | Alexander's Sun Country Pale Malt Extract | 8.26 | 1.033 | 4.2 | 0.13 | 0.17 | negative | 4.8 | 0.22 | 0.29 | 4.45 | 5.93 | 0.05 | 0.07 | 113.00 | 150.48 |
| 1017 | Ireks Munich Amber Unhopped | 8.98 | 1.036 | 5.0 | 0.17 | 0.21 | trace | 21.3 | 0.29 | 0.36 | 6.33 | 7.75 | 0.09 | 0.11 | 172.00 | 210.69 |
| 1014 | Ireks Munich Light Unhopped | 8.87 | 1.035 | 5.2 | 0.14 | 0.17 | trace | 10.0 | 0.13 | 0.16 | 6.31 | 7.83 | 0.08 | 0.10 | 158.00 | 195.94 |
| 1008 | Mountmellick Products, Ireland, Unhopped Light | 8.89 | 1.035 | 5.4 | 0.11 | 0.14 | negative | 7.5 | 0.12 | 0.15 | 7.09 | 8.77 | 0.07 | 0.09 | 140.00 | 173.23 |
| 1001 | Premier Reserve Gold Label Brewer's Wheat Malt Extract | 9.33 | 1.037 | 5.1 | 0.18 | 0.21 | negative | 8.1 | 0.07 | 0.08 | 6.48 | 7.64 | 0.11 | 0.13 | 205.00 | 241.69 |
| 1004 | Premier Reserve Gold Label Unhopped Malt Extract | 8.98 | 1.036 | 5.3 | 0.16 | 0.20 | negative | 9.7 | 0.16 | 0.20 | 6.56 | 8.04 | 0.09 | 0.11 | 225.00 | 275.61 |
| **Dry Malt Extract** | | | | | | | | | | | | | | | | |
| 1013 | Northwestern Extract Co. Amber Dry Malt Extract | 11.03 | 1.044 | 5.1 | 0.16 | 0.16 | negative | 24.7 | 0.13 | 0.13 | 7.68 | 7.66 | 0.11 | 0.11 | 223.00 | 222.89 |
| 1019 | Northwestern Extract Co. Dark Dry Malt Extract | 10.98 | 1.044 | 4.5 | 0.18 | 0.18 | negative | 54.5 | 0.51 | 0.51 | 6.89 | 6.90 | 0.10 | 0.10 | 203.00 | 203.37 |

| | | | | | | | | | | | | | | |
|---|---|---|---|---|---|---|---|---|---|---|---|---|---|---|
| 1011 | Northwestern Gold Dry Malt Extract | 11.54 | 1.046 | 5.3 | 0.18 | 0.17 | negative | 12.2 | 0.05 | 7.51 | 7.16 | 0.11 | 225.00 | 214.47 |
| 1002 | Premier Dia Malt Light Dry Malt Extract | 10.98 | 1.046 | 5.3 | 0.17 | 0.17 | negative | 10.0 | 0.01 | 7.20 | 7.21 | 0.09 | 163.00 | 163.30 |

**Liquid Malt Extract Kits**

| | | | | | | | | | | | | | | |
|---|---|---|---|---|---|---|---|---|---|---|---|---|---|---|
| 1006 | Armstrong New Zealand Premier Lager Kit | 9.18 | 1.037 | 5.0 | 0.14 | 0.17 | trace | 14.8 | 0.11 | 6.43 | 7.70 | 0.07 | 152.00 | 182.14 |
| 1016 | Black Rock New Zealand East India Pale Ale Kit | 9.01 | 1.036 | 5.1 | 0.14 | 0.17 | trace | 10.6 | 0.13 | 6.11 | 7.46 | 0.07 | 134.00 | 163.60 |
| 1005 | Cooper's Australian Brewery Bitter Kit | 8.84 | 1.035 | 5.2 | 0.10 | 0.12 | negative | 24.6 | 0.07 | 6.44 | 8.01 | 0.06 | 120.00 | 149.32 |
| 1012 | Cooper's Australian Brewery Stout Kit | 8.87 | 1.035 | 4.9 | 0.13 | 0.16 | negative | 86.5 | 0.13 | 5.91 | 7.33 | 0.07 | 153.00 | 189.74 |
| 1003 | Edme Superbrau Gold Weizen Lager Wheat Bier | 9.27 | 1.037 | 5.4 | 0.15 | 0.18 | negative | 14.8 | 0.40 | 6.70 | 7.95 | 0.11 | 191.00 | 226.65 |
| 1009 | Edme Superbrau Pale Lager | 8.90 | 1.035 | 5.3 | 0.10 | 0.12 | negative | 10.6 | 0.11 | 6.98 | 8.63 | 0.07 | 127.00 | 156.97 |
| 1010 | Mountmellick Products, Ireland, Brown Ale Kit | 8.82 | 1.035 | 5.1 | 0.13 | 0.16 | negative | 38.2 | 0.03 | 6.40 | 7.98 | 0.07 | 133.00 | 165.87 |
| 1018 | Munton's Gold Docklands Porter | 8.88 | 1.035 | 5.0 | 0.12 | 0.15 | negative | 54.2 | 0.09 | 5.73 | 7.10 | 0.08 | 142.00 | 175.90 |
| 1020 | Munton's Gold Imperial Stout | 8.78 | 1.035 | 4.9 | 0.14 | 0.18 | negative | 120.1 | 0.06 | 5.62 | 7.04 | 0.08 | 133.00 | 166.63 |

Note: Because the actual gravity for one pound of dry extract per gallon of water is very close to the reference of 11 °Plato, the compensated test results are nearly equal to the raw test results. This is just a coincidence.

Lodahl, Martin, "Malt Extracts: Cause for Caution," *BrewingTechniques*, vol. 1, no. 2, July/August 1993.

Manning, Martin P., "Understanding Specific Gravity and Extract," *BrewingTechniques*, vol. 1, no. 3, September/October 1993.

Paik, J., N. H. Low, and W. M. Ingledew, "Malt Extract: Relationship of Chemical Composition to Fermentability," *ASBC Journal*, vol. 49, no. 12, 1991.

Papazian, Charles N., *The New Complete Joy of Home Brewing*, Avon Books, 1991.

*Brew Free or Die: Beer and Brewing*, vol. 11, Brewers Publications, 1991.

Various authors, *Evaluating Beer*, Brewers Publications, 1993.

*Norman Farrell, a homebrewer since 1978, is a founding member of Just Brew It and a Certified BJCP Judge. Norman's other hobby, bicycling, goes great with brewing because both involve "drafting."*

# MALT—A SPECTRUM OF COLORS AND FLAVORS

## by Neil C. Gudmestad and Raymond J. Taylor

*This article originally appeared in Zymurgy, Special 1995 (vol. 18, no. 4)*

As homebrewers, many of us enjoy the brewing process as much as sampling the finished product. Mixing and matching malts, hops, and techniques with beer styles is not only challenging but fun. For many of us, making the switch to grain brewing meant expanding the versatility of beer-making. While it is true that brewing with raw product is less expensive than producing a beer from extract, we find that the real advantage is that it has increased our enjoyment of homebrewing.

The purpose of this article is to provide a basic reference on the variety of malts available to homebrewers today. We do not intend to provide a definitive work on malt usage, but rather an introduction to malts for anyone wanting to pursue grain brewing.

### Table 5-3. Malt Companies with Products Available in the United States:

*Argentina*
Malteria Pampa S. A.—Buenos Aires

*Belgium*
DeWolf-Cosyns Maltings—Aalst

*Canada*
Canada Malting Co.—Toronto, Ontario
Gambrinus Malting Co.—Armstrong, British Columbia
Prairie Malt Ltd.—Biggar, Saskatchewan
United Canada Malt—Petersborough, Ontario

*Germany*
Bamberg Malzerei—Bamberg
Brauerei Ayinger—Munich
Durst Malz—Bruchsal
Ireks Arkedy—Kulmbach
Weyermann Malzerei—Bamberg

*Great Britain*
Archer Daniels Midland Crisp Maltings of Great Ryeburg
Edme—Manningtree, Essex
Hugh Baird & Sons—Witham, Essex
Munton & Fison PLC—Stowmarket, Suffolk
Pauls Malt Ltd.—Newmarket, Suffolk

*Scotland*
Beeston Malt Co.—Arbroath
Brewing Products (U.K.)—Kirkliston
Telfords – Kirkliston

*United States*
ADM Malting Co.—Decatur, Ill.
Briess Malting Co.—Chilton, Wis.
Chilton Malting Co.—Chilton, Wis.
DVC Ultramalt—Manitowoc, Wis.
Froedtert Malt Co.—Milwaukee, Wis.
Great Western Malting Co.—Vancouver, Wash.
Ladish Malting Co.—Milwaukee, Wis.

Minnesota Malting Co.—Cannon Falls, Minn.
Premier Malt Products—Grosse Point, Mich.
Rahr Malting Co.—Minneapolis, Wis.
Schreier Malting Co.—Sheboygan, Wis.
Sunrise Milling—Johnson Creek, Wis.

## BASE MALTS

### Pale Malt

Compared to just ten years ago (see *Zymurgy*, Special Issue 1985, vol. 8, no. 4), the variety of pale malts available to homebrewers today is a smorgasbord. Maltsters in both the United States and Europe (Table 5-3) have made their products available to homebrewers through an expanded network of homebrew suppliers. Homebrewing as we know it definitely took a giant leap forward when the Belgian, British, and German malts finally invaded our shores in quantity and variety. Homebrewers have numerous pale malts from which to choose, and while they may appear similar, their subtle differences will lend a uniqueness to each beer in which they are used.

Pale barley malt is the basis for nearly every beer. It provides most of the enzymatic (diastatic) power to convert complex starches into fermentable sugars. In short, pale malts are the workhorses of the mash. Since pale malts usually compose greater than 50 percent of the total grain bill, they also provide the bulk of the fermentable sugars available in the wort for yeast fermentation. These malts are dried completely before they are kilned at relatively low temperatures to preserve enzyme capability and to minimize color development. Pale malts from America and Europe are readily available to homebrewers (Table 5-4).

U.S. and Canadian pale malt comes in either two-row or six-row varieties. Briefly, American six-row malt has less starch but greater husk weight than two-row malt. This means that for any given weight, more extract will be obtained from two-row than six-row. However, to many homebrewers, these differences are negligible. In cases where homebrewing equipment provides a shallow grain bed (less than 6 inches), six-row malt may actually provide higher extraction rates than two-row malt. It should be noted however, that six-row malt contains a higher proportion of proteinaceous substances. This is important because the higher protein content can result in greater break material (hot and cold). This increased protein can also result in increased problems with haze in the

## Table 5-4. Malts Available in the United States

| | Extract (SG per pound per gallon) | Color (in °Lovibond per pound per gallon) | Diastatic Power (relative rating based on common °Lintner values) |
|---|---|---|---|
| **Pale Malts** | | | |
| American two-row | 1.035–1.037 (1.037) | 1.4–2.0 (1.8) | very high |
| American six-row | 1.031–1.035 (1.035) | 1.5–2.1 (1.8) | very high |
| Belgian two-row | 1.036–1.037 (1.037) | 1.8–3.8 (3.0) | moderate |
| British two-row | —— (1.038) | 2.0–3.5 (2.5) | low |
| Canadian two-row | 1.034–1.037 (1.036) | 1.8–2.4 (2.1) | very high |
| British Pilsener | —— (1.036) | 1.0–2.5 (1.8) | moderate |
| German Pilsener | 1.035–1.038 (1.038) | 1.0–2.0 (1.6) | high |
| Belgian Pilsener | 1.035–1.037 (1.037) | 1.4–1.8 (1.8) | high |
| British Lager (two-row) | 1.035–1.038 (1.038) | 1.4–1.8 (1.4) | moderate |
| Lager malt (two-row) | —— (1.035) | —— (1.7) | high |
| Lager malt (six-row) | —— (1.031) | —— (1.7) | high |
| **Other Malts** | | | |
| American wheat | 1.038–1.039 (1.038) | 1.7–2.1 (2.0) | very high |
| American soft white wheat | 1.039–1.040 (1.040) | 2.5–3.5 (2.8) | very high |
| Belgian wheat | 1.038–1.039 (1.038) | 1.6–2.0 (1.8) | moderate |
| German wheat | 1.038–1.039 (1.039) | 1.0–2.0 (1.8) | moderate |
| German dark wheat | 1.038–1.039 (1.039) | 7.0–9.0 (8.0) | moderate |
| American rye | 1.029–1.030 (1.030) | 2.0–4.7 (3.5) | moderate |
| **Specialty Malts** | | | |
| LIGHT | | | |
| Sauer (acid) | —— (1.035)* | 1.0–2.0 (1.5) | high* |
| American Vienna | 1.030–1.035 (1.035) | 3.0–4.0 (4.0) | very high |
| German Vienna | 1.030–1.037 (1.037) | 2.0–4.0 (3.0) | moderate |
| British mild | 1.033–1.037 (1.037) | 3.0–4.2 (4.0) | moderate |
| German smoked (Bamberg) | —— (1.037) | 6.0–12.0 (9.0) | moderate* |
| British peated | —— (1.038)* | 1.6–6.0 (5.0) | low* |
| Scottish peated | —— (1.038)* | —— (5.0) | low* |
| DARK | | | |
| American victory | 1.029–1.034 (1.034) | 7.0–30.0 (25.0) | nil |
| British brown | —— (1.032) | 38.0–70.0 (70.0) | nil |
| Belgian biscuit | 1.030–1.035 (1.035) | 22.5–27.0 (24.0) | very low |
| Belgian aromatic | 1.030–1.036 (1.036) | 15.5–26.0 (25.0) | low |

## MALT

| | Extract (SG per pound per gallon) | Color (in °Lovibond per pound per gallon) | Diastatic Power (relative rating based on common °Lintner values) |
|---|---|---|---|
| British amber | —— (1.032) | 30.0–35.0 (35.0) | nil |
| Canadian honey | 1.030–1.035 (1.030) | 18.0–25.0 (18.0) | low |
| American two-row (toasted) | —— (1.033) | —— (30.0) | low* |
| American special roast | —— (1.033)* | —— (40.0) | low* |
| Melanoidin | —— (1.033)* | 30.0–40.0 (35.0) | low* |
| Belgian Munich | 1.032–1.038 (1.038) | 5.0–10.0 (7.8) | moderate |
| German Munich | 1.030–1.037 (1.037) | 5.0–8.0 (8.0) | moderate |
| American Munich (light) | 1.033–1.034 (1.033) | 8.0–12.0 (10.0) | moderate |
| American Munich (dark) | 1.032–1.033 (1.033) | 18.0–22.0 (20.0) | moderate |
| Canadian Munich (light) | —— (1.034)* | —— (15.0) | moderate* |
| Canadian Munich (dark) | —— (1.034)* | —— (30.0) | moderate* |
| British Munich | 1.036–1.037 (1.037) | 4.0–8.0 (6.0) | low |
| CARAMELIZED MALTS | | | |
| Dextrin malt (CaraPils) | 1.030–1.033 (1.033) | 1.5–3.0 (1.8) | nil |
| Belgian caramel Pils | 1.030–1.034 (1.034) | 4.0–8.0 (7.9) | nil |
| British CaraMalt | —— (1.035)* | 10.0–13.0 (12.0) | nil |
| American crystal 10 °L | 1.024–1.035 (1.035) | —— (10.0) | nil |
| American crystal 20 °L | 1.024–1.035 (1.035) | —— (20.0) | nil |
| American crystal 30 °L | 1.024–1.035 (1.035) | —— (30.0) | nil |
| American crystal 40 °L | 1.024–1.034 (1.034) | —— (40.0) | nil |
| American crystal 60 °L | 1.024–1.034 (1.034) | —— (60.0) | nil |
| American crystal 80 °L | 1.024–1.034 (1.034) | —— (80.0) | nil |
| American crystal 90 °L | 1.024–1.033 (1.033) | —— (90.0) | nil |
| American crystal 120 °L | 1.024–1.033 (1.033) | —— (120.0) | nil |
| British Light Carastan | —— (1.035)* | 12.0–19.0 (15.0) | nil |
| British Carastan | —— (1.035)* | 30.0–39.0 (34.0) | nil |
| British crystal 50–60 °L | —— (1.034) | 50.0–60.0 (55.0) | nil |
| British crystal 70–80 °L | —— (1.034) | 70.0–80.0 (75.0) | nil |
| British crystal 95–115 °L | —— (1.033) | 95.0–115.0 (105.0) | nil |
| British crystal 135–165 °L | —— (1.033) | 135.0–165.0 (150.0) | nil |
| German Carahell | —— (1.034)* | 10.0–15.0 (12.0) | nil |
| German light caramel | 1.035–1.037 (1.037) | 2.2–2.8 (2.5) | nil |
| German dark caramel | 1.035–1.037 (1.037) | 50.0–80.0 (65.0) | nil |
| German wheat caramel | 1.036–1.038 (1.038) | 50.0–60.0 (55.0) | nil |
| Belgian CaraVienne | 1.030–1.034 (1.034) | 15.0–30.0 (22.0) | nil |
| Belgian CaraMunich | 1.032–1.033 (1.033) | 53.0–80.0 (75.0) | nil |
| Belgian Special "B" | 1.029–1.030 (1.030) | 75.0–250.0 (220.0) | nil |

## Malt—A Spectrum of Colors and Flavors

| | Extract (SG per pound per gallon) | Color (in °Lovibond per pound per gallon) | Diastatic Power (relative rating based on common °Lintner values) |
|---|---|---|---|
| **ROASTED MALTS** | | | |
| American chocolate | 1.023–1.029 (1.029) | 325.0–400.0 (350.0) | nil |
| Belgian chocolate | 1.029–1.030 (1.030) | 375.0–500.0 (500.0) | nil |
| British chocolate | 1.029–1.034 (1.034) | 350.0–600.0 (475.0) | nil |
| German Carafa | —— (1.030)* | —— (400.0)* | nil |
| German Carafa special | —— (1.030)* | 500.0–750.0 (600.0) | nil |
| American black patent | 1.023–1.029 (1.028) | 475.0–530.0 (500.0) | nil |
| British black patent | 1.023–1.030 (1.027) | 500.0–750.0 (525.0) | nil |
| Belgian black | 1.028–1.030 (1.030) | 500.0–675.0 (600.0) | nil |
| **UNMALTED** | | | |
| American roasted barley | 1.024–1.029 (1.028) | 450.0–510.0 (450.0) | nil |
| American black barley | 1.023–1.028 (1.027) | 500.0–550.0 (530.0) | nil |
| Belgian roasted barley | 1.029–1.030 (1.030) | 450.0–650.0 (575.0) | nil |
| British roasted barley | —— (1.029)* | 500.0–600.0 (575.0) | nil |
| German roasted wheat | 1.028–1.030 (1.030) | 550.0–750.0 (650.0) | nil |
| Roasted rye | 1.028–1.029 (1.029) | 400.0–650.0 (500.0) | nil |
| German Carafa chocolate | —— (1.030)* | —— (525.0)* | nil |

---

*The information presented here was compiled from a wide variety of sources including Charlie Papazian's* The Home Brewer's Companion *(Avon, 1994), numerous homebrewing catalogs and technical data from maltsters' fact sheets. Ranges are given followed by the most commonly published value for the malt type. Note: This value will not necessarily represent the median value for the ranges obtained from all the sources. \*Indicates authors' estimate based on available information on related malts.*

finished beer. Six-row malt may also be slightly less expensive than two-row malt, depending on the supplier.

Six-row malts are generally regarded as possessing greater enzymatic power than two-row malts, although many brewers believe two-row malts lend a more mellow flavor to the beer. Six-row can be useful in brewing beers with a high proportion of adjuncts that lack enzymes, or wheat malt that lacks husk. The use of six-row malt can help the homebrewer achieve conversion of starch to sugar. Beyond this, color differences between six-row and two-row malts are slight.

Continental pale malts, primarily from Germany and Belgium, are also widely available to homebrewers. British pale malts are in ample supply, too. Although generally more expensive than domestic pale malts, there are benefits in their use for homebrewers who desire to make beer styles using authentic ingredients.

Because of differences in the malting procedures of European and American maltsters, pale malts from Europe generally have less diastatic power compared to those from the United States. These differences are not generally great enough to cause significant problems for the homebrewer. We believe British and continental pale malts provide a more complex malt palate and rounder flavors than their domestic counterparts when used to make beers that lack a significant proportion of specialty malt. This makes them worth the extra cost, especially when trying to duplicate your favorite English bitter, German Oktoberfest, or Bohemian Pilsener.

## *Other Malted Grains*

Although most homebrewers are familiar with malted barley, other grains can make significant contributions to the malt profile of a beer. The most common of these is wheat malt, which is essential in making American and German wheat beers and Kölsch. Also, the addition of 3 to 5 percent wheat malt in many other all-malt beers can aid in improving head retention without imparting any significant changes in flavor.

The amount of extractable sugar obtained from wheat malt is somewhat higher than can be achieved by using barley malt alone. Wheat malt flavor is different from barley malt, usually lighter, although this can vary with the mash procedure used. The color derived from wheat malt is quite light, typically in the same range as pale barley malt, with the exception of soft white wheat malt (see Table 5-4). Diastatic enzyme power is higher than that of barley malt.

Unfortunately for the homebrewing community, German dark wheat malt, wonderful in a dunkelweizen, and roasted wheat malt, which have been available in the past, are becoming more difficult to find. Apparently demand was insufficient for most homebrew suppliers to keep these in stock. We suggest you contact your local microbrewery or brewpub for information if you would like to experiment with these malts.

Rye malt imparts a very distinctive flavor to the final product. If you like the taste of rye bread, you are sure to enjoy the flavor of a beer in which a substantial proportion of the grain bill includes rye malt. Additions as low as 5 percent of the total grist can lend a nutty, rye finish to a beer, depending on the style. We recommend starting out with small quantities of rye malt to determine if you do indeed appreciate the flavor. More intense yet is roasted rye, which is very dark and a more acrid variant of rye malt. This malt, however, also is becoming difficult to find.

Rye malt, like wheat malt, is huskless. The barley that makes up the remainder of a grain bill including rye or wheat should be properly crushed with husks intact to reduce the chances of a stuck runoff during sparging. The extract obtained from rye malt is considerably less than that obtained from the malts previously discussed. The color of rye malt is slightly darker than barley or wheat malt.

### Specialty Malts

Now that we know something about pale malt, we are ready to begin discussing the types of malts that add character to the beer we are trying to brew. For us, this is where the fun really begins! We enjoy seeking bold new brews through experimentation, testing, and tasting. Creativity makes the homebrewing world go round, and specialty grains are what get that big beer ball turning. Most of these malts are produced in similar ways to the malts described above except they are kilned at slightly higher temperatures once they are completely dry or have very low moisture content. The result is a malt with darker color, sweeter flavors, and slightly fewer fermentables that still retains some diastatic capability. Examples of this type of malt are Vienna, Munich, amber, brown, Biscuit, aromatic, mild, and Victory, to name a few. Still others are kilned over open fires (German smoked malt) or peat (Scottish peat malt) to give a special "smoked" flavor characteristic of certain beers.

Specialty malts can be particularly enjoyable for the homebrewer to use. Each imparts characteristic flavors ranging from lightly toasted to a biscuity, malty sweetness. Some, such as aromatic, greatly enhance the malt aroma of a beer even when used in small quantities. The color they contribute is red to amber depending on the quantity used and other malts in the grist.

Most of the malts listed here can be used in either small or quite significant quantities. For example, using 10 percent Munich malt will lend a malty sweetness and slight toasty flavor to a number of beer styles such as German alts, California Common beers, or any beer style you fancy (including your own). It is important to remember that the enzymatic power of darker (unroasted) malts, such as Munich and aromatic, is quite low and therefore you should be careful not to overshoot your mash temperatures in brews made up mostly of these malts.

Amber and brown malts were used in British beers such as porters during the eighteenth and nineteenth centuries. Their use tapered off after 1817 when black malt became available, and they are now something of a rarity. Try using them to make a British brown or mild ale as well. Our own experience with these malts is that, if used in high proportion, the beer can take on harsh notes in the finish that are difficult to resolve, even with extended aging. The lesson with specialty malts such as these is simple: The use of small to moderate quantities (less than 10 percent) will greatly enhance the flavor characteristics and malt complexity of your beer. We suggest they be used judiciously until you are certain the flavor imparted to the beer is to your liking. This is especially true in the case of smoked malts. The key to the use of any specialty malt is controlled experimentation, keeping detailed notes on the flavors imparted by the malt and using only one at a time so you can define the flavor you obtain with each malt.

### Caramelized Malts

Crystal and caramel malts are fully modified during malting procedures and are kilned at relatively high temperatures when still moist. The result is a "stewing," rather than a roasting or toasting, in which the starches convert to sugars and then caramelize. These malts come in a wide range of colors from light to very dark amber. They provide a sweet maltiness to a beer and contribute considerable mouth feel and body to the final product even when used in 5 to 15 percent concentrations.

Because the starches are already converted, these malts need not be mashed, and many extract brewers steep crystal malts to add color and body to their beers. These malts also have no enzyme activity because of the kilning process. The wide color range of crystal and caramel malts makes them quite versatile and they can be used in a variety of beer styles with great success.

Belgian CaraVienne and CaraMunich are unique caramelized malts. While generally similar to domestic, British, and German crystal and caramel malts, the Belgian varieties produce a very subtle toasted flavor with some residual caramel sweetness that can enhance a beer such as bock or Oktoberfest, but work well in a wide variety of beer styles. CaraMunich is a darker and more robust version of CaraVienne. Both contribute a soft, smooth finish. Belgian Special "B" is an extremely dark caramel malt that imparts a distinctive but intense toasted (almost toffeelike) malt flavor. It should be used in very small quantities (2 to 4 ounces per 5-gallon batch) until you are sure you like its flavor contributions. This malt could be used in almost any beer style, but it would seem most at home in a toasty, roasty Scottish ale.

Dextrin and CaraPils malts are produced much like crystal malts except that kilning temperatures are considerably lower, which minimizes color development. When used in small quantities, say 5 to 10 percent, these malts greatly enhance body, mouth feel, and head retention while changing the color of the beer very little. Dextrin malts are commonly used in light-colored lagers in which pale malt alone, or the use of adjuncts (discussed below), may be insufficient to achieve a desired malt characteristic.

Finally, British carastan malts offer a unique experience for homebrewers. Although not widely available to homebrewers, they are used quite extensively by microbreweries and brewpubs, which is where we have obtained them. Apparently they were developed as a "crystal" malt for use in lagers in which mouth feel and body were desired without residual sweetness. In this regard, carastans are similar to dextrin malt, although, in our experience, the former do provide for some additional sweetness not obtained from dextrin or CaraPils. Belgian caramel Pils also fits into this category but is much lighter in color than any available carastan malt.

## ROASTED MALTS AND GRAINS

Malts and grains in this category are generally used in small quantities, so the potential extract yield is not particularly important to the homebrewer. Roasted malts and grains are produced by kilning at very high temperatures that carbonize the starch and sugars. Examples are chocolate malt, black patent (also called roasted malt or black malt), black barley, and roasted barley. The latter two differ in that the grain is not malted prior to kilning.

These malts impart a very dark color to beer and an intense nutty (in the case of chocolate malt) or anything from a chocolate-like to a burnt charcoal-like flavor. They are used to some degree in most porters and stouts. Chocolate and black malt can also be used to add color and malt complexity to a number of other beer styles when used in very small quantities. Roasted barley is used almost exclusively in dry and sweet stouts. All of these highly carbonized malts and grains have no diastatic power and need not be mashed in order to take advantage of color and flavor enhancement of beer. When used with caution (1 to 2 ounces per 5 gallons), these dark roasted malts can provide a hint of brownish red color as well as an enhanced background flavor to the overall malt profile of a pale beer.

A particularly nice variant of these malts is German Carafa chocolate malt. Although not widely available, it is certainly worth seeking out. It is darker than other domestic and imported chocolate malts, but slightly lighter than black malt and black barley. This malt gives a luscious but intense chocolate finish even when as little as 4 ounces in a 5-gallon batch is used. We use it often in our favorite beer style—porter!

There are several obscure extremely specialized malts such as white malt (very pale at 1 °L for lightening Pilseners, etc.), melanoidin malt (high in melanoidin content for fullness and flavor stability), and Sauer malt (an acidified German malt for pH adjustment and intensifying fermentation). Brumalt is a German green malt good for intensifying color, flavor, and aroma (see Narziss, *Zymurgy*, Winter 1993, vol. 16, no. 5). These malts have been marketed in this country at one time or another. However, most of them have been discontinued because of lack of demand. If you do happen to run across any of them in your malt-hunting expeditions, they may be worth a test batch or two. Gambrinus Malting Company of Canada recently began exporting to the United States their version of Brumalt called "honey malt" because of its intense, sweet honeylike aroma and flavor. Based

## Table 5-5. Additional Fermentable Grain Adjuncts

| Grain Adjunct | Extract (SG per pound per gallon) | Color (in °Lovibond) |
| --- | --- | --- |
| flaked corn | 1.039–1.040 (1.040) | 0.5–0.8 (0.5) |
| flaked oats | 1.025–1.033 (1.033) | 2.0–2.5 (2.2) |
| flaked rice | 1.038–1.040 (1.040) | 0.5–1.0 (0.5) |
| flaked rye | 1.033–1.036 (1.036) | 1.5–3.0 (2.8) |
| flaked wheat | 1.020–1.036 (1.034) | 1.5–2.5 (2.0) |
| flaked barley | 1.025–1.032 (1.032) | 1.5–2.5 (2.2) |

*The above unmalted grains can be incorporated into the mash to enhance mouth feel, head retention, and/or flavor. The same grains can be used in their raw (unflaked) form with similar results, but must be gelatinized first.*

on our limited experience, we believe honey malt probably could find a home in many different styles of beer, light or dark. It would be particularly useful where more malt aroma and/or a subtle malty sweet flavor is desired.

## GRAIN ADJUNCTS

There is really nothing sacred about using malted barley, wheat, and the other malts already discussed. Fermentable sugars can be derived from other sources of starch and can contribute desirable characteristics to beer. Examples of adjuncts that are routinely used in the production of beer include flaked barley, flaked corn, flaked wheat, flaked rice, flaked rye, and rolled oats, all of which are unmalted (Table 5–5). When using unmalted, unprocessed grains, the starches in them should be gelatinized prior to use for best results. Depending on the grain, gelatinization involves heating that renders the starch vulnerable to enzyme degradation into simple sugars. Since gelatinization occurs during the rolling process, flaked adjuncts can be incorporated directly into the mash. If all of this appears too daunting a task, refined corn starch, available at any supermarket, can be added without any processing prior to use. A number of the adjuncts listed here will result in a beer with lighter body than if it were brewed with all barley malt. This is particularly desirable when brewing a number of American lagers, cream ales, and the like.

Flaked rye imparts a distinct sharp flavor. Flaked barley not only contributes to foam stability (head retention) but can improve mouth feel. Rolled oats also will improve mouth feel but impart a slight oiliness to the beer. Both flaked barley and rolled oats can be used to improve the quality of any classic stout recipe already in your repertoire.

Homebrewers not experienced in the use of adjuncts should be aware of some potential problems that can result from their use. For example, because these unmalted grains generally lack enzyme power, it is sometimes advisable to use a domestic two-row or six-row malt in conjunction with them. The extra enzyme capability of these pale malts is generally sufficient to convert the starch available in the flaked grains with little noticeable delay. Flaked grains also can lead to slow runoffs during the sparging process. Therefore, many brewers prefer to use six-row pale malt with its high husk-to-endosperm ratio. It also is advisable to use 20 percent or less adjunct in the grain bills. Some adjuncts, such as flaked barley, can lead to chill haze problems, which is why they are more commonly used in darker beer styles where this is less a concern.

## NONTRADITIONAL GRAINS

Although this topic is really beyond the scope of this article, we believe it is important to mention that grains from a wide variety of other crop species could be, and have been, used as adjuncts in beer. Those that have definite brewing potential range from the common (buckwheat, millet, sorghum, triticale, and wild rice) to the exotic (adlay, amaranth, dinkel, fundi, kamut, kasha, quinoa, spelt, and teff). These can be gelatinized and used as is, or the ambitious homebrewer might attempt to malt them. This area is wide-open and ripe for experimentation. We take the time to mention these nonstandard fermentables because we have personal experience with amaranth and kasha, both of which contribute a subtle nuttiness to the overall malt flavor profile. Amaranth has the highest protein content per unit of weight of any basic grain in the world, and this attribute contributes to the formation of a rich, creamy head. Even with the high protein content, our amaranth beer was remarkably free of protein haze. Amaranth and kasha have worked very well for us in British mild and brown ales. Additional information on alternative fermentable grains can be found in *Zymurgy*, 1994 Special Issue (vol. 17, no. 4). As a note of caution, stick to grains that are known to be edible. Grains (seeds) of many plants are not eaten because they contain toxic substances. Exercise caution when breaking new ground.

## RECIPE GUIDELINES

Now that you have some familiarity with the available malts and grains that can be used by the homebrewer, it's time to give some thought to how you should use them. It is difficult to argue with the adage "Keep it simple, stupid!" especially because so many fine beers contain only one, two, or three malts. However, we believe one of the simple pleasures of homebrewing is to combine malts in small quantities to experience the diverse flavors and other characteristics imparted to the finished product.

An easy way to demonstrate the differences in malt quality achieved by using a variety of malts is by producing a simple pale ale. In a 5-gallon batch, try making a pale ale using 8 pounds of pale malt with 1 pound of 40 °L crystal malt. Next time, while using the same mashing procedures, hopping schedule, and yeast strain, make a similar pale ale but change the color of the crystal. Rather than using 1 pound of 40 °L, use ½ pound while blending in ¼ pound of 20 °L and ¼ pound of 80 °L crystal malt. While the color of both beers will be copper, you will find considerable differences in malt complexity and residual sweetness flavors. This little experiment may well start you off on an exploration of the mysteries of malt.

When using specialty and a number of caramelized and highly carbonized malts, we have a rule of thumb: Keep them at 15 percent or less of the total grist. When we experiment with a new malt, it rarely exceeds 5 percent of the total grain bill. That way if the flavors are intense, they usually will not overwhelm the malt profile and make the beer difficult to enjoy. Once you become accustomed to the flavor of a particular malt and determine that you like it, try blending small quantities of different malts to increase malt depth and complexity.

## ON THE GELATINIZATION PROCESS . . .

In the strictest sense, gelatinization is simply the release of starch into suspension by rupturing starch storage granules. A certain degree of gelatinization occurs whenever malted or unmalted grains are exposed to temperatures over 140 degrees F (60 degrees C). Gelatinization is therefore important because starches not liberated during the malting process are made available for enzymatic conversion during the mash, but the process is most important when incorporating unmalted grains or seeds in your brews.

Unmalted grains contain large quantities of starch that will be used as a source of energy by the developing sprout. This starch is basically in an "unmashable" form because it is tied up in cellular storage structures. In addition to activating the enzymes important for mashing, the malting process also liberates much of this stored starch. Soaking unmalted grain in hot water extracts this starch by helping break the hard seed coat, disrupting cell walls within the seed itself and eventually causing the starch storage particles to burst. The freed starch forms a thick colloidal suspension. At this point the starch can be used by the enzymes of the mash.

## ON GELATINIZING ADJUNCTS . . .

Flaked (rolled) adjuncts are produced from unmalted grains by softening them with steam, then passing them through pressure rollers. The heat associated with both of these processes almost completely gelatinizes the starches, so flaked materials can be added directly to the mash without further preparation.

Not so with unprocessed unmalted grains—they must first be gelatinized. Unfortunately, complete details on how to accomplish this at home are not readily available. The amount of time required for gelatinization varies depending on the grain adjunct used. Details are sparse because many of these adjuncts have had only limited use, and procedures for gelatinizing them are unknown.

To gelatinize an unmalted adjunct for the first time, mix it with water at a rate of 2 gallons (7.57 l.) of water per pound (0.45 kg.) of grain. Bring to a boil, reduce heat, and simmer the grain suspension for a minimum of 15 minutes (30- to 60-minute boils may be necessary for some grains). Use plenty of water, and additional water may be added at any time because as gelatinization proceeds and the grain breaks down, the slurry will become very thick, pasty, and sticky. Heat the mixture carefully because the potential for burning it is high. Gelatinization is complete when the suspension becomes uniform in texture and the adjunct grain is indistinguishable. The slurry will have the consistency of Cream of Wheat.

The gelatinized adjunct then may be added directly to the mash, but since it has not been modified by malting, include a protein rest in the mash schedule prior to raising the mash to saccharification temperatures. Allow the gelatinized suspension to cool to close to protein-rest temperature before mixing it into the grains. A final recommendation: Take good notes (especially details about quantities of water and boil times).

## SUMMARY

There are times when the abundance of available malts can be overwhelming. The key to making sense of all this is for homebrewers to take advantage of the situation and use the available malts to improve their beer. The extensive selection of malts lends great versatility and flexibility in the production of quality homebrew. These malts may be best used by homebrewers who choose to brew their own style of beer while refusing to be trapped by conventional recognized styles. For these brewers, this period in homebrewing is particularly exciting.

## REFERENCES

Blenkinsop, Peter, "The Manufacture, Characteristics and Uses of Specialty Malts," MBAA *Technical Journal* 28, 1991.

Fix, George, "Belgian Malts: Some Practical Observations," *BrewingTechniques* 1(1), 1993.

Foster, Terry, *Porter*, Brewers Publications, 1992.

Hayden, Rosannah, "Brewing with Rye," *BrewingTechniques* 1(3), 1993.

Narziss, Ludwig, "Specialty Malts for Greater Beer Type Variety," *Zymurgy*, Winter 1993 (vol. 16, no. 5).

Noonan, Gregory J., *Brewing Lager Beer*, Brewers Publications, 1986.

O'Rourke, Timothy, "Making the Most of Your Malt," *The New Brewer* 11(2), 1994.

Papazian, Charlie, *The New Complete Joy of Home Brewing*, Avon Books, 1991.

Papazian, Charlie, *The Home Brewer's Companion*, Avon Books, 1994.

Thomas, David, and Geoffrey Palmer, "Malt," *The New Brewer* 11(2), 1994.

*Zymurgy*, Special Issue 1985 on Grain Brewing (vol. 8, no. 4).

*Zymurgy*, Special Issue 1994 on Special Ingredients and Indigenous Beer (vol. 17, no. 4).

# six

# Hops

## HOP VARIETIES AND QUALITIES

### BY BERT GRANT

*This article originally appeared in* Zymurgy, *Special* 1990 *(vol.* 13, *no.* 4)

In 1990 the most popular bittering hops in the United States are Cluster, Galena, Nugget, and Chinook. Most widely used aroma hops are Cascade, Hersbrucker, Willamette, Hallertauer, and Mt. Hood.

Popularity and availability of some of the traditional varieties are declining, in some cases rapidly. Brewers Gold and Bullion, for instance, which were a major factor in the Oregon market up to ten years ago, have now all but disappeared. Several new "starts" on the hop variety scene a few years ago also have faded away (Comet, Columbia, and Olympic). While imported traditional aroma types such as Hallertauer and Mittelfrüh are sometimes available in very small amounts, the growing trend in the major breweries and microbreweries is to use the newer U.S. varieties, which in most cases produce beers of equal or superior quality.

### SAAZ

Traditional fine aroma hops from Czechoslovakia. High price, low yield, but one of the world standards for noble hop aroma in a lager beer.

## CHINOOK

U.S. Department of Agriculture cross from Petham Golding released in 1985. Good yields and high alpha-acid content make this an economic bittering hop. Mild, spicy aroma with a positive, lingering bitterness.

## GALENA

Developed in Idaho from Brewers Gold by open pollination. Released in 1978 and now rivals Cluster as the most popular U.S. bittering hop. Good yield and very high alpha content make this an excellent economic value. Aroma neutral if used early in the kettle boil. Excellent as early kettle addition when combined with Cascade or Willamette for late "aroma" addition.

## NUGGET

New high-alpha variety, based on Brewers Gold and released in 1982. Good aroma profile and storage stability. Interesting aroma character—heavily spiced and herbal. Good bitterness quality. Increasing production and demand.

## TETTNANGER

Traditional German variety from the Tettnang area. True noble aroma variety recently planted extensively in the United States. Low yields keep the costs high, but Tettnanger remains one of the finest hops in the world for aroma.

## MT. HOOD

This is a triploid seedling of Hallertauer with improved agronomic characteristics. Released in 1988 but acreage increasing rapidly because of acceptance as Hallertauer equivalent by major brewers.

Table 6-1. Hops

| Variety | % Alpha Acids (fresh) | Freshness Stability | Oz. per 5-gal. for average bitterness | Origin | Comments |
|---|---|---|---|---|---|
| Aquila | 7.0 | Fair | 1.5 | U.S. | New, aroma |
| Banner | 10.0 | Fair | 1.0 | U.S. | New, bitter |
| Brewers Gold | 9.0 | Poor | 1.1 | U.S./U.K. | Declining use |
| Bullion | 9.0 | Poor | 1.1 | U.S./U.K. | Declining use |
| Cascade | 5.5 | Poor | 2.0 | U.S. | Good aroma |
| Chinook | 12.0 | Very good | 0.9 | U.S. | Recent, bitter |
| Cluster | 7.0 | Excellent | 1.5 | U.S. | Standard |
| Columbia | 10.0 | Fair | 1.0 | U.S. | Obsolete |
| Comet | 10.0 | Fair | 1.0 | U.S. | Obsolete |
| Eroica | 10.0 | Fair | 1.0 | U.S. | Bittering |
| Fuggles | 4.0 | Fair | 2.2 | U.S./U.K. | Rapidly declining use |
| Galena | 12.0 | Very good | 0.9 | U.S. | Most popular bittering hop |
| Goldings | 5.0 | Fair | 2.0 | U.S./Canada | Declining use |
| Kent Goldings | 5.0 | Fair | 2.0 | U.K. | Declining use |
| Hallertauer | 5.0 | Poor | 2.0 | U.S./Germany | Declining, premium aroma |

| Variety | % Alpha Acids (fresh) | Freshness Stability | Oz. per 5-gal. for average bitterness | Origin | Comments |
| --- | --- | --- | --- | --- | --- |
| Hersbrucker | 5.0 | Poor | 2.0 | U.S./Germany | Aroma |
| Mt. Hood | 5.0 | Fair | 2.0 | U.S. | Good aroma |
| Northern Brewer | 8.0 | Fair | 1.4 | U.S./Germany | Bittering |
| Nugget | 12.0 | Good | 0.9 | U.S. | Bittering, aroma |
| Olympic | 10.0 | Good | 1.0 | U.S. | Obsolete |
| Perle | 8.0 | Good | 0.9 | U.S. | Bittering, aroma |
| Pride of Ringwood | 8.0 | Good | 1.4 | Australia | Bitter |
| Saaz | 5.0 | Fair | 2.0 | Czechoslovakia | Premium aroma |
| Spalt | 7.0 | Poor | 1.5 | Germany | Aroma |
| Styrian Goldings | 6.0 | Fair | 1.8 | U.S./Yugoslavia | Aroma |
| Talisman | 8.0 | Good | 1.4 | U.S. | Declining use |
| Tettnanger | 5.0 | Poor | 2.0 | U.S./Germany | Aroma |
| Willamette | 5.0 | Fair | 2.0 | U.S. | Aroma, popularity rapidly increasing |
| Wye Target | 10.0 | Fair | 1.0 | U.K. | Bitter |

## PERLE

Developed in Germany from English Northern Brewer. Moderate bittering potential (about 8% alpha) and some good aroma characteristics. Recently introduced into U.S. growing areas.

## HALLERTAUER (HERSBRUCKER)

Traditional German variety selected in the Hallertauer area. More properly called Hallertauer Mittelfrüh, as many other hops are also called Hallertau. Very good aroma hop, until recently the standard for lager beers. Declining production because of sensitivity to verticillium wilt and low yields. Recently largely replaced by Hersbrucker, another German "area" hop, not as highly regarded but much more readily available.

## CASCADE

Aroma hop developed in Oregon by open pollination of a Fuggle seedling. Released commercially in 1972. Rapid acceptance followed by rejection by major brewers and recently renewed interest. Still a major variety, well liked for its distinctive aroma derived in part from the relatively high content of linalool and geraniol in the oil fraction. Pleasant bitterness even at high bitterness unit levels.

## CLUSTER

The standard bittering hop in the United States and much of the world since the turn of the century. Parentage is lost in antiquity, but it is likely a cross between an English import and a wild American hop. Best aroma and bitterness stability of any variety. Good yields, vigorous plants, susceptible to mildew in rainy areas but reasonably resistant to other problems. Use declining in favor of much higher alpha-value types such as Galena.

## FUGGLE AND WILLAMETTE

Fuggle is an English variety selected in 1875. Widely used in the United States and England until the 1980s. Largely supplanted in

United States by the recently developed triploid progeny—Willamette.

Fuggle use declined in England because the Wye varieties were developed as replacements. Fuggle still is widely used as dry hops, but even here Willamette is an adequate replacement.

Acreage figures for 1989 give an estimate of the major varieties in commercial production:

| Variety | Acreage |
| --- | --- |
| Cluster | 6,862 |
| Galena | 6,451 |
| Willamette | 6,299 |
| Nugget | 3,519 |
| Tettnanger | 2,941 |
| Chinook | 1,560 |
| Cascade | 1,297 |
| Perle | 1,064 |

Some varieties already are nearly obsolete, not having survived competition from newer arrivals, or losing to disease and pests. These are Brewers Gold, Bullion, Columbia, Comet, Olympic, and Talisman.

A few new varieties, Mt. Hood, Aquila, Banner, are increasing in acreage, but still are considered experimental (see Table 6-1).

# CALCULATING HOP BITTERNESS IN BEER

### BY JACKIE RAGER

*This article originally appeared in Zymurgy, Special 1990 (vol. 13, no. 4)*

Has confusion over the use of AAU, BU, HBU, and IBU left you with a bitter taste in your mouth? Was that bitter taste left by your latest homebrew that should have emphasized the malt? If so, this article may be your ticket to better-balanced brews.

First, let's familiarize ourselves with the four terms for the two calculation methods used to determine the bitterness in beer. They are BU (Bitterness Units), HBU (Homebrew Bitterness Units), AAU (Alpha Acid Units), and IBU (International Bitterness Units). Why we need two names for each calculation is beyond me. That's an issue best left up to the HSC (Homebrewing Standards Committee), if one exists. If I understand what I've read, AAU and HBU are calculated by multiplying the weight of hops (in ounces) by their alpha acid. IBU and BU include boiling time in the calculation.

Which is best for determining bitterness in beer? What do you do if you want to decrease or increase the hop bitterness of a particular recipe, or if you want to develop your own recipe? What would happen if you boiled the hops 20 minutes less or more? To answer these questions, you need more data on the effect boiling has on hop utilization.

What homebrewers need is a formula for determining how much bitterness will be derived from a hop addition or how to determine how much hops to add to get the desired bitterness. The following utilization chart (Table 6-2) and formulas have been developed to satisfy that need. I must give credit to Fred Eckhardt, Dave Miller, and Byron Burch because information gleaned from their books forms the basis for this chart and formula. My contribution was to experiment with their data, fill in some of the gaps, and develop the formula to cover any batch size and wort density.

Keep in mind that Table 6-2 and the calculations are for bitterness only. Your aroma, flavoring, and bittering hop rates should be determined based on the style of beer being brewed. Hops boiled 5 minutes or less contribute more to aroma than to bitterness. Hops boiled 6 to 25 minutes contribute greatly to the flavor of the beer. Bitterness is extracted in greater amounts by longer active boiling.

### Abbreviations used in the following formulas:

| | |
|---|---|
| %U | Percent utilization |
| %A | Percent alpha acid |
| $W_{gr}$ | Weight in grams |
| $W_{oz}$ | Weight in ounces |
| $V_{gal}$ | Volume in U.S. gallons |
| $V_L$ | Volume in liters |
| GA | Gravity adjustment |
| GB | Gravity of boiling wort |
| IBU | (International) Bitterness Units |

## Table 6-2. IBU Chart

| Boiling Time | Percent Utilization |
|---|---|
| Less than 5 minutes | 5.0% |
| 6–10 minutes | 6.0% |
| 11–15 minutes | 8.0% |
| 16–20 minutes | 10.1% |
| 21–25 minutes | 12.1% |
| 26–30 minutes | 15.3% |
| 31–35 minutes | 18.8% |
| 36–40 minutes | 22.8% |
| 41–45 minutes | 26.9% |
| 46–50 minutes | 28.1% |
| 51–60 minutes | 30.0% |

Note: In formulas, percent utilization must be expressed as a decimal.

A gravity adjustment calculation is used for high-gravity beers or when the brew kettle is too small to boil the full batch. It is calculated by taking the larger of gravity of boil (GB) minus 1.050, divided by 0.2. If GB is less than 1.050, then (GA) equals 0.

$$GA = \frac{(GB)-1.050}{0.2}$$

All percentages should be expressed as decimal equivalents.

If using metric measurements of grams and liters, the equations for finding IBUs and grams of hops are:

$$IBU = \frac{\%U \times \%A \times W_{gr}}{V_L \times (1+GA)} \times 1000$$

$$W_{gr} = \frac{V_L \times (1+GA) \times IBU \times .001}{\%U \times \%A}$$

If using English measurements of ounces and U.S. gallons, the equations for finding IBUs and ounces of hops are:

$$\text{IBU} = \frac{W_{oz} \times \%U \times \%A \times 7462}{V_{gal} \times (1+GA)}$$

$$W_{oz} = \frac{V_{gal} \times (1+GA) \times \text{IBU}}{\%U \times \%A \times 7462}$$

Suppose you have 1½ ounces of Bullion hops with an alpha acid of 8.0. You want to know how many IBUs you would get by boiling them 45 minutes in 5 gallons of 1.045 OG wort.

$$\text{IBU} = \frac{W_{oz} \times \%U \times \%A \times 7462}{V_{gal} \times (1+GA)}$$

$$= \frac{1.5 \times .269 \times .08 \times 7462}{5 \times (1+0)}$$

$$\text{IBU} = 48.2$$

To determine how many hops to add to get a desired IBU; say 32 IBU:

$$W_{oz} = \frac{V_{gal} \times (1+GA) \times \text{IBU}}{\%U \times \%A \times 7462}$$

$$= \frac{5 \times (1+0) \times 32}{.269 \times .08 \times 7462}$$

$$W_{oz} = 1.0 \text{ oz}$$

## EXAMPLE:

How many Hallertau hops at 4.8 alpha acid boiled 40 minutes are required to give 32 IBUs in a 5-gallon batch with a specific gravity of 1.048? If the gravity is below 1.050, the GA is not calculated. But if my brewkettle will only boil 2.5 gallons (half the batch), then my gravity of boil (GB) would be 1.096, twice the 1.048 OG.

$$GA = \frac{1.096 - 1.050}{0.2} = 0.23$$

## Table 6-3. Beer Style

| Types of Beer | IBU Range |
|---|---|
| *Light Beers* | |
| American Light | 7.0–19.5 |
| American Standard Premium | 9.3–17.0 |
| International Style Lager | 18.0–40.0 |
| North German Lager and Pils | 28.5–40.0 |
| Cream Ale | 20.0–70.0 |
| | |
| *Amber Beers* | |
| Vienna Lager | 14.6–26.0 |
| Oktoberfest/Marzen | 17.0–34.0 |
| Steam | 40.0 |
| Bitter | 23.0–44.0 |
| Pale Ale | 19.0–54.0 |
| Kölsch and Alt | 21.0–31.0 |
| India Pale Ale | 19.0–87.0 |
| Trappist | 11.2–24.0 |
| | |
| *Dark Beers* | |
| Schwarzbier | 28.0–40.0 |
| Bock | 26.0–35.0 |
| Brown Ale and Mild | 31.0–38.0 |
| | |
| *Black Beers* | |
| Porter | 34.0–56.0 |
| Sweet Stout | 29.0 |
| Dry Stout | 35.0–90.0 |
| Belgian Triples/ Barley wines | 32.0–100.0 |
| | |
| *Wheat Beers* | |
| Weisse and Weizen | 10.5–20.0 |
| Berliner Wiesse | 4.0–5.0 |
| American Wheat | 15.0–27.0 |

IBU *figures were derived from* Essentials of Beer Styles, *by Fred Eckhardt.*

The hop weight calculation is:

$$W_{oz} = \frac{V_{gal} \times (1+GA) \times IBU}{\%U \times \%A \times 7462}$$

$$= \frac{5 \times (1+.24) \times 32}{.228 \times .048 \times 7462}$$

$$W_{oz} = 2.4 \text{ oz.}$$

*Jackie Roger has been a serious homebrewer for 18 years. He is a charter member of the Kansas City Bier Meisters. Roger is one of the managing partners of Bacchus & Barleycorn Ltd. of Merriam, Kansas, and has conducted homebrewing classes for the last 13 years. When asked what his favorite beer is, he replied, "The one in my stein."*

# seven

# Yeast

## ACTIVE DRY YEAST FOR THE HOMEBREWER

BY RODNEY MORRIS

*This article originally appeared in* Zymurgy, *Spring 1991 (vol. 14, no. 1)*

Many homebrewers have misconceptions on how dried yeast is made and how to use it correctly in starting a batch of beer.

Yeast grown in a homebrew fermentation produces a limited number of yeast cells and ethanol as a desired product. An expression of this type of fermentation is:

Maltose + Amino acid →
100 g.       0.5 g.

Yeast + Ethanol + Carbon Dioxide + Energy
5 g.    48.8 g.    46.8 g.         50 kCal

The producers of dried yeast wish to maximize the yield of yeast cells, so they recommend using a fermentation with vigorous aeration. The expression for this type of fermentation is:

Molasses + Ammonia + Oxygen →
100 g.      5 g.       51 g.

Yeast + Water + Carbon Dioxide + Energy
48 g.   35 g.    75 g.            194 kCal

The amounts of yeast cells produced are in dry weight. As you can see, the production of brewer's yeast under aerobic conditions produces much more yeast and energy than an anaerobic fermentation for beer production.

Some homebrewers believe that the dried yeast is a freeze-dried product. The dried yeast for brewing (and wine-making) is produced by removing water from the yeast with warm air to a carefully controlled level. Brewing yeast that has been freeze-dried has poor viability, so this process is not used commercially.

Yeast is produced under food-level sanitary conditions. Some bacterial and wild yeast contamination is almost inevitable, but producers try to keep such contamination low. The yeast is propagated in large open vats or closed tanks with vigorous aeration of the beet-cane molasses wort. Ammonium phosphate salt and other minor minerals and nutrients are added for growth. The concentration of the sugar-molasses in the wort is kept under 0.1 percent to prevent the fermentation from shifting to an anaerobic fermentation with production of alcohol with low yeast cell yield. Additional molasses is added during the fermentation as required.

After fermentation is complete, the yeast is removed by centrifugation, washed, and pressed into a cake to remove excess water. The crumbly yeast cake is forced through a perforated steel plate to form noodles that are broken up and dried to a final moisture level of about 5 to 8 percent. The yeast is dried by running the particles on a conveyor belt of stainless steel screening through a tunnel of warm air. Some producers place the yeast on trays and dry it in warm ovens. Others use rotary drums or air lift columns to tumble the yeast particles in warm air until dried to the desired degree. The dried yeast is then packed in small foil-plastic laminated packages and flushed with nitrogen before sealing.

The viable population of dried yeast when manufactured is typically 10 to 20 billion cells per gram. One manufacturer of dried beer and wine yeasts specifies that 70 percent by weight of the dried pellets consists of yeast cells. During the first month of storage at 70 degrees F (21 degrees C), the viability drops about 10 percent. Thereafter the viability drops at least 20 percent per year at this temperature and 5 percent or more per year at 38 degrees F (3.5 degrees C). Storage at temperatures around 100 degrees F (38 degrees C) can cause a considerable drop in yeast viability in a short time, so dried yeast held in a non-temperature-controlled warehouse during summertime in a place such as Arizona can result in sluggish starts in homebrews.

Packages of dried yeast obtained at homebrew supply stores and tested in my microbiology laboratory had widely varying viable counts, ranging from 200,000 to 12 billion cells per gram. Except for the Paul Arauner yeast, there was no consistent correlation between the brand of dried yeast and viability. The Arauner brand was packaged in paper envelopes that provide no protection from moisture or oxygen and was always found to be low in viable yeast.

I examined ten different brands of dried brewing yeast and found that only two of these had correct rehydration instructions printed on the packages. For example, the instructions on one package erroneously recommended that the homebrewer sprinkle the dried yeast on the surface of the cooled wort. Yeast dried to a low water content has good stability but is sensitive to the temperature of the water used to rehydrate it. If the yeast is rehydrated in cold water, up to 25 percent by weight of the yeast soluble solids is leached out. The resulting viability is typically 10 percent. Dried yeast produced experimentally with a moisture content of 2.5 percent has excellent keeping properties, but it must be rehydrated with water vapor rather than warm water to give good viability.

## REHYDRATING DRY YEAST

A package of dried yeast should be rehydrated in about ¼ cup of sterile water at a temperature of 95 to 105 degrees F (35 to 40.5 degrees C) for 5 to 10 minutes. The yeast is not harmed by this short period of warm rehydration, and the viability is improved. Rehydrating in warm wort or sugar solutions rather than sterile water will result in reduced viability and retard the yeast's initial growth. Yeast cells suspended in warm wort, 104 degrees F (40 degrees C), or a fermentable sugar solution will literally explode. Moreover, the dried yeast has a storage carbohydrate (trehalose) that provides sufficient energy for the initial growth, so any added sugar present during the rehydration phase is unnecessary.

## MAKING A STARTER

Prepare a starter by adding enough malt extract to water to give a gravity of 1.040 to 1.050. Do not make higher-gravity starters, because yeast growth may be sluggish. Rinse out a bottle with boiling water. Dip the bottle cap in the boiling water to sanitize it. Boil 1 quart of the malt extract solution for 10 minutes and pour it into the hot bottle. Recap and cool to room temperature of 75 degrees F (24 degrees C).

After rehydrating for 10 minutes, add it to the quart of starter. Avoid dumping the rehydrated yeast directly into wort chilled to 45 degrees F (7 degrees C), because this can increase the number of respiratory-deficient (petite) mutants. Respiratory-deficient mutants produce high levels of diacetyl and esters. Leave the yeast in this starter a few hours or overnight before pouring it into your fermenter. Rouse or aerate the wort when the starter is added to stimulate the initial growth of yeast cells. To ensure a rapid start of fermentation, the temperature of the wort in the primary fermenter should be about 60 to 65 degrees F (15.5 to 18 degrees C) when the yeast starter is added. If a lager beer is to be made, place the fermenter in a refrigerator a few hours after the yeast starter is added.

When making a starter for a 5- or 6-gallon batch of beer, use a minimum of one 7-gram package of yeast or two 7-gram packages if you feel the yeast is old or has a low viability. If the yeast is sluggish when dumped in, using two packages or purchasing a fresh package is recommended. For larger batches of 10 gallons of beer, begin with a larger volume of starter and an additional 7-gram package of yeast.

## PROBLEMS WITH DRY YEAST

In addition to having wide differences in yeast viability, some samples also have considerable contamination from bacteria and wild yeasts. Very few homebrewers have microscopes and culture equipment to examine dried brewing yeast for bacterial and wild yeast contamination, but they should not despair of producing good beer with dry yeast, because they can evaluate the quality of yeast before using it.

Here is how to do it: Purchase several different brands of dry yeast from your local homebrew supply store. Dissolve enough light dry malt extract in water to give a gravity of 1.040. Boil this wort for 30 minutes, then pour the boiling hot wort into hot sanitized quart bottles (one bottle for each yeast sample) until they are about three-fourths filled. Use no hops, because you wish no hop bitterness or aroma to obscure any off flavors or aromas from the test ferments. Fit each bottle with an air lock and cool to room temperature. Rehydrate about 2 grams of yeast with warm, sterile water as described and inoculate the quart bottles of wort. Ferment at 60 degrees F (15.5 degrees C) to minimize production of esters or fusel oils by the brewing yeast. You can estimate viability by noting how fast the air locks bubble during the first few days of fermentation. When fermentation stops, refrigerate the bottles at about 38 degrees F (3.5 degrees C) for one week to settle out the yeast cells. Invite several of your homebrewing friends over for a tasting session. It is not essential for the beer to be carbonated for tasting, because off flavors and aromas can be detected easily in the flat beer. Participants should note any sour, phenolic, solventlike, and diacetyl aromas or flavors characteristic of bacterial or "wild yeast" contamination.

I have found this test correlates very well with cultural and microscopic examination of the yeasts in my laboratory for contaminants. All yeast samples that had a mild, yeasty flavor by taste test had little or no detectable contamination by laboratory examination. Go to your homebrew supply store and buy several packages of the same lots of dried yeasts that you have found to have good viability and no off flavors. Keep the dried yeast packages in a refrigerator at 38 degrees F (3.5 degrees C) to preserve their viability until used.

## FINAL COMMENTS

Inform the homebrew supply store of dry yeast samples you feel have low viability or contamination so they can give feedback to their suppliers on the quality of dry yeast they receive. I am often asked which is the best dry yeast. I have not found a particular brand that was always of good viability and free of bacterial or wild yeast contamination from sample to sample. If pressed, I find Whitbread dry yeast usually to be of good quality for those brewers who refuse to make a taste test of yeasts.

I recommend that you do not save an opened package of dried yeast. The storage viability of the dried yeast is improved when oxygen is excluded by the nitrogen flushing of the package. An open package of yeast may also pick up moisture, resulting in rapid loss of viability.

*Rodney Morris has been brewing for more than twenty-two years. A Malthopper since the club was formed, he was previously a Maltose Falcon.*

# BECOME SACCHAROMYCES SAVVY

## BY PATRICK WEIX

*This article originally appeared in* Zymurgy, *Summer 1994 (vol. 17, no. 2)*

Yeast are unicellular fungi. Most brewing yeast belong to the genus Saccharomyces. Ale yeast is S. *cerevisiae*, and lager yeast is S. *uvarum* (formerly *carlsbergensis*, and sometimes considered to be a subspecies of S. *cerevisiae*). Another type of yeast you may hear mentioned, usually in conjunction with Weizens, is S. *delbrueckii*. Finally, lambicophiles will want me to say that Brettanomyces is also used in brewing; however, I can't think of anything that somebody somewhere hasn't used to brew a lambic! You may ask, "If ale and lager yeasts are basically the same species, why all the fuss?" The fuss has to do with strain variation. All dogs are the same species, yet no one will ever mistake a basset hound for a Doberman (at least not twice).

Using different strains can add fun and spice to brewing, especially if you have some idea of the differences. I originally collected the information in the accompanying table to catalog the flavor profiles of the various strains available. But first I would like to discuss some of the general characteristics of brewing yeast and try to answer some of the more frequently asked questions. You may ask yourself, "Why should I care? I just rip open that little packet on top of the Can-O-Malt and toss it in!" Well, you should care. The more informed you are about the different aspects of brewing, the better your beer will be—and isn't that what it's all about?

## TEMPERATURE

One of the most obvious differences between ale and lager yeasts is the different temperatures at which fermentation is carried out. The normal temperatures for ale yeast range from 60 to 75 degrees F (16 to 24 degrees C). A few strains ferment well down to 55 degrees F (13 degrees C), but 68 degrees F (20 degrees C) is a good average. Lager strains normally undergo primary fermentation at 50 to 55 degrees F (10 to 12 degrees C). Then a slow, steady reduction to the desired temperature for secondary fermentation (usually 32 to 45 degrees F or 0 to 7 degrees C) usually will give good results.

The fermentation rate also is closely related to temperature within a specific range. The reason is simply that fermentation is a sequence of chemical reactions facilitated by the enzymes in the yeast. The lower the temperature, the slower the rate of fermentation. At higher temperatures, yeast will produce more esters, diacetyl, and higher alcohols. Therefore, it is important to take into account that yeast are living organisms and do not thrive under rapidly varying conditions. They have evolved mechanisms to protect themselves from wild fluctuations in temperature, mechanisms that are not always in your beer's best interests! So try to keep your fermenting beer at a steady temperature, and when you reduce the temperature to lager, do so slowly.

## ATTENUATION

Attenuation refers to the percentage of sugars converted to alcohol. Each yeast strain ferments different sugars to varying degrees, so attenuation is determined by both the composition of the wort and the yeast strain used. The degree of attenuation also affects the sweetness and body because those sugars not broken down by fermentation remain in the wort. Larger, longer-chain sugars contribute mostly to the mouth feel and body of the beer while the shorter, smaller sugars (mono- and disaccharides) contribute more to the sweetness.

Apparent attenuation of yeast normally ranges from 67 to 77 percent, and is calculated by:

$$\text{Apparent Attenuation} = \frac{\text{(Original Gravity} - \text{Final Gravity)}}{\text{(Original Gravity} - 1.000)}$$

where 1.000 refers to the specific gravity of water.

For example, if the OG is 1.040 and the FG is 1.010, then:

$$\text{Apparent Attenuation} = \frac{(1.040 - 1.010)}{(1.040 - 1.000)} = \frac{0.03}{0.04} = 75\%$$

Actually, it's slightly more complex than that (isn't everything?). There's "apparent attenuation" and "real attenuation." The difference comes about because alcohol has a specific gravity of less than 1 (about 0.8). The calculation of real attenuation takes into account the changing gravity of the wort caused by the increasing amount of alcohol present. Most attenuation figures, however, are given in terms of apparent attenuation.

## PITCHING RATES AND METHODS

So is the attenuation of the yeast strain the only aspect that determines the final percentage of alcohol? No, of course not! Two other factors are very important: pitching rate and flocculation (see below). The pitching rate refers to how many viable yeast cells are added to your wort. Pitching rate is especially important for those wanting to make barley wines, bocks, Scotch ales, or other high-alcohol beers. Underpitching can cause a number of problems, such as long lag times and/or incomplete fermentations. Again, if you are happy with the beer you make, it should not be necessary to make any changes, but if you are just starting out, are having problems with long lag times, or are interested in making a stronger style of beer than usual, you might want to try the following methods and pitching rates.

**Hydration Procedure for Dry Yeast:** Use 14 grams of dry yeast (usually two packets) per 5 gallons of brew. Rigorously sanitize everything used in the hydration procedure. This includes boiling and cooling the water for rehydration so chlorine is boiled off and the water is sanitized. I find it easiest to do this by heating a Mason jar containing one cup of water and covered with either a plastic lid or plastic wrap in the microwave. Heat the water to boiling, then let it cool in the microwave until the jar can be handled but is still warm. The temperature of the water should be about 90 to 100 degrees F (32 to 38 degrees C). Carefully open the yeast packages, add them to the water, and let stand for 15 minutes.

Once the wort has been chilled and aerated (shaking the carboy works well), pitch the yeast and water slurry. Shake or swirl the carboy to disperse the yeast. Attach the blowoff tube or fermentation lock.

The two essential rules are to sanitize everything in sight and aerate your wort to ensure rapid initial yeast growth—your best defense against bacterial or wild yeast infection.

**Preparation of Liquid Yeast:** Liquid yeast typically requires more preparation than dry yeast. You must buy it a day or two in advance so you can activate it and start its growth. Some yeasts are packaged with a starter, but others require you to make your own. Either follow the distributor's directions or use the following recipe.

## Recipe for Starter Wort

> 5 tbsp. dry malt or 6 tbsp. liquid malt
> 2 c. water
> optional—a hop pellet or equivalent amount of loose hops (hops have a natural antibacterial effect, and may help keep your starter free from contamination)

Boil and add to a sanitized wine or quart beer bottle. Allow both the wort and liquid yeast to come to room temperature to avoid shocking the yeast. Aerate the starter well after it has cooled. Carefully sanitize the neck of the bottle and the outside of the yeast package. Add the yeast to the wort. Attach an air lock and leave the bottle at room temperature. When the starter activity subsides, it is ready to pitch. It's that easy!

If all this piques your interest and you want to go the extra step and start your own yeast ranch, I recommend purchasing a kit. Several manufacturers offer full-featured kits for beginners and advanced brewers. Remember to keep yeast notes along with your beer notes so that you can learn from experience. (See Table 7-1 on pages 220 to 237 for a summary of different yeast strains and their characteristics.)

## FLOCCULATION

Flocculation refers to the tendency of yeast to clump together and settle out of suspension. The primary determinant of how well a strain flocculates appears to be the "stickiness" of the carbohydrates in the cell wall. The degree and type of flocculation vary for different yeasts. Some strains clump, producing a firm, stable yeast cake. Some flocculate very little, giving a more granular consistency. Most yeast strains clump and flocculate to a moderate degree. A yeast that is more flocculant will fall out of suspension better. How does that affect the final clarity of your brew? Because it will be in the bottle at least a week or two before you drink it, it really doesn't seem to matter so much. However, it does matter for other characteristics of the beer; namely, attenuation and diacetyl. If the yeast settles out too quickly, some chemical reactions may remain unfinished. These strains may not be as attenuative because of shorter contact time of yeast and sugars and may not finish reducing all the diacetyl, leaving a butterscotch flavor.

In short, only the extremes of flocculation are likely to be noticeable to the homebrewer. Strains with a very low flocculency are likely to need some sort of finings, such as isinglass, while highly flocculent strains are suitable for those beers in which diacetyl is desirable, such as certain British ales.

## ALCOHOL TOLERANCES

The alcohol tolerance of most brewing yeast is at least 8 percent alcohol by volume. Barley wines up to 12 percent can be produced by most ale strains. Pitching rates need to be increased proportionally to higher gravities. Alternately, champagne or wine yeast can be used for high-gravity beers, sometimes resulting in alcohol levels up to 18 percent. To get the characteristics of particular beer yeast strains in barley wines or imperial stouts, some brewers start with the desired strain of beer yeast, ferment to 5 to 8 percent, then finish with a champagne or wine yeast.

## SMELL AND TASTE

Although the principal flavors present in a beer result from the malts and hops, the strain of yeast can add important flavors, good and/or bad. Yeasts that add little in the way of extra flavors are usually described as having a clean taste. These are especially useful for beginners because they permit experimentation with different malts and hops without worrying about yeast influence.

Yeast produce three main classes of metabolic by-products that affect beer flavor: phenols, esters, and diacetyl. Phenols can give a spicy or clovelike taste, but can also result in medicinal tastes. Esters can lend a fruity taste to beer. Diacetyl can give beer a butterscotch or woody taste. The desirability of any one of these components depends largely on the style of beer being brewed. In addition, there are certain by-products in these families that are more noxious than the others. A lot depends on the individual palate and the effect you're aiming for. A final note: Some yeast, especially lager yeast during lagering, can produce a rotten-egg smell. This is the result of hydrogen sulfide production. Although this scent bubbling out of the air lock is enough to make the strongest homebrewmeister blanch, fear not! The good news is that this will usually pass, leaving the beer unaffected. Relax, etc.

## ACKNOWLEDGMENTS

Special thanks to David Adams, George Fix, Al Korzonas, and Doug O'Brien for providing information and insight.

Patrick Weix is an M.D./Ph.D. student in the genetics and development program at the University of Texas Southwestern at Dallas. He became interested in homebrewing when, within the span of one week, his friend Chuck Hodge suggested it, he drove past a store called Homebrew Headquarters, and he found the Homebrew Digest on the Internet. Patrick took this as a sign from above and has been hooked ever since. He currently brews with the Lakewood Grain Co-op.

# YEAST STOCK MAINTENANCE AND STARTER CULTURE PRODUCTION

### BY PAUL FARNSWORTH

*This article originally appeared in Zymurgy, Special 1989 (vol. 12, no. 4)*

To make clean beer, you must add at least 1 cup of actively growing yeast to your wort as soon as it is cool. The yeast then quickly takes over the whole 5 gallons and suppresses any other "bugs." Commercial yeasts may be contaminated or are available in such small volumes that they will not add enough to protect 5 gallons.

The only answer is to make your own cultures. One way of doing this is to store pure yeast strains and turn these into starter cultures to pitch into the wort. Once you have prepared slants and plates, you can store them for many months. Routinely you only have to do steps 5, 6, and 7. If your last petri dish of working culture is contaminated, you need to repeat step 4. Here is a simple general scheme for yeast management.

## Table 7-1. Yeast Strains

### Part 1: Ale Yeast (*Saccharomyces Cerevisiae*)

| Dry Strains | Characteristics and Styles[1] | Attenuation[2] | Flocculation[2] | Notes |
|---|---|---|---|---|
| Coopers Ale Yeast | Very clean fruitiness. | — | — | — |
| Doric Ale Yeast | General ale yeast. | — | — | Good reputation. |
| Edme Ale Yeast | Some fruity esters. | High | — | Starts quick. Good reputation. |
| Lallemand Nottingham Yeast | Nutty tastes/smells. | — | High | Very good reputation. It is a fast starter with quick fermentation at 62 degrees F (17 degrees C). |
| Lallemand Windsor Yeast | Estery to both palate and nose with a slight fresh yeast flavor. | — | Medium | Produces a beer that is clean and well-balanced. Not as quick as the Nottingham. |
| Munton & Fison Ale Yeast | Some fruity esters. | High | — | — |
| Red Star Ale Yeast | Excellent general purpose ale yeast with a clean taste. | 76-78% | — | Fast, reliable starter. New strain—nothing like the Red Star from 5 years ago. |
| Whitbread Ale Yeast | Pale ales, and other ale styles. | — | — | Sometimes seems to have odd aftertaste in finished ale. |

| Liquid Strains | Characteristics and Styles[1] | Attenuation[2] | Flocculation[2] | Notes |
|---|---|---|---|---|
| BrewTek CL–10 Microbrewery 1, American | A smooth, clean, strong fermenting ale yeast. | High | — | Works well down to 56 degrees F (13 degrees C). |
| BrewTek CL–20 Microbrewery 2, American | Creamy malt profile with hints of diacetyl. | — | — | — |
| BrewTek CL–120 Pale Ale, British | Bold, woody and dry character. | — | — | Accentuates mineral and hop flavors. |
| BrewTek CL–130 British Pale Ale 2 | A smooth, full-flavored, well-rounded ale yeast. Mildly estery. | High | — | — |
| BrewTek CL–160 British Draft Ale | Well-rounded flavor with a buttery rich diacetyl. | — | — | — |
| BrewTek CL–170 Classic British Ale | Complex ale with very British tones and fruit like esters. | — | — | — |
| BrewTek CL–240 Irish Dry Stout | Leaves a very recognizable character to dry stouts with roasted malts coming through well. | — | — | — |

Table 7-1. Yeast Strains (cont.)

## Part 1: Ale Yeast (*Saccharomyces Cerevisiae*)

| Liquid Strains | Characteristics and Styles[1] | Attenuation[2] | Flocculation[2] | Notes |
|---|---|---|---|---|
| BrewTek CL-260 Canadian Ale | Pleasant, lightly fruity and complex finish. | — | — | — |
| BrewTek CL-300 Belgian Ale | Robust and estery with notes of clove and plum. Flanders-style yeast. | — | — | Produces a classic Belgian ale flavor. |
| BrewTek CL-320 Belgian Ale 2 | Makes a terrific strong brown. | — | — | A good base brew for fruit-flavored beers. |
| BrewTek CL-340 Belgian Ale 3 | Classic Trappist character with esters of spice and fruit. | — | — | — |
| BrewTek CL-380 Saison | Mild yet pleasant esters and apple pie spices. | — | — | — |
| BrewTek CL-400 Old German | Mildly estery flavor. Good for traditional altbiers. | High | — | — |
| BrewTek CL-450 Kölsch | Clean, lightly yeasty flavor in the finish. | — | — | Mineral and malt characters come through well. |
| Wyeast 1007 | Ferments dry and crisp, | 73–77% | High | Produces an extremely rocky |

| | | | |
|---|---|---|---|
| Ale Yeast German | leaving a complex yet mild flavor. | | | head and ferments well down to 55 degrees F (12 degrees C). This is actually a Kölsch yeast. |
| Wyeast 1214 Belgian Ale Yeast | With both clovelike phenolics and alcohol spice. Banana estery flavor. Good for abbey beers. | — | — | Ferment warm or with inadequate aeration and you're likely to get a bubble-gum note. Reported to be the Chimay strain. |
| Wyeast 1028 London Ale Yeast | Rich minerally profile, bold woody slight diacetyl production. | 73–77% | Medium | Optimum fermentation temperature: 68 degrees F (20 degrees C). |
| Wyeast 1056 American/Chico Ale Yeast | Ferments dry, finishes soft, smooth and clean. Very well-balanced. | 73–77% | Low to Medium | Optimum fermentation temperature: 68 degrees F (20 degrees C). The cleanest of the bunch, this is Sierra Nevada's yeast. Probably the best available all-around yeast. |
| Wyeast 1084 Irish Ale Yeast | Slight residual diacetyl is great for stouts or Scotch ales. It is clean, smooth, soft and full-bodied. | 71–75% | Medium | Optimum fermentation temperature: 68 degrees F (20 degrees C). Reputed to be the yeast Guinness uses. |

## Table 7-1. Yeast Strains (cont.)

### Part 1: Ale Yeast (*Saccharomyces Cerevisiae*)

| Liquid Strains | Characteristics and Styles[1] | Attenuation[2] | Flocculation[2] | Notes |
|---|---|---|---|---|
| Wyeast 1087 Wyeast Ale Blend | General purpose ale strain. | — | — | Yeast blends are created to ensure a quick start, good flavor and good flocculation. (They come in the new 80-gram packages.) |
| Wyeast 1098 British Ale Yeast | Great in pale ales and bitters, good in porters. Tart, crisp, clean. | 73–75% | Medium | Ale yeast from Whitbread. Ferments well down to 55 degrees F (12 degrees C). |
| Wyeast 1338 European Ale Yeast | Especially well-suited to altbier. A full-bodied complex strain finishes very malty. | 67–71% | High | Alt yeast from Wissenschaftliche (#338) in Munich. Produces a dense rocky head during fermentation. Optimum fermentation temperature: 70 degrees F (21 degrees C). |
| Wyeast 1728 Scottish Ale Yeast | Scottish-style ales, smoked beers and high-gravity ales. | — | — | — |

| | | | |
|---|---|---|---|
| Wyeast 1968 Special London Ale Yeast | Rich malty character and balanced fruitiness. | High | Possibly Young's yeast strain. |
| Wyeast 1565 Kölsh Yeast | Develops excellent maltiness and subdued fruitiness with a crisp finish. | — | A hybrid of ale and lager characteristics. Ferments well at moderate temperatures. |
| Yeast Culture Kit A01 | Barley wine, brown ale, pale ale, India pale ale, cream ale, porter, stout. | — | From California. |
| Yeast Culture Kit A04 | Düsseldorf Altbier, Kölsch. | — | From Oregon. |
| Yeast Culture Kit A06 | Porter, stout, imperial stout. | — | From Oregon. |
| Yeast Culture Kit A08 | Barley wine. | — | From Dorchester, England. High residual sweetness. |
| Yeast Culture Kit A13 | Porter, stout, imperial stout. | — | From Ireland. |
| Yeast Culture Kit A15 | Brown ale, pale ale, India pale ale, cream ale, bitters and milds. | — | From England. Strong yeast flavors. |
| Yeast Culture Kit A16 | Trappist ales (abbeys, doubles, trippels). | — | From Belgium. |
| Yeast Culture Kit A17 | Brown ale, pale ale, India pale ale, cream ale, bitters and milds. | — | From London. |

## Table 7-1. Yeast Strains (cont.)

### Part 1: Ale Yeast (*Saccharomyces Cerevisiae*)

| Liquid Strains | Characteristics and Styles[1] | Attenuation[2] | Flocculation[2] | Notes |
|---|---|---|---|---|
| Yeast Culture Kit A34 | Barley wines, Scotch ale, Scottish bitters, strong ale. | — | — | From Edinburgh, Scotland. |
| Yeast Culture Kit A35 | Belgian whites. | — | — | From central Belgium. |
| Yeast Culture Kit A36 | Belgian ales. | — | — | From Houffalize, Belgium. |
| Yeast Culture Kit A37 | Altbier, Kölsch. | — | — | From Bavaria, Germany. |
| Yeast Lab A01 Australian Ale Yeast | Brown ales and porters. Produces a very complex woody and flavorful beer. | 74–75% | Medium | Australian origin. |
| Yeast Lab A02 American Ale Yeast | Produces a very fruity aroma. | 74–75% | Low | Clean strain with soft and smooth flavor when fermented cool. |
| Yeast Lab A03 | Classic pale ale strain. | 74–75% | Medium | A powdery yeast. |

| | | | |
|---|---|---|---|
| London Ale Yeast | A hint of diacetyl and rich minerally profile, crisp and clean. | | |
| Yeast Lab A04 British Ale Yeast | Pale ales and brown ales. A complex estery flavor. Ferments dry with a sharp finish. | 74% | Medium | This strain produces a great light-bodied ale. |
| Yeast Lab A05 Irish Ale Yeast | Stouts and porters. A hint of butterscotch in the finish, soft and full-bodied. | 73% | High | Slightly acidic. |
| Yeast Lab A06 Düsseldorf Ale Yeast | German Altbier. Finishes with full body, complex flavor and spicy sweetness. | 75% | High | — |
| Yeast Lab A07 Canadian Ale Yeast | Light and cream ales. Light-bodied, clean and flavorful beer. | 76% | Medium | Very fruity when fermented cool. |
| Yeast Lab A08 Trappist Ale Yeast | Trappist strain, producing a malty flavor. | 76% | High | Alcohol tolerant. A balance of fruity, phenolic overtones when fermented warm. |

## Table 7-1. Yeast Strains (cont.)

### Part 2: Lager Yeast (*Saccharomyces Uvarum*)

| Dry Strains | Characteristics and Styles[1] | Attenuation[2] | Flocculation[2] | Notes |
|---|---|---|---|---|
| Yeast Lab Amsterdam Lager or Yeast Lab European Lager | Both are supposed to produce a clean smooth lager. | Medium | High | Ferment at 60 to 70 degrees F (16 to 21 degrees C). |
| **Liquid Strains** | | | | |
| BrewTek CL–600 Original Pilsener | Sweet, underattenuated finish with a subdued diacetyl character. | — | — | Makes a full-bodied lager. |
| BrewTek CL–620 American Megabrewery | Leaves a light, crisp, almost dry finish to lagers. | — | — | A strong fermenter. |
| BrewTek CL–640 American Microbrewery | A clean, full-flavored, malty finish. | — | — | A strong fermenter. |

| | | | |
|---|---|---|---|
| BrewTek CL-660 North German Lager | German Pilseners, Mexican and Canadian lagers. Exhibits a clean, crisp, traditional lager character. | High | — | A clean, crisp traditional lager. A strong fermenting and forgiving yeast. |
| BrewTek CL-680 East European Lager | Imparts a smooth, rich, almost creamy character, emphasizing a big malt flavor and clean finish. | — | — | — |
| BrewTek CL-690 California Esteem | Steam-style beers. Leaves a slightly estery, well-attenuated finish. | — | — | Use to create "California common beers." |
| Wyeast 2007 Pilsen Lager Yeast | Specific for Pilsener-style beers. Ferments dry, crisp, clean and light. | 71–75% | — | Optimum fermentation temperature: 52 degrees F (11 degrees C). Leaves some residual green-apple notes. (It is worth mentioning that this yeast strain is reportedly used quite a bit in St. Louis, if you know what I mean.) |

## Table 7-1. Yeast Strains (cont.)

### Part 2: Lager Yeast (*Saccharomyces Uvarum*)

| Liquid Strain | Characteristics and Styles[1] | Attenuation[2] | Flocculation[2] | Notes |
|---|---|---|---|---|
| Wyeast 2035 American Lager Yeast | Unlike American Pilsener styles. It is bold, complex and woody. Produces slight diacetyl. | 73–77% | Medium | Optimum fermentation temperature: 50 degrees F (10 degrees C). |
| Wyeast 2042 Danish Lager Yeast | Rich, yet crisp and dry. Soft, light profile that accentuates hop characteristics. | 73–77% | Medium | Optimum fermentation temperature: 48 degrees F (9 degrees C). |
| Wyeast 2112 Lager Yeast, California | Steam-style beers. Malty profile. | 72–76% | Low | Warm fermenting bottom cropping strain, ferments well to 62 degrees F (17 degrees C) while keeping lager characteristics. |
| Wyeast 2124 Bohemian Lager Yeast | Ferments clean and malty. | 69–73% | High | Optimum fermentation temperature: 48 degrees F (9 degrees C). |

| | | | | |
|---|---|---|---|---|
| Wyeast 2178 Wyeast Lager Blend | Supposed to make a good clean lager. | — | — | Yeast blends are created to ensure a quick start, good flavor and good flocculation. (They come in the new 80-gram packages.) |
| Wyeast 2206 Bavarian Lager Yeast | Rich flavor, full-bodied, malty and clean. Bocks, lagers | 73–77% | Medium | Lager yeast strain used by many German breweries. Optimum fermentation temperature: 48 degrees F (9 degrees C). It is reported to be a slow starter. Weihenstephan 206. |
| Wyeast 2278 Pils Yeast, Czech | Pilseners and bock beer. Classic dry finish with rich maltiness. | — | Very Low | Sulfur produced during fermentation dissipates with conditioning. (My recommendation, and that of a commercial brewer using it: use some sort of mechanism for clearing the beer. The commercial brewer said they always use finings with this strain.) |

## Table 7-1. Yeast Strains (cont.)

### Part 2: Lager Yeast (*Saccharomyces Uvarum*)

| Liquid Strains | Characteristics and Styles[1] | Attenuation[2] | Flocculation[2] | Notes |
|---|---|---|---|---|
| Wyeast 2308 Munich Lager Yeast | Smooth, soft, well-rounded and full-bodied. | 73–77% | Medium | Optimum fermentation temperature: 50 degrees F (10 degrees C). More likely to bring out hop flavor than Wyeast 2206. Weihenstephan 308. |
| Yeast Culture Kit L09 | American dark lager, American lager, Bavarian dark, Doppelbock, Dortmund/Export, Eisbock, German Bock, German Lagers, German Schwarzbier, Hellesbock, Munich Helles, Marzen/Octoberfest, Pilsener. | — | — | From Bavaria, Germany. That long list is the distributor's suggested uses. |

| | | | | |
|---|---|---|---|---|
| Yeast Culture Kit L17 | American lagers, Bohemian Pilsener. | — | — | From Plzenv, Czech Republic. |
| Yeast Lab L31 Pilsener Lager Yeast | Ferments dry and clean. | 73% | Medium | — |
| Yeast Lab L32 Bavarian Lager Yeast | Rich in flavor with a clean, malty sweetness. Medium-bodied lagers and bocks, Vienna and Märzen styles. | 75% | — | — |
| Yeast Lab L33 Munich Lager Yeast | Medium-bodied lagers and bocks. Wissenschaftliche strain with subtle and complex flavors. | 75% | Medium | A hint of sulfur when fresh. |
| Yeast Lab L34 St. Louis Lager Yeast | American-style lagers. Produces a round, very crisp and clean fruity flavor with medium body. | 74–76% | Medium to High | — |
| Yeast Lab L35 California Lager Yeast | California common beer strain. Malty with a sweet woody flavor and subtle fruitiness. | 74–75% | Medium | — |

## Table 7-1. Yeast Strains (cont.)

### Part 3: Weissen, Lambic, Mead and Barley Wine Styles

| Strains | Characteristics and Styles[1] | Attenuation[2] | Flocculation[2] | Notes |
|---|---|---|---|---|
| BrewTek CL-900 Belgian Wheat | Leaves a sweet, mildly estery finish. | — | — | — |
| BrewTek CL-920 German Wheat | Spicy, clovy and estery. | High | — | — |
| BrewTek CL-930 German Weiss | Spicy, clovy and estery. | — | — | Still produces the sought-after clove and phenols, but not as intense as CL-920. |
| BrewTek CL-980 White Ale, America | A smooth, slightly sweet wheat beer. | — | — | Underattenuated malt flavor. |
| BrewTek CL-5200 *Brettanomyces lambicus* | Belgian lambic beers. | — | — | — |

| | | | |
|---|---|---|---|
| BrewTek CL–5600 *Pediococcus damnosus* (a bacteria) | Used in lambics. | — | Produces acid and diacetyl. |
| Wyeast 3056 Bavarian Weissen Yeast | Produces a South German-style wheat beer with cloying sweetness when the beer is fresh. | 73–77% | A 50/50 blend of S. *cerevisiae* and *delbrueckii*. Problematic to get the right flavor, often produces relatively unattenuated beer without the clove-like aroma/flavor. |
| Wyeast 3068 Weihenstephan Wheat Yeast | Bavarian Weizen. A very consistent and clean cloviness. | — | *Saccharomyces delbrueckii* single strain for German wheat beers (especially Bavarian Weizen). |
| Wyeast 3273 *Brettanomyces bruxellensis* | Rich, earthy, odiferous character. | — | *B. bruxellensis* is the dominant Brettanomyces strain in the Brussels area. |
| Wyeast 3944 White Beer Yeast, Belgian | Belgian ales, wit beers, Grand Cru, abbey biers. A very estery strain without the usual battering of bananas. | High | Rich, phenolic character for classic Belgian styles. |

## Table 7-1. Yeast Strains (cont.)

**Part 3: Weissen, Lambic, Mead and Barley Wine Styles (cont.)**

| Strains | Characteristics and Styles[1] | Attenuation[2] | Flocculation[2] | Notes |
|---|---|---|---|---|
| Yeast Culture Kit M01 | American wheat, Dunkel Weizen, German Weizen, Weizenbock. | — | — | From Bavaria. Although the vendor lists American wheat as a suggested style, it appears to produce too much clove taste for that; however, that does make it excellent for the Bavarian Weizens! After all, it is a Bavarian yeast. |
| Yeast Lab W51 Bavarian Weizen | Moderately high, spicy phenolic overtones reminiscent of cloves. | Medium | Medium to Low | This strain produces a classic German-style wheat beer. |
| Yeast Lab M61 Dry Mead | Ferments dry, fruity and clean, yet leaves noticeable honey flavor and aroma. | Tolerance to 14–15% | Medium to Low | Very alcohol tolerant. |
| Yeast Lab M62 Sweet Mead | A very fruity, sweet mead with tremendous honey aromas. | Tolerance to 12–13% | Medium to Low | This strain has reduced alcohol tolerance. |

## Wine Yeast

| Strains | Characteristics and Styles[1] | Attenuation[2] | Flocculation[2] | Notes |
|---|---|---|---|---|
| Red Star Pasteur Champagne Yeast | Imperial stouts and barley wines, mead. | High | — | Good reputation. High tolerance for alcohol. Some use it by itself, others pitch Pasteur after their chosen beer yeast poops out. From Montreal, Canada. |
| Yeast Culture Kit M06 | Barley wine (Champagne). | — | — | |
| Wyeast 3021 Pasteur Champagne Yeast | Imperial stouts and barley wine, mead. | High | — | Good reputation. High tolerance for alcohol. Some use it by itself, others pitch after their chosen yeast poops out. |

[1] Characteristics refer to the scents and flavors produced; styles refer to the beer styles for which the particular yeast is especially suited.
[2] Information not available for all strains. Attenuation as given by distributor.

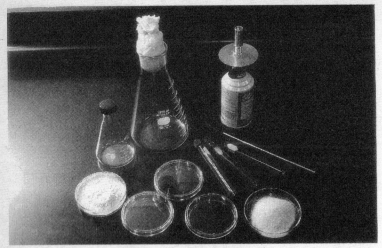

Photo 1: Equipment Required
PAUL FARNSWORTH

## EQUIPMENT AND SUPPLIES

1 small (250 ml.) conical flask with cap
1 large (1,000 ml.) conical flask with either a screw cap or a plug made from cotton wrapped in gauze
small test tubes with caps and rack
petri dishes
inoculating loop
aluminum foil
electrician's tape
agar
dry malt extract
a source of open flame: gas burner, portable stove, propane torch or mini welding torch
pressure cooker, or large boiling pot (not shown)

## CULTURE PREPARATION

**1. Make Wort**

Add 8 rounded tablespoons of dry malt extract, or 10 tablespoons of syrup, to 2½ cups of water. Boil, covered, for 20 minutes and let cool. Pour off the liquid from the junk in the bottom. If you are not going to use it straightaway, pour it into sanitized bottles, cap, and put in fridge (see Photo 2).

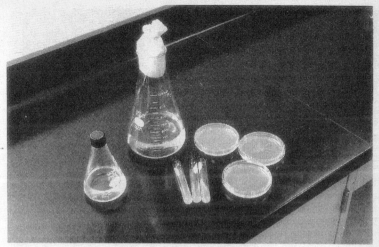

**Photo 2: Liquid and solid wort**
PAUL FARNSWORTH

An easy way to do this is to save some wort from a prior brew. An extremely lazy way is to dissolve the malt extract in hot water and not boil and cool it. This will work but will look ugly because when you sterilize it later on, you will get brown clumps of trub that look nasty but do no harm.

**2. Make Solidified Wort**

This is just like making Jell-O. When stirred into boiling liquid, the dry agar powder will dissolve. When it cools it sets into a solid.

Add 2 rounded tablespoons of powdered agar to 2½ cups of boiling wort and stir till dissolved.

If you are making slants, pour the mixture into each tube so they are one-third full and cap loosely. If you are making plates, pour the mixture into two small jars (baby food jars or canning jars) and put lids on loosely.

You must now sterilize the tubes and jars of mixture. This is just like home canning. Stand them on a rack in a pan and add water until they are about half covered. Boil for 20 minutes, tighten lids, and let cool for 30 minutes. An even better idea is to use a pressure cooker. Cook for 10 minutes and leave to cool for 30 minutes. When the tubes are almost too hot to touch you must do the next step quickly, before the agar sets.

If you are making slants, tighten the tube caps and tilt all the

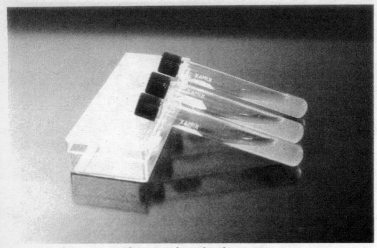

**Photo 3: Tubes of molten wort**
PAUL FARNSWORTH

tubes at a 60-degree angle, prop them up, and let them set (see Photo 3). If you are making petri plates, put the sterile dishes out, unopened, in a row along the edge of a bleach- or alcohol-cleaned kitchen counter. Take one of the small jars of hot agar/wort mixture in one hand. With the other hand lift the lid of one of the petri plates just far enough to pour in the mixture. Add enough to half fill the dish. Replace the lid and go on to the next dish.

Work quickly, leave the lids open for as short a time as possible, leave the jar in your other hand tilted over all the time so you don't slop it about a lot, and do not stop until you have run out of mixture or burned your fingers. Two cups of mixture should make twenty plates or forty or more tubes, so don't worry, you will have plenty. Any leftovers can be stored in the jar in the fridge for next time.

Leave the plates and tubes at room temperature for 5 days. If they are contaminated, you will see things growing. Anything that shows any change should be discarded. Those that pass inspection should be sealed with electrician's tape and stored in the cold. I know this all sounds like a pain, but you only have to do it occasionally!

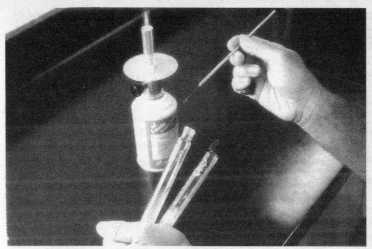

Photo 4: Transferring yeast from an old slant to a new slant
PAUL FARNSWORTH

## MASTER CULTURE

You must start with a clean, pure culture of yeast. You can buy these from the author (see address at end of article). You may also purchase them from J.E. Siebel & Sons in Chicago. Alternative sources are to beg some from a brewery, steal some from the sediment in a bottle-conditioned commercial beer, or culture your own from one of the dried or liquid yeasts available commercially. If you use one of these sources, you must first plate out the yeast onto solidified wort to check for purity and then brew a batch of beer with it to make sure you have what you think you have. If the beer is good, use the culture on solidified wort to make your master culture.

3. Inoculate Some Slants for Storage

Light the flame of your gas stove, torch, or burner. Hold the tube with the yeast, and a new slant tube of solidified wort, both in your left hand and with the tops pointed away from you and spread a couple of inches apart. Hold the loop in your right hand (see Photo 4). Loosen but do not remove the tube caps. Heat the loop till it is glowing red and hold it in your right hand like a pen.

With the crook of your little finger, remove the cap from the yeast tube, pass the neck of the tube through the flame twice, then poke the loop into the tube and scrape up a little of the yeast—not much, there are billions of cells and you only need a few! Withdraw the loop and pass the neck of the tube through the flame twice, then recap it. Uncap the second tube, pass the neck twice through the flame, squiggle the loop carrying the yeast onto the surface of the agar in the "slant" tube, then flame the tube neck and replace the cap.

Put the tubes down and heat the loop to sterilize it. Work quickly, leaving the tubes open for the minimum time. This idea of holding both tubes in one hand and removing the cap with the little finger of the other hand, though it feels awkward at first, will speed things up.

Inoculate several slants and leave them at room temperature for 5 days. If the surface is covered with smooth, pale cream or white yeast, label, seal, and store the tubes in the cold. If it looks weird, dump it and try again.

## WORKING CULTURE

**4.** Inoculate from a Slant to a Petri Plate
Light your flame.

Hold a slant tube of yeast in your left hand, the loop like a pen in your right hand, and place a plate of solidified wort, unopened, in front of you. Loosen but do not remove the tube cap.

Heat the loop till it's red. Remove the tube cap in the crook of your little finger, pressing it against your palm. Pass the tube neck twice through the flame, then put the loop into the tube and scrape up a little of the yeast. Withdraw the loop, pass the tube neck twice through the flame, and replace the cap.

Lift the petri plate lid halfway and squiggle the loop over the plate surface. Replace the lid. Resterilize your loop. Leave the plate in the warm for 4 or 5 days. You should see white or cream-colored dots up to ¼-inch across (usually about ⅛ inch). They should all look the same. It's OK if they are fused together into a streak, but next time pick up a little less on the loop. If the dots, which are colonies of yeast cells, are colored or there is a hairy or cottonlike growth on the plate, discard it—it is contaminated (see Photo 5). If it's good, seal it with tape and store it in the cold.

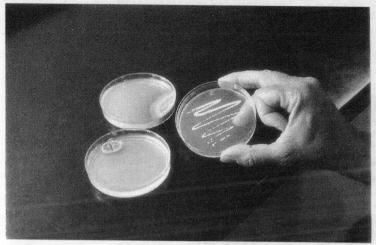

Photo 5: Petri dish with yeast growing in it (in hand) and two showing contaminants
PAUL FARNSWORTH

## STARTER PREPARATION

**5. Make Flasks of Sterile Wort**

Dissolve 8 rounded tablespoons of dry malt extract in 2½ cups of water and boil, covered, for 20 minutes, cool, and pour the liquid off the sediment.

Put about ¼ cup into the small flask (up to the 75-ml. mark), and 2 cups (up to the 500-ml. mark) into the large flask. Cap them loosely and sterilize them, as you would for home canning. Put the flasks on a rack in a pot and add water, boil for 20 minutes, tighten the caps, and let cool. Alternatively, pressure-cook them for 10 minutes and let them cool in the cooker. Store in the cold.

**6. Make a Small Starter Culture**

Three days before brew time, you need a petri plate of yeast and two flasks of sterile wort. Take the small flask of wort, the petri plate, and your inoculating loop to the burner.

Heat the loop till red-hot, open the plate lid halfway, and touch the loop to the surface of the agar. You will hear a *pssst* as the loop cools. Now pick up one of the nice smooth-looking colonies with the loop, and close the dish lid (see Photo 6). This dish can be reused until it has visible contamination.

Keeping the loop in your right hand held like a pen, pick up the

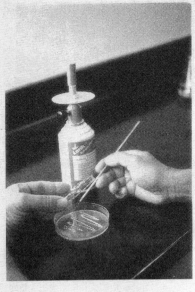

Photo 6: Picking colonies from a culture on a petri plate

PAUL FARNSWORTH

flask with your left. Remove the cap using the crook of the little finger on your right hand, pressing it against your palm. Pass the flask neck through the flame. Then put the end of the loop into the juice in the flask and shake it to dislodge the yeast. Don't worry about getting it all in there; there are millions of cells in a loopful and you only need a few. Withdraw the loop, pass the flask neck through the flame, and replace the cap.

Leave the flask in the warm for two days until the wort has become opaque. It's a sign you have waited too long when the yeast begins to settle down the bottom.

### 7. Make a Large Starter Culture

Though this is the simplest step (you simply pour the contents of the small flask into the large flask of sterile wort), it is the most prone to contamination, so you must do the flame-passing ritual again.

Hold the small flask toward the base in your right hand and remove the cap with your left. Put the cap down. Pick up the large flask in your left hand and remove its stopper using the crook of your little finger. Do not let the end of the stopper that goes in the flask touch anything, especially you. Pass the necks of both the large and small flasks through the flame several times each, twisting them as you do so, to heat at least half of the circumference. Then pour the contents of the small flask into the large flask, trying

**Photo 7:** Starter culture at optimum growth stage for transfer
PAUL FARNSWORTH

not to get any juice on the inside neck of the large flask. Pass the large flask neck through the flame twice and replace the stopper.

Leave the large flask in the warm until the wort is opaque. A head may rise and the culture is at its optimum when this just begins to fall. It is past its prime but still usable as the yeast begins to settle out (see Photos 7 and 8). You cannot store it in the cold and have it work correctly because the yeast cells will begin to die. If you do forget and put it in the refrigerator, take ¼ cup of the mixture and use it as the small starter culture.

"Feeding" the yeast in an old culture by adding sugar or malt is not a good idea. This greatly increases the risk of infection. You also will be feeding any bacteria that are already present. Instead, you should prepare another flask with 2 cups of sterile wort, add ¼ cup from the old culture to this new culture flask, and then wait for 24 hours for the new one to grow up.

A *Quick Review of Culturing Yeast*

---

**MASTER CULTURE**
*Kept on solidified wort in sealed tubes ("slants") in fridge. Stock culture of pure yeast that you rarely use so it stays pure.*

**WORKING CULTURE**
*Kept on solidified wort in petri dishes sealed and stored in fridge. Cul-*

Photo 8: Starter culture past its peak, the yeast has settled. Make a new culture before using.
PAUL FARNSWORTH

tured yeast spread out so you can see whether it is pure. Used routinely to make starter cultures.

**SMALL STARTER CULTURE**
*One colony of yeast transferred to ¼ cup of sterile wort in a flask.* Yeasts grown 36 to 48 hours initially in a small volume of liquid until they are healthy enough to dominate a large volume.

**LARGE STARTER CULTURE**
*Contents of small flask transferred to 2 cups of sterile wort in large flask.* Yeasts grown for 24 hours in large enough quantity to start 5 gallons.

**FERMENTER**
*Contents of large flask added to 5 gallons as soon as wort is below 80 degrees F.*

---

Paul Farnsworth was born and raised in Burton-upon-Trent, England, where he thought fresh air was supposed to smell like boiling wort and the only jobs available were at the breweries. He now lives in the U.S. where he teaches fermentation science, helps out at competitions as a certified beer judge and still seeks the perfect ale recipe.

# eight

# Water

## WATER TREATMENT: HOW TO CALCULATE SALT ADJUSTMENT
### A For the Beginner Column
#### BY DARRYL RICHMAN

*This article originally appeared in Zymurgy, Winter 1989 (vol. 12, no. 5)*

Do you blindly follow recipes that call for gypsum? Have you ever wondered what it does for your pale ales, and why it's not needed for stouts? Was your curiosity piqued by other recipes that call for bizarre minerals such as chalk and Epsom salts? What do these chemicals do for your beer?

Different beer styles have grown up in different areas. The locals will claim that the same style of beer brewed elsewhere doesn't taste the same and is therefore inferior. These claims can be laid to a variety of ingredients—different strains of barley or hops, soil conditions, weather, yeast—and different water.

Water has been called the "universal solvent." This name reflects the fact that water can dissolve more compounds than any other solution known. For one example, just look at what it does to the landscape: one can hardly name a more dramatic feature of the earth than the Grand Canyon. It was created solely by running water.

When water runs across or underneath the land, it dissolves some geologic features. Some of these features erode more quickly than others. For example, limestone is composed chiefly of calcium carbonate, which goes into solution easily. The Carlsbad Caverns were created by water running through a limestone formation.

The water used in each major brewing center flows through its own unique set of geologic formations. It stands to reason that each center's water has its own unique mineral combination. Over time, the breweries using this water, and their customers, have decided what beer styles best make use of the water's mineral makeup.

When we try to reproduce famous beer styles we should try to understand the water used as well as understanding the other ingredients, such as barley and hops.

The minerals dissolved in our water affect each stage of the brewing process. Free calcium is necessary in lighter beer mashes to form a weak acid with the barley, enhancing conversion of starch to sugar. Calcium and magnesium also work with hop resins in the boil to form the hot break, aiding beer clarification. Trace elements are required by our yeast so it can follow normal metabolic pathways and produce a clean-tasting beer. Sulfates and chlorides, although not distinguishable by themselves, add a certain dryness to beer that makes you want to have another, while carbonates can amplify the effect of bittering hops.

We can't understand what adding salts to beer will do if we don't know what is already present in the water we use. Finding out is very easy: call up your water supplier and ask for a water analysis. The Los Angeles Department of Water and Power and a local independent water supplier, Arrowhead, sent them to me free. The quantities they quote are given in parts per million, or milligrams per liter, which are equivalent measures.

The next difficulty lies in determining what minerals are present in the target water supply. Dave Line, in his *Big Book of Brewing*, shows a table of approximate mineral composition for English-style ales (but beware—his recipes are for *English* gallons and pints); Greg Noonan in *Brewing Lager Beers* does likewise for many lager styles; and Charlie Papazian has a table of brewing water compositions in *The Complete Joy of Home Brewing*. All have long, detailed discussions about mineral components. If after reading this article you want to know more, here is the start of your reading list.

Once you know what your water has in it and what you want it to have, all you need do is subtract the former from the latter and add to make up the difference, right? Well, maybe. First of all, you must realize that all those tables I mentioned are approximate.

Even the legendary Samuel Smith's, whose Old Brewery in Tadcaster lies atop an artesian well in a limestone deposit and which they claim supplies perfect brewing water, adds other minerals such as calcium chloride (I have photographic proof!). So you must take those numbers with (*ahem*) a grain of salt.

Also, you couldn't match those numbers if you wanted to. The salts we have available are made of complementary ions. The gypsum, which is the calcium sulfate you add to your brewpot, is locked into a fixed ratio of calcium and sulfate ions. If your water has plenty of sulfate but needs more calcium, you are out of luck. Selecting the proper salts is a matter of compromise.

The final problem is that you may already have too much of some or all of the ions you want. Using a home water softener may actually increase this problem because most of them work by trading the calcium for sodium ions. You want some calcium, but you rarely need to add any sodium. Carbonates (by combining with calcium) can be removed by boiling your water vigorously and racking it off the sediment (which is precipitated calcium carbonate). Dave Line says that all but the last 40 parts per million (ppm) of carbonate can be removed in this way. Other minerals are not easily removed unless you have a reverse-osmosis water purifier at your disposal, so you may have to make do or find another water source. (That's why I asked both water suppliers.)

## ON TO THE MINERALS

As brewers, we are most concerned with the following ions: calcium, magnesium, chloride, carbonate, and sulfate. These are available as compounds: calcium chloride ($CaCl_2$), calcium carbonate ($CaCO_3$, chalk), calcium sulfate ($CaSO_4$, gypsum), and magnesium sulfate ($MgSO_4$, Epsom salts). As you will see, the quantities involved are minuscule, so a small bottle will work fine.

When working with these materials it is important to realize that a little goes a long way. If the hardest water in the driest pale ale has 250 ppm of calcium, it only takes 16 grams (slightly more than ½ ounce) of gypsum in 5 gallons of completely soft water to achieve this level.

Also notice that I've just quoted you a figure *by weight*, not by volume. Like measuring flour, you must be aware of whether you have packed it tightly into the cup. This is because the physical nature of flour is such that it doesn't naturally compress into the

## Table 8-1. Breakdown of Mineral Compounds into Constituent Ions

| Compound | Ions | % Composition by weight | 1 g./5 gal. (ppm) |
|---|---|---|---|
| Calcium carbonate | Ca | 40.0 % | 21.2 |
| (chalk) | $CO_3$ | 60.0 | 31.7 |
| Calcium chloride | Ca | 36.1 | 19.1 |
|  | $Cl_2$ | 63.9 | 33.8 |
| Calcium sulfate | Ca | 23.3 | 12.3 |
| (gypsum) | $SO_4$ | 55.8 | 29.5 |
| Magnesium sulfate | Mg | 9.9 | 5.2 |
| (Epsom salts) | $SO_4$ | 39.0 | 20.6 |

smallest space it can occupy. Mineral salts behave the same way. Dave Line does give volume approximations, but you should be aware that they can vary dramatically in actual content. Greg Noonan gives an example of an experiment he carried out to compare how much salt he could get into a particular volume. If you are going to use volumes, you must try to be consistent.

Table 8-1 shows the breakdown of each compound into its constituent ions. For simplicity, the next column shows how many ppm of each ion are added when one gram of the compound is added to 5 gallons of water.

You may have noticed that the percentages of the last two compounds don't add up to 100 percent. When you buy these salts as gypsum and Epsom salts, they contain some water in their crystals (two molecules for gypsum and seven for Epsom salts), but we already have a lot of water.

Now you can construct a table like Table 8-2 to figure the salts you need to add to achieve the water you want. Begin by writing out the ions and the values from your water report. Write your target water's values next (I targeted the water of Munich, Germany) and subtract them from your water. Experiment by adding salts to achieve a total that matches this result.

Our goal here is to add salts to closely match the "Difference" column. The most important goal is a close match on calcium, because that is what drives many of the reactions in mashing, boiling, and fermenting.

To get this needed calcium, you can choose from three salts. But there is a problem with using calcium chloride because we already have too much chloride. We're not too high in terms of absolute quantity, so as long as we don't add more, we'll be all

## Table 8-2. Figuring the Salts Needed to Match Target Water

| Ion | My Water | Target Water (Munich) | Difference | Add CaCO₃ 3 g. | CaCl₂ 0 g. | CaSO₄ 0 g. | MgO₄ 1 g. | Total Ions |
|-----|----------|-----------------------|------------|----------------|------------|------------|-----------|------------|
| Ca  | 17.7     | 75                    | 57.3       | 63.6           |            |            |           | 81.3       |
| Mg  | 4.7      | 18                    | 13.3       |                |            |            | 5.2       | 9.9        |
| CO₃ | 47.4     | 150                   | 102.6      | 95.7           |            |            |           | 143.1      |
| Cl₂ | 8.1      | 2                     | -6.1       |                |            |            |           | 8.1        |
| SO₄ | 0.0      | 10                    | 10         |                |            |            | 20.6      | 20.6       |

right. This eliminates one possible salt. If we add gypsum (calcium sulfate) to get our calcium, we would be adding a great deal of the sulfate ion as well—much more than we need. So we choose three grams of chalk (calcium carbonate), which also happen to neatly fill our requirements for the carbonate ion. A gram of Epsom salts gives us both the magnesium, although a bit under, and sulfate, a bit over.

As you can see, these numbers are somewhat of a compromise, although for the minerals present in large quantities, we are very close. Our calcium is within 9 percent and the carbonates are within 5 percent of the targets. We could have added more magnesium sulfate to more closely match the magnesium, but this would have thrown the sulfates further off.

So why is this mythical water, which we have spent much space and time calculating, any better than just going with what comes out of the tap?

First of all, just looking at the numbers explains a lot about the kinds of beer produced in the Munich area. Their beers are colorful—from bocks and doppelbocks through Märzens and dunkels to the dunkelweizens, they are all darker beers. Perhaps this stems from the fact that, to obtain the proper level of acidity, more dark malt is needed to balance out the carbonate water. Carbonates, as mentioned above, combine with calcium and precipitate out when heated. Without calcium, the barley malt does not form a weak acidic environment; the mash is left strongly alkaline, which is unfavorable for the enzymes. Dark malts come to the rescue because they add their own acidity and stabilize this tendency toward alkalinity.

As you can see, understanding an area's water can help explain the local brewing style. Walking a mile in the local brewer's galoshes can help you produce stylistically correct beers and understand why they developed.

*Adapted from an article originally appearing in* Brews & News, *newsletter of The Maltose Falcons Home Brewing Society, Woodland Hills, California.*

# BEER FROM WATER—MODIFY MINERALS TO MATCH BEER STYLES

## BY JON RODIN AND GLENN COLON-BONET

*This article originally appeared in* Zymurgy, *Winter 1991 (vol. 14, no. 5)*

The single largest component of any beer is water, but how many of us actually pay attention to the water we use to brew our beer and how suitable it is for the style we are making? The water of the different brewing regions of the world played a significant part in defining some of the classic beer styles—the soft water of Plzeň gives its Pilsener a smooth, delicate flavor; the hard, carbonate waters of Dublin, suitable only for ales with large amounts of dark roasted malt, gave rise to dry stouts; and the complex minerals of Burton-on-Trent give its bitter and pale ales a unique dryness and hop character. Many other examples of styles evolved from the characteristics of the brewing water, and to fully match these styles, duplicating some of the characteristics of their brewing water helps.

The brewing water you use contains many dissolved minerals that affect the flavor of the finished beer. By looking at the mineral content of some of the classic beer styles and looking at our own brewing water, we can determine how to modify our water to match the style. Before we adjust the mineral content of our water, it is important to understand which minerals are important and how each of them affects the finished beer. Below is a list of some of the more important minerals or *ions* that appear in brewing water and what effect they have on beer.

### CALCIUM (Ca)

Increases mash acidity, assists enzyme action, helps gelatinize starch, helps to extract hop bitterness, reduces haze, and decreases wort color.

## MAGNESIUM (Mg)

Assists enzyme reactions, acts as a yeast nutrient, and accentuates beer flavor. In high concentrations, imparts an astringent bitterness.

## SODIUM (Na)

Accentuates beer flavor through its sour, salty taste. In excess, adds harshness to beer and is harmful to yeast.

## IRON (Fe)

In concentrations above 0.05 ppm, imparts bloodlike flavor. Weakens yeast, increases haze and oxidation of tannins.

## ZINC (Zn)

In low concentrations (0.1–0.2 ppm), acts as a yeast nutrient. Weakens yeast and inhibits enzymes in higher concentrations.

## CARBONATE ($CO_3$)

Raises mash pH, hinders starch gelatinization, impedes flocculation, increases risk of infection, and contributes harsh, bitter flavor. High concentrations (above 200 ppm) acceptable only when balanced with acidity of dark roasted malts.

## SULFATE ($SO_4$)

Gives beer a dry, fuller flavor, enhances hop bitterness. Above 500 ppm, sulfate is strongly bitter. Levels of less than 150 ppm are recommended.

## CHLORIDE (Cl)

Smooths bitterness, improves clarification, produces palate fullness, and enhances beer sweetness.

In order to adjust your brewing water, you'll need some information about the concentrations of these ions in your water supply. If you are on a municipal water system, you can ask the city for a water analysis report. If you are using well water, you can either have the water tested or use a water hardness test kit to get the information you need. Look over your water analysis and check the concentrations of the different ions, also noting the pH, alkalinity, and hardness of the water. Most of the concentrations are specified in mg./l. or parts per million (ppm), which are equivalent in this case. Likewise, the calcium concentration is given in mg./l. as $CaCO_3$. To get just the calcium concentration, multiply by the percentage weight of calcium in calcium carbonate—40 percent. For example, if the $CaCO_3$ is 100 mg./l., then Ca is 40 mg./l. If the water is very hard or contains large concentrations of iron or magnesium, it may not be very suitable for brewing certain styles. In that case you may want to use an alternate source or blend your tap water with distilled water to get the concentration below the taste threshold. If your water is hard, some of the carbonate hardness can be precipitated out by simply boiling the brewing water about 10 minutes and racking the water off the magnesium and calcium carbonate sediment. Also, you can use small amounts (1 to 2 teaspoons for 5 gallons) of lactic or citric acid to help buffer the alkalinity, although these may affect the flavor. Adding lactic or citric acid lowers the pH toward the acidic side, causing magnesium and calcium carbonate to settle out of the solution. A better option would be to add an acid rest step at 95 degrees F (35 degrees C) to your mash. Table 8-3 is a summary of some of the different brewing cities and how they compare against Fort Collins, Colorado. This list can be used to determine how to adjust the mineral content.

### Table 8-3. Mineral Content in Various Brewing Cities

| Brewing Center | Ion Concentrations (mg./l.) | | | | | |
|---|---|---|---|---|---|---|
| | $Ca^{++}$ | $Mg^{++}$ | $Na^+$ | $Co_3^{--}$ | $SO_4^{--}$ | $Cl^-$ |
| Fort Collins | 13 | 2 | 7 | 18 | 13 | 2 |
| Plzeň | 7 | 2 | 2 | 14 | 5 | 5 |
| Munich | 76 | 18 | 2 | 152 | 10 | 2 |
| Vienna | 200 | 60 | 8 | 120 | 125 | 12 |
| Dortmund | 225 | 40 | 60 | 180 | 120 | 60 |
| London | 52 | 16 | 99 | 156 | 77 | ? |
| Dublin | 118 | 4 | 12 | 319 | 54 | 19 |
| Burton | 268 | 62 | 54 | 200 | 638 | 36 |

The water in Fort Collins is relatively soft, making it well suited to brewing Pilseners without any adjustments. If we wanted to brew a stout or bitter, though, we should adjust the water to be a closer match for the style. To adjust the ion concentrations, it would be great if we could just go to the homebrew shop and buy some calcium or sulphate tablets, but these chemicals aren't very safe or stable in that form, so typically mineral salts are used for water treatment. These salts are used because they are safe and commonly available in pure (USP) form, making them suitable for brewing. They are compounds involving the various minerals and ions we described earlier, so when you add these salts, they will contribute differing amounts to each ion in the ratios shown below. Table 8-4 lists the salts used and shows how adding them affects the ion concentrations of your water. The descriptions show the contribution by weight with a volume conversion to teaspoons. Because these compounds are sold as powders or crystals, the contribution of 1 teaspoon will vary depending on how fine the powder is and how tightly you pack the teaspoon. Unfortunately, though most of us have teaspoons, few of us have scales accurate to 1 gram! The weight measurements are much more accurate, but if you are consistent, using volume measurements should be okay.

Now that we know what the water composition is, we can adjust our water based on the style of beer we are trying to match. We won't always be able to exactly match the water of a certain location given the salts we have to work with, but we can usually get close enough. Many times you may not even want to match the water exactly! You may find certain flavors undesirable, such as the effects of Epsom salts, in which case, feel free to omit them.

Now we know all our water data and we understand how to use the brewing salts, so we should be all set to start figuring out the adjustments, right? Not quite. The information so far is a little incomplete. We know that dry stouts originated in Dublin, so we can match that easily. We subtract the ions in our water, and given how much each salt contributes, we can figure out how much to add. But what if we wanted to make a sweet stout? Or how about an Altbier? Let's break down the ion concentration information by styles, so they are more directly usable. Table 8-5 shows the ion concentration for some different beer styles based on the water of some of the major brewing cities combined with the information from Gary Bauer's article in *Zymurgy* (vol. 8, no. 4).

Given this information, we can go through the same procedure described in Darryl Richman's article in *Zymurgy* (vol. 12, no. 5) on water adjustments and come up with a table showing how much

## Table 8-4. Salts

| | | 1 Teaspoon Weighs †† | 1 Gram Per 1 U.S. Gallon Adds | 1 Teaspoon Per 5 U.S. Gallons Adds |
|---|---|---|---|---|
| Table Salt* | NaCl | 5.3 g. | Na 104 ppm<br>Cl 160 ppm | Na 110 ppm<br>Cl 170 ppm |
| Gypsum | $CaSO_4 \cdot 2H_2O$ | 4.8 g. | Ca 62 ppm<br>$SO_4$ 148 ppm | Ca 59 ppm<br>$SO_4$ 142 ppm |
| Chalk | $CaCO_3$** | 1.8 g. | Ca 107 ppm<br>$CO_3$ 159 ppm | Ca 39 ppm<br>$CO_3$ 57 ppm |
| Epsom salts | $MgSO_4 \cdot 7H_2O$ | 3.4 g. | Mg 26 ppm<br>$SO_4$ 103 ppm | Mg 18 ppm<br>$SO_4$ 70 ppm |

* Use only noniodized salt without additives; read the label.
** $CaCO_3$ is insoluble in neutral or alkaline water. It must be added to either the mash or kettle.
†† Accuracy in determining weights by volume teaspoon measurements is somewhat variable, especially with gypsum and chalk, which are light powders. For best accuracy the chemicals should be weighed.

## Table 8-5. Recommended Ion Concentrations for Beer Styles (mg./l.)

| Beer Style | $Ca^{++}$ | $Mg^{++}$ | $Na^+$ | $CO_3^{--}$ | $SO_4^{--}$ | $Cl^{-1}$ |
|---|---|---|---|---|---|---|
| Pale ale | 100–150 | 20 | 20–30 | 0 | 300–425 | 30–50 |
| Bitter | 60–120 | 10 | 15–40 | 0 | 180–300 | 25–50 |
| Mild | 25–50 | 10 | 30–40 | 0 | 95–170 | 50–60 |
| Brown ale | 15–30 | 0 | 40–60 | 0 | 35–70 | 60–90 |
| Scottish ale | 20–30 | 0 | 12–20 | 0 | 50–70 | 18–30 |
| Porter | 60–70 | 0 | 40 | 60 | 50–70 | 60 |
| Sweet stout | 55–75 | 0 | 10–20 | 60–80 | 35–55 | 18–30 |
| Dry stout | 60–120 | 10 | 10–20 | 60–200 | 35–110 | 18–30 |
| Pilsener | 7 | 2–8 | 2 | 15 | 5–6 | 5 |
| Light lager | 35–55 | 0 | 20–35 | 0 | 85–130 | 35–55 |
| Dark lager | 75–90 | 0 | 40–60 | 90 | 35–70 | 60–90 |
| Munich dark | 50–75 | 0 | 5–15 | 60 | 20–35 | 5–20 |
| Maerzen | 30–60 | 0 | 30–40 | 0 | 70–140 | 45–60 |
| Bock | 55–65 | 0 | 40–60 | 60 | 35–55 | 60–90 |
| Doppelbock | 75–85 | 0 | 40–70 | 90 | 35–55 | 60–110 |
| Alt | 30–45 | 0 | 25–30 | 0 | 70–110 | 40–50 |
| Weizen | 15–30 | 0 | 5–15 | 0 | 35–70 | 10–20 |
| Dortmunder | 60–90 | 0 | 45–60 | 0 | 140–210 | 70–90 |

of each salt to add for each of the beer styles. Table 8-6 shows my computations for adjusting the Fort Collins water for brewing these different styles. These values should work if your water is less than 100 ppm hardness. The teaspoon measures are provided for convenience, but it is much more accurate to measure the additives by weight. The table is a general guideline, and you may want to change some values as you gain experience, but it is a great starting point for trying to match that authentic taste.

*Glenn Colon-Bonet is a member and former President of the Fort Collins homebrew club,* The Mash-Tongues. *A recognized Judge in the BJCP, he works as an electrical engineer for Hewlett-Packard and spends much of his free time brewing and enjoying beer. Jon Rodin recently moved to Andover, Massachusetts, from Colorado in search of fall foliage. He occasionally takes time away from reading the* Homebrew Digest *to play with his children and develop software.*

Table 8-6. Water Treatment for 5 gallons

| Beer Style | Table Salt NaCl | Gypsum CaSO$_4$ | Chalk CaCO$_3$ | Epsom salts MgSO$_4$ |
|---|---|---|---|---|
| Pale ale | 1.0 g. (¼ tsp.) | 10.5 g. (2 ¼ tsp.) | — | 2.0 g. (⅝ tsp.) |
| Bitter | 1.0 g. (¼ tsp.) | 8.0 g. (1⅔ tsp.) | — | 2.0 g. (⅝ tsp.) |
| Mild ale | 1.5 g. (⅓ tsp.) | 3.0 g. (⅝ tsp.) | — | 1.0 g. (¼ tsp.) |
| Brown ale | 2.5 g. (½ tsp.) | 2.0 g. (½ tsp.) | — | — |
| Porter | 3.0 g. (½ tsp.) | — | 3.5 g. (2 tsp.) | 1.5 g. (½ tsp.) |
| Sweet stout | 2.0 g. (⅜ tsp.) | 1.5 g. (⅓ tsp.) | 2.0 g. (1⅛ tsp.) | — |
| Dry stout | 0.5 g. (⅛ tsp.) | 1.0 g. (¼ tsp.) | 4.0 g. (2 ¼ tsp.) | 0.5 g. (⅛ tsp.) |
| Pilsener | — | — | — | — |
| Light lager | 1.0 g. (¼ tsp.) | 2.0 g. (⅜ tsp.) | 3.0 g. (1⅔ tsp.) | 1.0 g. (⅓ tsp.) |
| Dark lager | 2.0 g. (⅜ tsp.) | 2.0 g. (⅜ tsp.) | 3.0 g. (1⅔ tsp.) | 1.0 g. (⅓ tsp.) |
| Munich dark | — | — | 3.0 g. (1⅔ tsp.) | 1.0 g. (⅓ tsp.) |
| Vienna | — | 3.0 g. (⅝ tsp.) | 2.0 g. (1⅛ tsp.) | 1.0 g. (⅓ tsp.) |
| Bock | 2.0 g. (⅜ tsp.) | 1.5 g. (⅓ tsp.) | 3.0 g. (1⅔ tsp.) | 1.0 g. (⅓ tsp.) |
| Doppelbock | 2.5 g. (½ tsp.) | 1.5 g. (⅓ tsp.) | 3.0 g. (1⅔ tsp.) | 1.0 g. (⅓ tsp.) |
| Alt | 1.5 g. (⅓ tsp.) | 3.0 g. (⅝ tsp.) | — | — |
| Weizen | — | 2.0 g. (⅜ tsp.) | — | — |
| Dortmunder | 2.5 g. (½ tsp.) | 4.0 g. (⅞ tsp.) | 4.0 g. (2 ¼ tsp.) | 2.0 g. (⅝ tsp.) |

# nine

# Unusual Ingredients

## OPTIONS FOR ADDING FRUIT
### BY AL KORZONAS

*This article originally appeared in* Zymurgy, Special, 1994 *(vol. 17, no. 4)*

Some fruit beer recipes call for adding the fruit at the end of the boil, some to the primary, and others to the secondary. Many recipes suggest the fruit be pulverized before use (actually this will make racking more difficult) when cutting the fruit into small pieces is sufficient, and others suggest using whole fruit. Which procedures are best and what are the factors brewers should consider when adding fruit to beer?

Three main things must be considered when making fruit beers: (1) blowoff, (2) maintaining aromatics, and (3) sanitation.

### BLOWOFF

A 5/16-inch outside-diameter blowoff tube, as is often recommended in brewing texts, can clog even when used for fermenting a regular beer without all the fruit pulp that will be carried up in the kraeusen. An oversized 3/4-inch outside-diameter blowoff tube won't clog with regular beers, but has been known to clog when used with fruit beers. The safest alternative is to use a 1 1/4-inch outside-diameter blowoff tube, which is nearly certain to remain clear regardless of the amount of blowoff that is created.

## MAINTAINING AROMATICS

Fruit flavors are mostly just aromas. Our tongues sense only sweet, salty, sour, and bitter, but our noses can distinguish thousands of aromas. What identifies most fruits is their aroma. Adding fruit at the end of the boil will reduce the aromatics because most will be boiled away. Adding fruit to the primary also will reduce the potential aromatics because the carbon dioxide emitted carries with it much of the aromatics. To maximize the aromatics you gain from fruit, they should be added to the secondary after the primary ferment has subsided, minimizing the amount of scrubbing of fruit (hop, herb, or spice for that matter) aromatics caused by evolving $CO_2$. Some $CO_2$ will be produced from the fermentation of the fruit sugars, but much less than from the primary ferment.

## SANITATION

Adding fruit at the end of the boil solves the problems of sanitation, but also sets the pectins that make the beer permanently cloudy and adds a cooked-fruit flavor. The cloudiness often is referred to as *pectic haze*. Pectic enzyme can be used in the fermenter to reduce this haze, but the cooked fruit or jamlike flavors will not be removed.

Adding fruit to the primary or secondary means that you will have to sanitize it somehow. Again, boiling is a possibility, but you have the same problems as noted before. One alternative is to put the fruit in water, heat to 150 to 160 degrees F (66 to 71 degrees C), rest for 15 to 20 minutes, and then drain the water.

Another way to sanitize the fruit is to use Campden tablets or metabisulfites. Typically one tablet per gallon of fruit juice is used. Sanitation occurs as a result of the sulfur dioxide gas that is produced. This will dissipate in about 24 hours, at which time the juice is ready to be added to the fermenting beer. The disadvantage with using Campden tablets and metabisulfites is that they add sulfites to the finished beer and cause allergic reactions in some people. Some brewers have reported elevated sulfur aromas in beers made with metabisulfites—it could be a matter of varying sensitivity.

Blanching is another possibility, but the user must walk a fine line between the risk of infection (too short a time, too cool a temperature) and the risk of setting pectins (too long a time, too hot a temperature). The fruit is first frozen (which helps break open juice "sacks" in the fruit) and then dipped for a few seconds in boiling water. Unblemished fruit (no cuts or insect marks) is best because you are only sanitizing the outside of the fruit. Blanching fruits like raspberries and blackberries is problematic because dipping the frozen fruit in the boiling water releases the juice, so brewers should be prepared to add the blanching water to the beer or lose much of the juice.

Commercially made fruit juices are a possibility and don't have to be sanitized, but make sure the juice is preservative-free. Some preservatives can kill your yeast. Read the label for juice content also. Many cherry juices, for example, are mostly white grape juice with only a small amount of cherry juice.

Finally, fruit extracts recently have become available to homebrewers. Many of these, despite claiming to be made from 100 percent real fruit, add medicinal or unpleasant bitter flavors to the finished beer. Try adding a few drops of the extract with an eyedropper to a glass of relatively neutral-tasting finished beer to see if the flavors added are what you want before you commit a full batch.

©1994 AL KORZONAS

# POTIONS!

## BY RANDY MOSHER

*This article originally appeared in* Zymurgy, Special 1994 (vol. 17, no. 4)

You are no doubt familiar with the common ways to add herb and spice flavors to your beer. Adding flavorings to the boil and/or during fermentation is relatively straightforward and these methods generally work well, but there are a couple of weak points.

Herbs and spices added during wort boiling are sanitized by the heat, but that same heat can modify and drive off some of their precious aromas. Primary fermentation generates prodigious quantities of $CO_2$ that carry spice aromas out of the beer along with it. Consequently the amount of flavor persisting into the finished beer is difficult to predict. And, of course, there is no way to go back to the boil to add more if the quantity is not sufficient. This is especially critical in beers such as Christmas ales wherein having a perfectly balanced blend of ingredients makes all the difference. Herbs and spices added during fermentation may not solubilize well, and if added during the primary, are affected by the same problem of aroma loss as those added in the boil. Plus there is always the potential for contamination by anything added to the fermenting beer.

There is another method that I have been using for several years now—potions.

Potions are nothing more than the use of a friendly solvent, alcohol, to sterilize and solubilize the flavorful components of spices and herbs. Fortunately, most of these aromatic compounds are soluble in alcohol, often more so than in water. When the flavors have been dissolved, the mixture is filtered and added to the beer during the secondary fermentation or prior to bottling.

The first step in the process is to buy a big jug of the least expensive vodka you can find. Not feeling shameful is actually the most difficult part of this flavoring method. More expensive brands may make you feel better (and your wallet lighter), but they add nothing in terms of quality, from my experience.

Once obtained, curtains drawn, you can put this cheap-yet-magical substance to work in the service of brewing. Put the herbs and spices you wish to use into a beaker or wide-mouthed jar. Use a little more than you think you'll actually need for the batch of beer. Pour a quantity of vodka into the same container, about double the volume of the seasonings. Tightly cover and allow this to soak for at least a week. Shake it every day or two if you like. The actual soaking time can be much longer, but the mixture matures pretty well in a few weeks and doesn't change much after that. Be sure to carefully measure the quantities of herbs and spices added, so you can repeat the recipe in case you win best of show. A gram scale works best, but dry measures serve well if there is no alternative.

After a week, taste it, or better, add a few drops to a beer. Scrutinize the mixture for balance, flavor, and depth. Now is the time to add more cloves, pepper, ginger, or whatever you think it will take to finesse the mixture. Add these as desired and allow to sit for a few more days. You can fine-tune the potion endlessly if you want.

The next step is filtration. The simplest method is to use a funnel and coffee filter. This removes nearly all of the spices, leaving just a bit of dusty stuff that settles right out with the yeast in the bottle. I recommend a funnel with shallow ribs on the inside designed for use with filter paper (or coffee filters). The ribs raise the paper and allow the liquid to flow more freely than in a smooth funnel. If you need finer filtration for some reason, various grades of filter paper and more sophisticated gizmos are available from scientific supply houses.

I usually add the potion to the beer at bottling, although you can just as well add it toward the end of secondary fermentation. If the beer is to be kegged rather than bottled, it is best added when the beer is racked into the keg.

At some point you need to do a test to decide how much of the potion to add for the best flavor. Many chemicals taste very different depending on the specific concentration present. Small changes in quantity can translate into large changes in taste. I have found during these tests that there is a sharp transition between not enough and too much, with only a very small range of just right in the middle.

The dose-testing procedure is as follows: Get a pipette or small syringe graduated in some small amount such as $1/10$ milliliter. You also will need a small measure, such as a shot glass, and calibrate it with a line indicating one ounce. (This assumes you are keeping track of your batches by gallons; if you are using liters, use a similar metric quantity—25 or 50 milliliters.)

You can use either a small amount of the intended beer for the test or one that is similar. Although the stand-in beer won't be the real thing, it will be fully carbonated, greatly affecting the flavor. I have done it both ways with good results. Pour an ounce of the beer into the shot glass. Withdraw a small amount of the potion into the pipette, taking a wild guess about how much to start with, and add it to the shot glass. Stir well with an inert stirrer and taste. Too much? Not enough? You just have to tinker until you get it right.

Once you have determined the tenths-of-a-milliliter-per-ounce ratio, all that remains is to scale it up. Get out the pocket calculator and do the arithmetic: 1 ounce $\times$ 128 (ounces per gallon) $\times$ 5 (gallons in batch) = 640 ounces per batch. If the dosing test determined that 0.2 milliliter was the correct amount for 1 ounce, multiply 0.2 milliliter $\times$ 640 and you get 128 milliliters of potion that must be added to match the small-scale test. If the beer is a strong one intended for long aging, you might bump up the quantity a bit to compensate for the inevitable fading of flavor that comes with extended time in the cellar.

One obvious question is how much alcohol gets added to the beer with this method. The answer depends on the quantity and strength added, but it really doesn't amount to much. In a 5-gallon batch, 16 ounces of 80-proof vodka will add 1 percent of alcohol. Some alcohol is lost to evaporation during the soaking period, so it may end up being a little less. This shouldn't affect things too much, but be aware that in beers with higher than 7 percent alcohol, priming may be slowed down a bit. If in doubt, you can always cut back the priming and add champagne yeast when you bottle. [Editor's Note: Adding champagne yeast, which is more alcohol-tolerant than most yeasts, to a high-alcohol beer at bottling time can result in overcarbonation.]

In addition to homemade potions, commercial liqueurs may be used as a source of exotic flavorings in beer. I have had very good results with triple sec (orange liqueur) and crème de cacao (chocolate). Spices may be added to these using the same method as with the vodka. Fruit-flavored brandies often have rather elegant fruit character and also work well, especially when some real fruit is used along with the brandy. I have used as much as a full 750-milliliter bottle of liqueur in a 5-gallon batch of beer.

Keep in mind that the liqueurs contain a certain amount of sugar that must be considered as priming material. You can determine the quantity of sugar present by measuring the specific gravity of the liqueur. (Alcohol is lighter than water, so the specific gravity will be lower than if you had a solution of sugar in water.) By subtracting the effect of the alcohol, which is a clearly labeled quantity, you can determine the amount of sugar present. These numbers will get you in the ballpark: for 80-proof, add 20 °Plato to the measured gravity; for 60-proof, add 15 °Plato; for 40-proof, add 10 °Plato. Once you've added the appropriate number of degrees Plato to compensate the alcohol, it is simple to calculate the amount of sugar present. Because degrees Plato is a measure of the percentage of sugar, just multiply the degrees Plato (as a decimal: 10 °Plato = 0.10) times the weight of the liqueur. You can weigh the stuff or calculate it. A 750-milliliter bottle of pure water weighs 26.7 ounces, which is probably a good figure to use since the lighter weight of the alcohol about cancels out the added weight of the sugar. Multiply the weight by the degrees Plato (10 °Plato = 0.10), $26.7 \times 0.10 = 2.7$ ounces (net weight) of sugar present in the liqueur. Subtract this from the weight of the sugar with which you intended to prime. It may be useful to know that a cup of corn sugar is roughly equal to 6.7 ounces by weight.

There are a number of commercially prepared flavoring extracts meant for preparing homemade liqueurs. Home wine and beer shops often carry these products in all common liqueur flavors. Many, like hazelnut and crème de cacao, are difficult to extract on your own and are well worth using. Some brands and flavors are better than others and some contain oil that may cause head-retention problems. Be sure to check out the one you plan to use with the small-scale test discussed above. Watch carefully for deleterious effects on the head, especially with orange-flavored varieties.

## 🍺 Christmas Ale

This beer is a moderately hopped dark "white" beer with complex malt character and brown color. Spice flavor is very soft and round with nothing sticking out. I used East Indian coriander, which is paler and more oblong than the common grocery-store variety and has a milder, less resiny taste.

*Ingredients for 5 gallons*
- 7 lbs. U.S. two-row lager malt
- 1½ lbs. U.S. six-row lager malt
- 1½ lbs. German or Belgian two-row Munich malt
- 1½ lbs. six-row pale malt, roasted 70 minutes at 350 degrees F (177 degrees C)
- 2½ lbs. unprocessed oats, in husk
- 4 lbs. wheat (malted or unmalted)
- 1 lb. imported pale crystal malt
- 1 oz. chocolate malt
- ¼ oz. Styrian Golding hops (90 min.)
- ¼ oz. Northern Brewer hops (90 min.)
- ½ oz. Styrian Golding hops (30 min.)
- ½ oz. Northern Brewer hops (30 min.)
- 1 oz. crushed coriander (30 min.)
- 1 oz. East Kent Goldings hops (10 min.)
- 1 oz. East Kent Goldings hops (end of boil)
- liquid Altbier yeast
- ½ c. corn sugar (to prime)

- Original specific gravity: 1.083
- Bitterness: 45 IBU (estimate)
- Color: 30 °SRM (estimated)

Cook oats and wheat as you would rice, until tender. If oatmeal and flaked wheat are substituted, they may be added directly to the mash without precooking. Mash at 152 to 155 degrees F (67 to 68 degrees C) for 1 hour. Be sure to mash out at 170 degrees F (77 degrees C) before sparging. This will help keep the rather gooey mash flowing smoothly. Do not let the sparge bed get below 165 degrees F (74 degrees C) during sparging or it will stiffen up.

*Spice Potion*
12½ oz. of the following mixture added at bottling:

- 250 ml. crème de cacao
- 600 ml. Curaçao or triple sec
- 50 ml. Benedictine liqueur
- ¼ tsp. black pepper, cracked
- 1 tsp. vanilla extract
- 2 tsp. cassia buds (try a Chinese market; use cinnamon if cassia buds are not available)
- ¼ oz. orange blossom petals or orange blossom water*
- ¼ tsp. aniseed
- ½ star of star anise
- 3 oz. cracked coriander

Mix all ingredients for the spice potion and allow to sit two weeks or more before filtering. This spice mixture will work with any kind of brown beer with 1.050 or higher original gravity.

The possibilities for potion concocting are endless. The best thing is that they let you "try before you buy," reducing the risks associated with such bold new territory. You can even dose a bottle or two and continue to nurture the seasoning elixir, which will last indefinitely. And what to make? How about a dark and dangerous Pirate Stout, filled with such exotic ingredients as galingale root and grains of paradise? Or a fresh, ethereal springtime ale with honey, heather, sweet woodruff, and basil. Maybe your taste leans toward a medieval gruitbeer with wild rosemary and sweet gale. And if you come up with something really weird, save a bottle for me.

*Randy Mosher is a beer author/lecturer and freelance graphic designer. A National BJCP Judge, Randy began brewing in 1984. He has made presentations to a number of national and regional homebrew conferences, including three engagements at AHA National Homebrewers Conferences. He is the author of* The Brewer's Companion (Alephenalia *Publications, 1994).*

---

*Try Middle Eastern or Mexican markets to locate orange blossom petals or blossom water.

# A BREWER'S HERBAL—RETURNING TO FORGOTTEN FLAVORS

### BY GARY CARLIN

*This article originally appeared in* Zymurgy, *Summer 1987 (vol. 10, no. 2)*

Homebrewing, very simply, is a matter of taste. As homebrewers, we are not bound by commercial tastes or trends. We are free to experiment, and hopefully create new tastes, by varying the amounts and types of ingredients we choose to include in our brew.

Early brewers experimented with a great variety of ingredients, some of which are still in use today, while others have long since been forgotten. Among the forgotten ingredients were brewing herbs and spices.

Unfortunately, few of the many herbs and spices used in brewing over the centuries remain in use today. The majority used were for bittering, for to the early brewers, the combination of malt's sweetness with a bitter flavor produced quite a pleasing drink.

Yet the addition of herbs and spices was not limited to producing a countertaste to the malt's sweetness. Throughout the evolution of ales and beers, herbs and spices have performed a variety of functions.

As homebrewers of the twentieth century, we are blessed with many luxuries not afforded our predecessors. Just having consistent quality of malt, hops, and yeast is something we tend to take for granted today. Even our scientific knowledge of fermentation and the necessary sanitary conditions are relatively new to the brewing industry. In comparison, early homebrewing at its best was a primitive business. The results were, for the most part, a gamble.

To account for the uncertainty of their efforts, early homebrewers learned to heavily flavor their brews with herbs and spices. This would mask undesirable flavors, make a brew drinkable, or perhaps even make a brew pleasant-tasting. Without the knowledge of proper sanitary conditions and yeast cultures, these early homebrewers were forced to deal with "sick" beers and ales, which they learned to cure with various herbs. When it became fashionable to clarify beers, many herbs with varying tastes were available.

At one time, both beers and ales were made with various herbs and spices to prevent and remedy sickness. Some used specific herbs and spices to purge the body, prevent scurvy, and cure "consumption." Beers and ales throughout history have served as a disease preventative by serving as a substitute to impure drinking water. They also unknowingly provided vitamins and minerals to prevent deficiency diseases. In some cultures they have even represented up to 50 percent of the population's food sources.

If flavoring was considered to be the primary use of herbs and spices in brewing, then their preservative abilities would have rated as the secondary function. Through the use of many herbs and spices over the centuries, certain ones were found to have strong preservative qualities.

Of all those used, hops were found to be the best preservative of beer and ale. In the sixteenth century, it became the brewing standard. Hops for the most part eliminated the use of most herbs and spices in brewing, unless necessity or personal taste advocated their use. It is interesting to note that at one time in brewing, the addition of hops made the difference between beer and ale. Beer was hopped and ale was not, although ale was generally of a much higher alcohol content, if it was to be stored!

Yet the use of herbs and spices in brewing was in part caused by the scarcity and high costs of traditional brewing ingredients at different times and places in history. Unfortunately, some of these herbs and spices have been found to be dangerous or unhealthy, while others create interesting flavors and aromas. It will be up to you to decide your own personal taste, and remember that the flavors you re-create are part of the past, and they will be different today.

The tastes and flavors of beer and ale have undergone, and will continue to undergo, some interesting and curious evolutions. Perhaps like any other aspect of history, the pendulum may once again swing back, reviving the herbal tradition of brewing.

## BREWING HERBS AND SPICES

*Caution: Be sure only pure, food-grade herbs are used, free of herbicides and pesticides.*

Agrimony, *Agrimonia eupatoria* (leaves)—very delicate aroma with a faint sweet smell. One to 2 ounces dry-hopped for one week per 5 gallons.

Balm, *Melissa officinalis* (leaves)—minty, lemon flavor with a bitter taste. Imparts a pleasant aroma with a subtle flavor. Used to clarify ale. Only use fresh, 2 to 4 ounces per 5 gallons.

Betony, *Stachys officinalis* (leaves)—salty, bitter flavor. Leaves must be dried for use—do not use fresh leaves. One half to 1 ounce dried herb per 5 gallons.

Bogbean, Buckbean, *Menyanthes trifoliata* (leaves)—an aquatic, perennial herb with a pleasant bitter taste. Used to improve the taste of beer and ale. Provides an excellent flavor for unhopped ale. As a flavoring agent, use ½ ounce herb per 5 gallons.

Bog Myrtle, Sweet Gale, *Myrica gale* (leaves and branches)—used in England in place of hops to brew Gale Beer. A small deciduous shrub with a "resinously aromatic" principal.

Cardamom Seed, *Elettaria cardamomum* (seed)—sweet aromatic odor, characteristic of ginger family. Used to give "strength" to beer, its pungent taste seems to provide warmth. Boil five to eight crushed seeds ½ hour per 5 gallons.

Chamomile, *Anthemis nobilis* (entire herb)—has an applelike fragrance: "apple on the ground." The flowers upon blooming are bitter and warming. Especially good in lighter beers and ales. Any combination of leaves, stalks, and flowers—fresh or dried. Try 1 ounce dried flowers or 3 ounces fresh per 5 gallons (hopped or unhopped).

Clary, *Salvia sclarea* (leaves)—the bitter and aromatic leaves are used to flavor ale. Said to increase "intoxicating quality," headiness, and body. The oil of clary is used in the production of muscatel.

Costmany, Alecost, *Chrysanthemum balsomita* (leaves)—a spicy, mint, and camphor flavor that is also bitter. Most popular flavoring for ale during the Middle Ages. Also gives body and head to ale. Is famous for its strong influence in the "parturition of ales and beers." Fresh or dried in place of hops.

Dandelion, *Taraxacum officinale* (entire plant)—all parts of the dandelion have been used in the production of beer. The leaves alone are used to brew Dandelion Beer. Dandelion provides a unique bitterness, which is increased if the sepals are not removed when the flowers are used. Simmer 1 gallon loosely packed flowers with sepals for 20 minutes per 5 gallons.

Elecampane, *Inula helenium* (rootstock)—bitter, yet with a strong violet and camphor odor. It is strongly antibacterial, preventing "sick" beer and ale. Boil the fresh rootstock, 1 to 2 ounces for 30 to 40 minutes per 5 gallons.

Garden Sage, *Salvia officinales* (leaves, flowers, and seeds)—a warm, bitter, and spicy taste. Used to dry-hop beers and ales to increase "inebriating quality." Used to produce a popular brew called Sage Ale. Dry-hop with ½ to 1 ounce of herb per 5 gallons.

Gentian Root, *Gentiana lutea* (root)—perhaps our most bitter plant material. Use with discretion. Boil ⅛ to ¼ ounce rootstock per 5 gallons. Be careful.

Ginger, *Zingiber officinale* (rootstock)—one of the few brewing spices still in use today. Try boiling ½ ounce to 2 ounces fresh root 20 minutes per 5 gallons.

Ground Ivy, Alehoof, Cat's foot, Gill over the Ground—*Glechoma hederacea*—the leaves and stems were used to flavor, clarify, and improve the keeping quality of ale. Ground ivy is a small creeping mint with a bitter principal. It is used dry, as you would use hops.

Hyssop, *Hyssopus officinalis* (leaves and young shoots)—has a mint-camphor odor, and is extremely bitter. Use in place of or with hops.

Indian Borage, *Coleus amboinicus* (leaves)— strong bitter flavor and aroma. Substituted for hops in India, also used to flavor wine.

Licorice, *Glycyrrhiza glabra* (rootstock)—used in the flavoring of beer and to promote headiness. Rootstock contains glycyrrhizin—fifty times sweeter than sucrose. A "small" piece of root boiled 20 minutes per 5 gallons. (Break root into small pieces). Excellent countertaste to malt's sweetness.

Meadowsweet, *Filipendula ulmaria* (leaves)—commonly called "meadwort," because of its use as a flavor in mead. The leaves have a pleasant wintergreen flavor. Found to contain salicylic acid (aspirin). Use as flavoring agent in beers and ales. Dry-hop with ½ ounce dried leaves per 5 gallons.

Mugwort, *Artemisia vulgaris* (leaves)—once used to flavor and clear beer. Contains absinthin, a bitter principal, which helps digestion. It will impart a pleasing bittersweet taste and aroma. Used in place of hops.

Southernwood, *Artemisia abratanum* (young shoots)—sweet lemon fragrance with a strong bitter flavor. Added to beers and ales for its flavor.

Spruce, *Picea* (young twigs or new tips)—only the Norway, red, and black species are used in brewing. Provide resins that preserve beers and ales. Commercially available as essence of spruce—1 tablespoon per 5 gallons. Fresh stems and needles—1 to 4 cups per 5 gallons.

Valerian, *Valerian officinalis* (rootstock)—has a sharp, bitter taste that brings out and livens apple flavors. Try with chamomile. Use sparingly, ¼ to ½ ounce rootstock boiled 25 minutes per 5 gallons.

- Wild Marjoram, *Origanum vulgare* (entire herb)—"balsamic fragrance." It was used to flavor beers and ales. Try ½ ounce hopped or dry-hopped per 5 gallons.
- Wintergreen, Checkerberry, *Gaultheria procumbens* (leaves)—wintergreen flavor is an excellent flavoring agent. Oil of wintergreen—Menthyl salicylate for the most part—is made synthetically or taken from young birch trees, *Betula lenta*. To get the flavor, wintergreen leaves must be allowed to sit 3 to 4 days until they naturally begin to ferment, then added to the brew. Boil ½ gallon of water, pour over 1 to 2 ounces wintergreen leaves, cover, and allow to stand until fermentation begins. Strain and use to begin 5 gallons of brew.
- Wood Sage, *Teucrium scorodonia* (leaves and flowers)—very similar to hops; therefore, it was a common substitute for hops, used in the same manner.
- Yarrow, *Achillea millefolium* (leaves and flowers)—has a bitter, astringent taste with a mild aroma, used in place of hops. Flowers have a stronger aroma and bitterness than the leaves. Reported to improve taste, headiness, and "intoxicating quality." Try 2 ounces dried herb boiled for 30 minutes per 5 gallons.

*Editor's Note: The author provided suggested amounts for the herbs he has experimented with. For the others, estimated amounts will have to be used.*

*The following recipes were inspired by Old English formulas, then modified for twentieth-century brewing by Ralph Bucca, Fort Washington, Maryland.*

## Herb Beer

This beer contains four different dried herbs that can be found at a health food store or herb shop. It has an interesting flavor that grows on you.

Ingredients for 1 gallon:
- 2 oz. dried meadowsweet
- 2 oz. dried betony
- 2 oz. dried agrimony
- 2 oz. dried raspberry leaves
- 2 lbs. corn sugar
- ½ lb. light dried malt extract
- 1 tsp. citric acid
- 1 tsp. yeast energizer
- 1 gal. water

Pour the boiling water over the ingredients, cool to 75 degrees F, and add 1 teaspoon ale yeast. Ferment in a primary for a week, rack and ferment in a secondary for three months, and then bottle. Carbonate with corn sugar at bottling time, if interested.

## 🍺 Dandelion Beer

Go out in your yard and dig up some dandelion plants, flowers not required. **[Editor's Note: Caution—do not use any plants treated with herbicides, pesticides, or fertilizers.]**

Ingredients *for* 1 *gallon*:
- ½ lb. young dandelion plants with taproot
- 1 lb. brown sugar
- ½ oz. crushed ginger root
- juice of 2 lemons
- 1 gallon water
- 1 tsp. ale yeast

Boil the washed roots in half the water, cool, and add the rest of the ingredients. Ferment for one week, then bottle.

## 🍺 Another Herb Beer

Take 1 pound of fresh nettle tops, 1 pound of young parsley leaves, four large potatoes, 3 pounds of dried malt extract, five or six chilis, and 1 ounce of fresh hops. Wash the nettles and the parsley very carefully and tie them in a bunch. Scrub the potatoes and prick them all over with a knitting needle or fork. Simmer the nettles, parsley, hops, potatoes, malt extract, and chilis together for 1 hour in 2 gallons of water. Strain through cheesecloth. When the wort has cooled, add 1 gallon of water and 2 teaspoons of yeast. After fermentation is complete, carbonate with ½ cup of corn sugar.

## Damikola Beer

This brew is reputed to have aphrodisiac qualities. I cannot honestly say that I have noticed any effects, good or bad. However, I have had two children since I have made this unusual brew.

Ingredients for 3 gallons:
- 1 oz. damiana leaves
- 2 oz. powdered kola nuts
- 1 lb. raisins
- 1 lb. corn grits
- 3 lbs. malt extract
- 1 oz. hops
- 3 gal. water
- 1 packet ale yeast

Boil the corn grits and malt extract in 1 gallon of water for 30 minutes. Then add the raisins, damiana leaves, kola nut, and hops, and boil for 30 minutes. Strain this liquid into 1 gallon of cold water and add the yeast when cool. Ferment as usual and bottle after two months; age for three months and then get a hot date.

---

*A homebrewer for the better part of eight years, Gary spends most of his brewing time experimenting with as many unusual ingredients as he can dream up.*

# ten

# Better Brewing

## STEEPING: THE EASY STEP

### BY SHAWN BOSCH

*This article originally appeared in Zymurgy, Special 1995 (vol. 18, no. 4)*

One of the simplest ways to incorporate grain into your very next batch is by steeping some specialty grains. This process only requires a thermometer and a way to remove the grains from the resulting extract.

Steeping specialty malts will enable you to control more carefully the character of your beer. With specialty malts, you'll be contributing color, sweetness, body, and aroma to your brew. Just like when you spice up a can of soup or jar of pasta sauce, the small effort makes a definite difference.

The kinds of malts you should consider adding include black or black patent, chocolate, crystal, caramel malts, or roasted barley. These malts don't require mashing, so their contributions can be made to your wort via the steeping process.

Adding the steeping step to your brewing procedure involves soaking the grains in 155-degree F (68-degree C) water for about 30 minutes. You have two alternatives: adding the grain to the water loose or containing the grain in a grain bag. If you choose to add them loose, you'll have to consider how to get them back out. A screened kitchen strainer, a colander with fine holes, a bucket with

# Steeping: The Easy Step

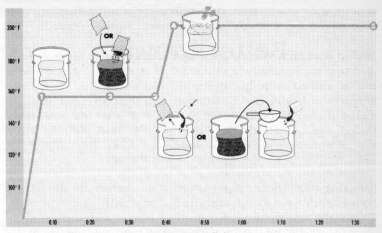

**Steeping Graph**

holes drilled in the bottom, or a nylon or cloth bag could serve this purpose. Once loose grain is captured, you can sparge, or rinse, the grain with additional water to get the most extract from your effort.

Steeping the grain in a bag in the brewing water means the bag can be removed easily, rinsed, and discarded with no extra mess. For a quick brew, it is hard to beat as far as convenience and time are concerned.

## STRAINING OPTIONS

**Kitchen strainer:** The strainer must be large enough to hold at least ½ pound of grain. At best it will hold almost 4 pounds. Most kitchen strainers are made of stainless steel, aluminum, or plastic. Stainless is most durable, plastic the most difficult to clean. Strainers generally require two people for ease of operation: one to hold the strainer and one to pour the extract and grains through.

**Colander:** Colanders are available in a wide variety of capacities and materials. Once again, stainless steel is the most durable, but aluminum, copper, and plastic are also popular. The determining factor in using a colander is the size of its perforated holes. If the holes are too large, grain will pass into the boiling kettle, which you don't want. Lining the inside of the colander with cheesecloth is an easy way to adapt a standard piece of kitchen apparatus to your brewing needs. Find a colander that can rest securely in your brewing pot and you won't need an extra pair of hands.

**Drilled bucket:** A very practical and inexpensive strainer that can lead to great versatility is a drilled bucket. Any food-grade plastic bucket drilled with hundreds of 1/8-inch holes in its bottom becomes a large-capacity strainer. The bucket's large opening makes it easy to clean. If the strainer is placed inside another bucket, one person will be able to handle the operation. Otherwise, one person will need to hold the bucket above the boiler while another pours the grain and water through.

**Straining bag:** These large bags function similarly to the drilled bucket. They have a large capacity and are relatively inexpensive. They work best when placed inside a bucket with the mouth of the bag folded over the lip of the bucket. (Securing the bag to the bucket with clothespins can prevent the bag from sagging into the bucket as you pour.) This setup allows one person to separate the extract from the grain.

Grain also can be tied into the bag and the whole thing put directly into the boiling pot, steeping it like a big tea bag. If the bag is made of nylon, don't let it rest on the bottom of a pot on a hot burner. The heat can melt the bag and release the grain.

To clean the bag after use, remove the spent grain, rinse it well, and send it through the washing machine; a true convenience!

**Disposable cheesecloth steeping bag:** Also known as muslin hop bags, these cheesecloth bags are the easiest, quickest way to steep grain. Each bag usually can contain up to 1 pound of grain. Simply tie the grain in the bag, steep, remove the bag, and dispose of the grain properly. For first-time grain users, this is an excellent way to experiment without investing much money. Steeping and straining are performed in one vessel, and as long as the bag isn't dropped on the floor, cleanup is a snap.

## THE PROCESS

For a 5-gallon batch of beer, put 1½ gallons (5.68 l.) of water in your boiling pot. If you've chosen to use a grain bag, tie the crushed grains in the bag and add it to the water. If you decided to add the grain loose, pour it directly into the water. Slowly raise the temperature of the water to about 155 degrees F (68 degrees C), being careful not to allow the mixture to boil. Boiling the grain will leach tannins and astringency from the grain husks. Maintain this temperature for about 30 minutes, then remove the grain bag, discard it, add your malt extract, and proceed with your boil. If you steeped loose grains, carefully pour the extract through the strainer and collect it in a brewing pot. If you want to maximize your extraction, pour an additional ½ gallon of 170-degree F (77-degree C) water over the grain in your strainer to rinse it thoroughly.

## STEEPING AT A GLANCE

1. Bring water temperature to 155°F.

2. Add grain bag and steep 30 minutes.
   OR
   Add loose grain and steep 30 minutes.

3. Remove and discard grain bag, add malt extract syrup.
   OR
   Pour "grain tea" through strainer to remove grain and add malt extract syrup.

4. Raise temperature to boiling and add hops.

5. Boil 60 minutes.

Here are some of my favorite extract specialty grain recipes.

### Never-Let-Me-Down Porter

*Ingredients for 5 gallons (19 l.)*
- ½ lb. chocolate malt (0.23 kg.)
- ½ lb. medium or dark crystal malt (0.23 kg.)
- 6½ lbs. unhopped amber malt extract (3 kg.)
- 1½ oz. Northern Brewer hops, 7.5% alpha acid (43 g.) (60 min.)
- ½ oz. Northern Brewer hops, 7.5% alpha acid (14 g.) (20 min.)
- 1 oz. Northern Brewer hops, 7.5% alpha acid (28 g.) (5 min.)
- English ale yeast (Wyeast No. 1968 is my favorite)
- ¾ c. dextrose (177 ml.) (to prime)

- Original specific gravity: 1.044
- Final specific gravity: 1.012

## 🍺 Georgia Stream Common Beer

Ingredients for 5 gallons (19 *l.*)
- ½ lb. Belgian Biscuit malt (0.23 kg.)
- ½ lb. medium crystal malt (0.23 kg.)
- 2 oz. Belgian Special "B" malt (57 g.)
- 3½ lbs. Munton and Fison Premium kit (1.59 kg.)
- 3 lbs. light dry malt extract (1.36 kg.)
- ½ oz. Northern Brewer hops, 7.5% alpha acid (14 g.) (45 min.)
- ½ oz. Cascade hops, 7.5% alpha acid (14 g.) (30 min.)
- ½ oz. Northern Brewer hops, 7.5% alpha acid (14 g.) (15 min.)
- Wyeast California lager No. 2112
- ¾ c. dextrose (177 ml.) (to prime)

- Original specific gravity: 1.045
- Final specific gravity: 1.014

This is a great full-flavored beer. It is hopped aggressively and tends to mature well in the bottle.

## 🍺 Double Dipper (Belgian Double)

Ingredients for 5 gallons (19 *l.*)
- ½ lb. Belgian Special "B" malt (0.23 kg.)
- 1 lb. light crystal malt (0.45 kg.)
- ½ lb. Belgian aromatic malt (0.23 kg.)
- 8 lbs. light dry extract (3.63 kg.)
- 2 oz. Mt. Hood hops, 4.5% alpha acid (57 g.) (45 min.)
- 1 oz. Saaz hops, 3% alpha acid (28 g.) (5 min.)
- Wyeast Belgian ale No. 1214
- ¾ c. dextrose (177 ml.) (to prime)

- Original specific gravity: 1.068
- Final specific gravity: 1.018

This is a satisfying beer to brew. Loaded with flavor from the darker grain and esters only a real Belgian brewing yeast strain can produce. Let it age.

Remember, what goes into your beer is what goes into your glass. Pick and choose your ingredients intelligently. Whether you stay with extracts and specialty grains or move on to mashing, using fresh grain will make a difference in your homebrew.

*Shawn Bosch, a homebrewer for over seven years, is a wine and beer making consultant at Brews Brothers at KEDCO Beer and Wine Supply Store in Farmingdale, New York. He has won various local and regional awards for brewing including the gold medal in the Belgian and French category in the AHA 1994 National Homebrew Competition.*

# START MASHING!
## BY CHARLIE PAPAZIAN

This article originally appeared in Zymurgy, Fall 1984 (vol. 17, no. 3)

All-grain brewing intrigues the hell out of me. I do it perhaps four times a year, not because it necessarily makes better beer, but because of the wacky satisfaction I get from getting involved with the process. If all my equipment is clean and ready to go, it usually takes 4 to 5 hours of brewing and cleanup to brew an all-grain batch. It does something for my sense of accomplishment. My extract brews are superb, and most of the time my all-grain brews simply are considered something special because I become a "father" of sorts.

## CONQUISTADOR

Now, don't forget, I am a kitchen brewer. No fancy cellar or basement setup. All the equipment comes out of storage and I play conquistador and assault the kitchen for a day. Forget about cooking dinner. Who feels like doing anything more in the kitchen when you're done brewing up an all-grain batch? Whew, not I.

And there the brew will remain, quietly inhabiting a space among the other carboys under my kitchen table. Like I said, I'm a kitchen homebrewer. I've learned a few tricks that work with my kitchen counters, sink, stove, tables, etc., and I'm sure you, too, have worked or will work out a game plan for your brewery.

What about all-grain brewing? Mashing is mashing—or is it? Frankly, there are hundreds of ways to combine ingredients, fabricate and use equipment, and process the ingredients. What I'd like to do in this burst of literation is discuss some of the basic principles of mashing without getting into the pros and cons of various equipment. At the same time I want to suggest some simple techniques that work for mashing from 1 pound to 15 pounds of grain.

## AMERICAN MALT

Virtually all American-made malted barley is undermodified. Undermodification means that germinating barley is halted at a stage just short of converting all of the raw starch to soluble starch. The degree of modification determines what process of mashing you as a brewer choose. Infusion, decoction, step-mashing—these are the choices, and it doesn't matter a goat's foot whether you are using two-row, six-row, high-enzyme, or low-enzyme malt. Remember, modification of your malt is the determining factor.

Why? During the malting process, modification develops enzymes, soluble starch (more suitable for mashing), and some sugar (not much). The malting process also develops yeast nutrients—amino acids for fermentation from the nitrogenous constituent of the barley. These amino acids are essential for healthy fermentation. Fully modified malts (often English two-row varieties) contain a proper spectrum of amino acids for healthy fermentation because they are fully modified. Undermodified malt lacks proper yeast nutrients.

## MAKING YEAST NUTRIENTS

By holding a mash of malted barley at 122 degrees F for a half hour, you can develop those missing yeast nutrients in undermodified malt. This part of the mashing process is called the "protein rest," because amino acid proteins are developed for good fermentation. Does everything begin to make sense? This entire mashing process is often referred to as a "step mash," or an "upwards infusion," because from here the mash is stepped up to diastatic activity at higher temperatures. The "decoction" method of mashing is seldom used by commercial brewers since the invention of the thermometer. Step mashing is what virtually all commercial brewers do. And that's what you should consider doing for healthier fermentation if you use undermodified malt.

The one-temperature infusion mashing process bypasses the 122-degree "protein rest" because it is unnecessary when using fully modified malt. That is why most British homebrew books emphasize infusion mashes—they use fully modified malt. It is not because they use two-row malt.

## GETTING HIGH YIELDS

What about extracting the most from your malt?

Let's take the hypothetical situation of mashing one type of malt using a step-mash process (protein rest and conversion at, let's say, 154 degrees F) and then mashing an equal amount of that same malt using an infusion process (conversion again at 154 degrees F). Will there be any difference in the yield? The answer is no! You will get the same yield and the same spectrum of sugars and nonfermentable dextrins. The only significant difference will be the development of yeast nutrients if undermodified malt is used.

What about the difference between two-row and six-row? Did you think that you get a better yield from two-row? You're right. But not because of the mashing process you used. Physiologically, two-row is generally a plumper grain with a lot less husk than six-row barley; therefore, in a pound-for-pound comparison, there is more good stuff for brewers in one pound of two-row, especially the British varieties.

So why do brewers bother with six-row? Well, for two good reasons. First, you can grow more six-row barley per acre, and second, six-row malted barley has more enzyme potential, important for quicker conversion or adjunct mashing when in excess of 10 percent.

## GRAINS AND ENZYMES

All brewer's malt, often referred to as pale malt or simply malted barley, has enough enzymes for converting its own soluble starches to fermentable sugars and nonfermentable dextrins. In the case of high-enzyme six-row malts, the enzyme potential is adequate for converting up to 30 to 40 percent adjuncts, such as corn, rice, potato, etc. Low-enzyme two-row may convert 10 percent adjuncts, while some high-enzyme two-row will have the ability to convert over 20 percent adjuncts. But as a homebrewer, don't try anything in excess of 10 to 15 percent unless you consider yourself a master brewer. There are too many variables that enter into the brewing process.

Now, what about all those specialty grains? Do they have enzymes? Do they need converting? Following is a table of ingredients that are most often available to homebrewers in America.

### Table 10-1. Enzymes in Grains

| Grain | Enzymes? | Conversion Required? |
|---|---|---|
| Pale malt | yes | yes |
| Munich malt | some | yes |
| Lightly toasted malt (homemade) | no | yes |
| Dextrine malt | no | yes |
| Crystal malt | no | no |
| Chocolate malt | no | no |
| Black Malt | no | no |
| Roasted barley | no | no |

## STOVE-TOP MASHING

Whether you are mashing 1 pound (for a combination malt extract–mash recipe) or 15 pounds for 5 gallons of strong doppelbock, the preceding information should be helpful for those of you contemplating taking the plunge into the wacky, wonderful world of mashing.

One last hint: I have found it possible to get complete conversion with a simple procedure of using my brewpot for my mash tun. I mash in with 1 quart of water for each pound of grain and boost the temperature to 122 degrees F and hold for ½ hour, then by adding boiling water (at a rate of an additional ½ quart of water per 1 pound of grain), raise the temperature to 150 degrees F and hold for 15 minutes. Finally, by turning on the heat and holding the temperature at 158 degrees F, I get complete conversion every time. Two quarts of 170-degree F sparge water per 1 pound of grains mashed is appropriate for most low- and medium-strength brews. While some may argue that this is only one way to mash, I couldn't agree more. For me it is painless. For the beginner it is painless and successful. If you want to vary your spectrum of sugars, then keep brewing and keep learning.

The next step is to get your hands on some good combination malt extract–mash recipes, then on to the guts and glory of all-grain brewing.

Of course, don't forget to relax, not worry, and have a homebrew.

**Decoction Equipment**
GREG NOONAN

# DECOCTION MASHING

## BY GREGORY J. NOONAN

*This article originally appeared in* Zymurgy, *Special 1985 (vol. 8, no. 4)*

For the growing body of whole-grain brewers, there has been a serious lack of information available on the subject of mashing. Largely because of this fact, even dedicated brewers of crushed malt may be unfamiliar with decoction mashing. They may also be unaware of the measurable extract increases and appreciable quality improvement decoction mashing makes possible.

Most homebrewers are at least passingly familiar with the infusion mash process, whereby boiling water is worked into crushed malt until it reaches the saccharification temperature of 149 degrees F. The mash is steeped at this temperature to extract malt starch and convert it to fermentable sugar.

The very word *decoction*, on the other hand, means to extract by boiling. Boiling does solubilize and disperse malt particles, but in actual fact, conversion of the malt to simpler fractions during decoction mashing is made during periods of steeping as well. The traditional decoction mash sequence is composed of four rests, and three "decoctions" of part of the mash. These boiled portions are returned to the mash, progressively elevating its temperature.

Regardless of which mashing method is employed, extraction of the malt is not accomplished by the effects of an elevated temperature alone. A hot, acidic solution only gelatinizes starch and "softens up" protein and hemicellulose. Extraction and fractionalization of the malt can be brought about only by the action of particular enzymes, each of which induces a specific reaction and is only active within a limited temperature range. Saccharification, for example, is by malt diastase, an enzyme group that converts malt starch to sugar.

At a superficial glance, the decoction sequence may seem to be no more than a means by which to raise progressively the temperature of the mash toward saccharification, as an alternative to the application of direct heat to the mash tun, or to the infusion of an unreasonable amount of boiling water.

## COMPLETE CONVERSION

In fact, decoction mashing is constructed upon a series of conditions that complete the conversion of the barley kernel that was begun during malting. The decoction mash rests at 95 and 122 degrees F are not merely incidental to raising the mash temperature toward saccharification, but capitalize on the effects of certain enzymes, enzymes that are never even activated in traditional infusion mashes.

The advantages of decoction mashing begin with doughing in the malt with cold water. It allows the water to evenly permeate the mash, whereas doughing in with boiling water invariably causes some of the malt flour to be encapsulated by paste.

The traditional decoction mash begins with the kneading of a conservative amount of cold water into the crushed malt, followed by an infusion of boiling water, raising the temperature to 95 degrees F. This is the "acid rest," during which the enzyme phytase is activated. Its temperature range is 86 to 128 degrees F. It acidifies the mash by inverting the insoluble malt phosphate phytin to phytic acid (and releases a B vitamin necessary for yeast growth as well). Because it liberates twice the amount of phytic acid released by infusion mashing, a decoction mash needs to rely less upon the brewing water being naturally acidic (sulfate) than does the infusion mash.

$$Ca_5Mg(C_6H_{12}O_{24}P_6 \times 3H_2O)_2$$

Phytin
+
$7H_2O$
Phytase
↓
$C_6H_6[OPO(OH)_2]_6$
Phytic Acid
$C_6H_{12}O_6$
myo-Inositol (B vitamin)
$5CaHPO_4 \times 2H_2O$
Calcium phosphate
↓ (secondary: precipitated) ↓
$MgHPO_4 \times 3H_2O$
Magnesium phosphate
↓ (seondary: precipitated) ↓

## FIGURE 1. PHYTASE ACTIVITY

Treatment with calcium sulfate should not be presumed to be necessary for all brewing water sources. Moreover, it is less necessary when decoction mashing than it is for an infusion mash because the release of phytic acid during the 95-degree F rest enables the pH to drop from 5.5 to 5.8 at mashing in to the 5.2 to 5.3 necessary at saccharification.

Very generally speaking, so long as the pH of the water source does not exceed 7.2, it need not be boiled or treated with an acid salt such as gypsum before use. Where the alkalinity of the brewing water is greater, carbonate mineral salts must be precipitated out of solution, either by boiling the water and decanting it off the sediment of calcium (and magnesium) carbonate, or by the addition of an acid salt such as calcium sulfate (gypsum). Gypsum lowers the pH of any mash because its acidity overcomes the buffering strength of carbonate slats, and because an excess of calcium inverts the organic malt phosphate phytate to its insoluble second form, releasing acidifying hydrogen ions into solution.

$$CaH_4(PO_4)_2 \times H_2O$$
Phytate
+
Calcium (Ca)
+
$H_2O$
↓
2H
Hydrogen ions
+
↓ $2CaHPO_4 \times 2H_2O$ ↓

Calcium phosphate
(secondary: precipitated)

## FIGURE 2. PHYTASE ACTIVITY WITH CALCIUM

This reaction is altogether a less desirable program for acidifying the mash than is the release of phytic acid, because it robs the mash and its extract of phosphorous that may be necessary for yeast growth.

Boiling part of the mash also contributes to lowering the pH of the decoction mash. More important, each time a decoction is withdrawn, that heavy part of the mash passes through the diastatic enzyme range while it is being heated to boiling, so that saccharification temperatures are reached three times in the course of a traditional decoction mash. Boiling ruptures balled starch particles and dissolves protein gum. When the boiled mash is returned to the mash tun, the dissolved protein is exposed to the enzyme activity in the albumin rest. Without benefit of having been boiled, protein gum is little affected by enzyme activity and passes through mashing largely unconverted, making the extract prone to haze and subject to oxidation. Undissolved paste and gum increase the risk of a set mash during sparging.

Boiling also deoxygenates the mash, so that it settles more densely in the lauter tun. Only the absence of residual protein gum makes the denser and more effective filter bed possible. Where an infusion mash has been employed, such a thick filter bed would likely result in a set mash.

Protein is dispersed by boiling, but the resulting complex protein polymers can only be reduced to manageable fractions by the

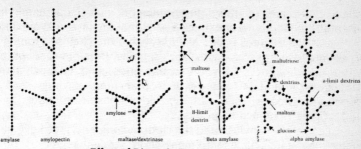

**Effect of Diastatic Enzymes on Starch**
GREG NOONAN

proteolytic enzyme group. This is the purpose of the decoction mash rest at 122 degrees F.

During the rest, protease, peptase, and peptidase progressively dissolve the pepticle links within the protein cells, liberating albuminous fractions. These coagulate with the hop resins in the kettle, providing a vehicle for the transmission of hop bitterness to the finished beer. It is also albumin, not protein, that gives beer its body and enables it to support a frothy, foamy head.

## DISSOLVES INTO AMINO ACIDS

During the course of the albumin rest, proteinase dissolves some of the albuminous matter to individual amino acids, which fuel yeast growth in the early stages of fermentation. The acidifying enzyme phytase remains active through the end of the protein rest, lowering the mash pH further. Extract efficiency is enhanced by the enzymatic dissolution of membranous proteins encasing starch and hemicellulose, and by the effect of debranching enzymes upon amylopectin during this rest.

Finally, after three rests and two decoctions, the traditional continental mashing comes to the saccharification rest. It reaches the diastatic enzyme range better solubilized, with its nitrogen complement more manageable and its starch more accessible than is otherwise possible.

The sole aim of the saccharification rest is to reduce malt starch to sugars. The art of it is in manipulating the mash temperature to yield a given proportion of fermentable to nonfermentable sugars.

Starch is made up of glucose molecules linked in straight and branched chains, respectively termed *amylose* and *amylopectin*. How these chains are separated determines the malt character and fermentability of the mash extract. The alpha-amylase enzyme, most active at 158 degrees F, breaks links within the chains to form glucose, maltose, trisaccharides, and dextrins. Beta-amylase, on the other hand, attacks only the chain ends, and is most active between 140 and 149 degrees F. Debranching enzymes, such as alpha-glucosidase (maltase) and dextrinase, separate amylopectin at its branching points.

Unless alpha-amylase and the debranching enzymes have separated the starch chains into many short, straight segments, beta-amylase hasn't many chains available and operates very ineffectively. The minimum saccharification temperature, therefore, is 149 degrees F, because this is the temperature at which starch becomes gelatinized, and because alpha-amylase is active. It yields the most fermentable wort possible.

Higher temperatures retard beta-amylase activity, generating more dextrinous extracts and consequently more slowly fermentable wort than is achievable at a lower temperature. Saccharification temperatures, then, may range between 149 degrees and 158 degrees F, to promote one enzyme or the other, regardless of which mashing technique is employed.

Again, the decoction mash enjoys an advantage yet, because its low-temperature rests give the heat-susceptible debranching enzymes opportunity to diminish the number and complexity of amylopectin fractions, an accomplishment that resting in the saccharification range alone cannot duplicate.

The decoction mash is the culmination of centuries of trial-and-error and discovery-by-chance brewing. The traditional sequence of rests and decoctions predates even the discovery of the enzymes that make it so effective. The decoction mash was not formulated theoretically; it evolved because it works.

A decoction mash program should be employed whenever "undermodified" malt is to be crushed, and whenever that crushing yields a coarse or uneven grist. Only well-crushed malt of British origin, such as Munton and Fison two-row pale malt, should ever be infusion-mashed.

## IDENTIFYING MALT

Malt is identified as undermodified or well modified only by British convention. The terms of distinction are appropriate to homebrewing nomenclature, however, given our indebtedness to

**Acrospire Growth**
GREG NOONAN

### Table 10-2. Malt Required for Worts of Various Specific Gravity

| Extract efficiency | Pounds of malt per gallon of wort at: | | | |
|---|---|---|---|---|
| | 1.040 | 1.044 | 1.049 | 1.061 |
| 80% | 1.08 | 1.2 | 1.31 | 1.66 |
| 75% | 1.16 | 1.28 | 1.4 | 1.77 |
| 70% | 1.24 | 1.37 | 1.5 | 1.9 |
| 65% | 1.33 | 1.48 | 1.61 | 2.05 |
| 60% | 1.45 | 1.6 | 1.75 | 2.22 |
| 50% | 1.73 | 1.92 | 2.1 | 2.66 |

the British homebrewing tradition. It is well to note that to most of the world's brewers, British malt is overmodified.

Identifying malt as being of either type is not difficult. To assess its modification, taste a sampling of twenty or so kernels. Identify the dorsal and ventral sides: the ventral side is longitudinally creased by the ventral furrow; the dorsal side is more rounded. The acrospire, or plant growth, of the germinated barley kernel lies beneath the dorsal husk. It is a conical, white spear growing from the base of the grain. Its growth is indicative of the extent of modification of the starchy endosperm.

If the malt is a thin-husked variety, the acrospire may be visible through the husk. Otherwise, the husk will have to be broken away above the acrospire, either by rubbing with the thumb or by cutting and lifting free with a razor blade.

Examine the malt. If the acrospire growth exceeds three quarters the length of the kernel in the majority of the grains, it is well modified. Its taste is very sweet with little starchy character. The endosperm will be soft and may even be granular, almost like table sugar.

Decoction mashing is indicated when the acrospire length of a significant number of the kernels is less than three quarters the length of the grains or when sprouting has been very inconsistent. The endosperm of grains that are cut through will have an opaque or "steely" appearance. Chewing a few grains will reveal a more mealy texture and grainy flavor, and very likely hard kernel tips. It is less well modified than British malt so that the endosperm will not have been depleted by fueling rootlet and acrospire growth, and in the interest of preserving enzyme strength, it must be decoction-mashed.

Let's look at the traditional decoction mash, crushing enough malt to yield 5½ gallons of cooled wort at a specific gravity of 1.049. This volume should be adequate for a 5-gallon fermentation, including wort for yeast starters or bottle priming.

Using either two-row or six-row pale malt of American origin, we can safely assume an extract efficiency of 70 percent. Actually, by using two-row malt, 80 percent is possible, and for six-row, 75 percent recovery of the dry weight as extract is possible. Even for the first-time brewer, then, 70 percent extract yield is not an unreasonable expectation.

Assuming a 70 percent extract efficiency, for 5½ gallons of wort at SG 1.049 we require 8¼ pounds of crushed malt. The malt can be conveniently measured by volume; a 1-quart measuring cup filled 8¾ times (or thirty-five 1-cup measures) will give us 8¼ pounds.

Crushing may be accomplished by using a countertop grain mill, set so that a feeler gauge of $25/1,000$ to $35/1,000$ inch thickness (of the type used for setting the spark plug gap on your automobile) just passes between the faces of the grinding disks. This range yields as fine a milling as can be tolerated without overly pulverizing the hulls.

For a mash tun, nothing performs as satisfactorily as an insulated polypropylene picnic chest of at least 24-quart capacity, except perhaps a glazed stoneware crock. The first satisfies the critical feature of the mash tun, the ability to maintain mash temperature, better than anything else. Its only drawback is that the inner surface will become heat-distorted with use. Stoneware doesn't suffer this malady, and holds heat well, but is dangerously subject to fracture if carelessly handled.

## Table 10-3. Water Required per Pound of Malt

|   | Very thick mash | Very thin mash |
|---|---|---|
| Doughing in | .75 quarts | 1.25 quarts |
| Mashing in | .35 | .625 |
| Temp. maintenance Sparging | 2.375 | 1.625 |

Total water requirement remains constant (3.5 quarts per pound of malt) regardless of the consistency of the mash.

To effect a uniform mix, the crushed malt goes into the mash tun before the brewing water. Sprinkle and knead cold water into the grist so that it is evenly and gradually moistened. For a very thick mash, the 8¼ pounds of malt are doughed in with 6 quarts of brewing water. The grain should be thoroughly wetted, but with little or no free-standing liquid left at the bottom of the tun. This is important, because flooding any part of the mash before the water has permeated every part of it destines dry particles to become encapsulated by nearly impenetrable paste.

The doughing in is with cold water, except where the malt is reasonably well modified and evenly crushed. In this case, an abbreviated decoction mashing may be made by doughing in the grist with 5 quarts of boiling water, bringing the mash to rest at 95 to 105 degrees F. Infusion of another 3 quarts of boiling water, raising the temperature of the mash to 122 degrees F, will allow the first decoction to be eliminated. The only profit in this method is expediency, however, and for most undermodified malt is inappropriate.

Usually the rest is made cold, allowing it to stand for 15 minutes or so before being fully mashed in. To raise the temperature to 95 to 105 degrees F for the acid rest, sprinkle and knead three quarts of boiling water into it. Immediately after establishing the strike temperature, measure the acidity of the mash by blotting a strip of narrow-range pH paper (4.6 to 6.2 or 5.2 to 6.8) with a drop of the liquid mash, and gauge it against the pH color scale. The pH should be between 5.5 and 5.8. If it exceeds this range, your brewing water is too alkaline, and requires correction.

It is a little late to discover this fact, but the situation is not unmanageable. It can be corrected by decanting all the free liquid from the mash tun and mixing into it a solution of gypsum dissolved in cold water, until the pH drops to 5.5. Carefully heat the liquid to above 100 degrees F and return it to the grist. Dispersion of the calcium salt will not be perfect, but it is the best that can be hoped for in what at this point is frankly a rescue effort.

For the very particular, where the water source is soft (relatively mineral-free: hardness as $CaCO_2$, less than 100 grams per milliliter, indicated by a pH of 7.0 or less), acidulation of the mash may be augmented by a lactic acid mash made several days before brewing begins.

## MAKING THE MASH

The mash is made by completely saccharifying a pound or so of crushed malt, cooling it to below 131 degrees F, and enclosing it in an insulated 2-quart container with a handful of crushed malt.

The acidulation is by *Lactobacillus delbruekii*, a heat-tolerant bacteria that metabolizes glucose to lactic acid. Normally present as a contaminant of the malt, it acidulates the mash without the harshness associated with sulfate treatment.

Prepare to hold this mash at 95 to 131 degrees F for at least two days, either by periodic infusions of boiling water, or by enclosing it in an insulated, aluminum-foil-lined box heated by a light bulb or other heat source.

There is no guarantee that this mash will not be contaminated by other thermophilic bacteria, but keeping it closely covered will at least discourage infection by airborne bacteria. A distinctly rancid odor is an indication of spoilage caused by the bacterium *Clostridium butyricum*.

## NO LACTIC ACIDS

Should the lactic acid mash be off in any way, it should not be used. Where successful, its pH will drop to 5.0 or below and be of an acidic but pleasant taste and normal appearance. It is used to establish an acceptable mash acidity at the start of the 95-degree F rest.

Whether any treatment of the mash or brewing water has been made or not, after 15 minutes of the 95-degree F rest, prepare the first decoction.

Using a 1-quart glass measuring cup, pull the grist to one side of the mash tun so that the thickest part of the mash can be withdrawn to a kettle of at least 2-gallon capacity. Lift the mash from the tun while pressing the rim of the cup against the side, so that most of the liquid runs free. Deposit this dense mash in the pan, and repeat the procedure until the thickest one-third part, or approximately 3½ quarts, are collected.

The thickest part of the mash is taken for the decoction not because its specific heat is greater than that of an equal volume of the mash liquid, but because it contains the greater part of any poorly solubilized or balled malt starch and protein gum. The more liquid part of the mash will have absorbed all of the soluble malt fractions (which need no further dissolution) and most of the malt enzymes. If this cold extract were to be boiled, the enzyme community would be devastated.

## MASH RETAINS ENZYMES

The heavy part of the mash retains adequate enzyme reserve to ensure sufficient reaction during the heating of the boiler mash. These are made more effective by its density. In fact, in the thick decoction alpha-amylase can more successfully decompose complex starch than it can during the later and thinner saccharification rest. The thickest part of the mash, then, is boiled, and enzyme strength preserved in the cold settlement left in the mash tun.

Although the tradition specifies that one third of the mash make up the decoction, in practice that fraction is subject to the density of the mash from which it is taken. The thinner the mash, the proportionally greater the volume of the decoction will have to be in order for it to affect the temperature of the next mash rest.

For our thick mash, however, the decoction need only be one third, and should be very thick as well. Any liquid standing above the settled decoction can be drained back into the tun.

Cover the mash tun and maintain its temperature at 95 to 105 degrees F. If necessary, restore its temperature by exchanging a part of it with some of the hotter boiled mash.

## APPLY HEAT

Heat the decoction to 155 degrees F, in 10 to 15 minutes, stirring and lifting the mash constantly so that it doesn't scorch on the bottom of the kettle. In the thick mash, 155 degrees F will produce the most rapid reduction of complex starch chains by the alpha-amylase enzyme, which at above pH 5.5 is operating at maximum efficiency. Better reduction of malt starch to small dextrins can be

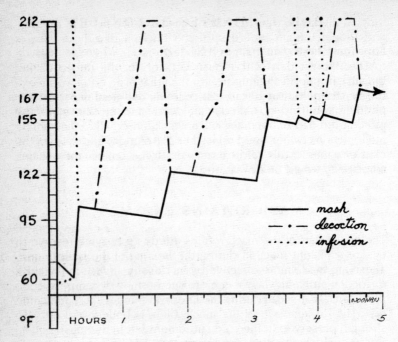

**Mash Decoction Infusion Graph**
GREG NOONAN

expected now than later, when the whole mash is brought into the diastatic enzyme range at a lower and less optimal pH.

The higher pH and mash-temperature levels also retard the activity of beta-amylase (pH optimum 4.7). Thus we are able to focus on the simplification of complex starch in decoction, so that later, during the saccharification rest of the whole mash, we can concentrate on controlling the maltose-to-dextrin ratio of the extract.

The decoction is held at the dextrinification temperature for 10 minutes, heated to 167 degrees F over the next 15, and then boiled. As the mash temperature rises above 170 degrees F, starch particles expanded by the heat will burst, exposing otherwise lost extract potential.

Diligently scrape the mash free from the bottom of the kettle. You cannot rest now—stir demoniacally. Stirring becomes less critical only when the mash comes to a boil.

## BOIL 15 MINUTES OR LONGER

For a mash from pale malt, boil for 15 minutes. Where using dark-roasted, very old, or enzyme-poor malt, the boil must be held longer, even up to 45 minutes.

Use the measuring cup to return the decoction to the mash tun. Mix them gradually and thoroughly so that the temperature dispersal will be even. The temperature will stabilize at 122 to 125 degrees F.

Generally 20 minutes' rest at this temperature is sufficient to break down most of the complex protein and yield satisfactory albumin-to-amino acid ratio. The heavy part of the mash can then be withdrawn again, for the second decoction.

For the second decoction, more than one third of the mash should be taken, so that upon return to the tun, it will raise the temperature to the desired point in the diastatic enzyme range. If the decoction were unable to raise the temperature of the mash to above 152 degrees F, a lager beer from it would be disappointingly thin. For our mash, a decoction of 4 to 4½ quarts should ensure that the proper saccharification temperature will be reached. Again, the decoction should be very thick, and any liquid that pools above it is ladled back into the mash tun.

Cover the mash tun and heat the decoction to 150 degrees F in 10 minutes, through 167 degrees F in 15 to 20 minutes, and then to boiling. Boil for 20 minutes.

## METHOD DEPENDS ON BEER TYPE

How the second decoction should be returned to the main mash depends on the type of beer being brewed. For light, dry beer and any beer to be bottled in less than one month's time, the decoction is remixed with the main mash slowly, over a period of 15 to 30 minutes, and finally allowed to rest at 149 to 151 degrees F. This pattern is mimicked by a temperature-programmed (upward infusion) mash. It gives the beta-amylase ample opportunity to hydrolize maltose, by eliminating most of the amylose liberated by alpha-amylase activity during the two decoctions.

The low strike temperature preserves beta-amylase activity, but still allows more complex starch to be dismantled by alpha amylase. Given a thick mash, at this temperature the enzymes will remain strongly active for up to 2 hours.

For lager beers this maltose-rich extract is inappropriate. Aging requires a slowly fermentable extract, predominantly composed of dextrinous sugars. Unlike maltose and nonsaccharides, these require extracellular enzyme reduction before the yeast can metabolize them, thereby providing a substrate for sustained fermentation.

A saccharification strike temperature of 152 to 155 degrees F must be reached quickly to discourage beta-amylase activity. The remixing of the mashes should be accomplished within 10 minutes, and the temperature dispersal must be absolutely uniform. Use the measuring cup to mix the mash. Temperature uniformity will be more readily achieved than by using a spoon.

When the strike temperature has been established, cover the resting mash. After 15 minutes, and every 10 minutes thereafter, uncover and remix the mash. Check its temperature, and if necessary restore it by infusions of boiling water. Whereas temperature stability was not a serious problem during the earlier mash rests, at the elevated temperature of the diastatic enzyme rest, heat loss is more pronounced. A temperature drop of even 2 degrees can have irreversible consequences. Happily, we have made a very thick mash, and can make liberal infusions of boiling water without jeopardizing enzyme activity. Although a dense mash favors overall extract efficiency, the thinner mash improves its fermentability. Up to 2 gallons of brewing water can be infused, by degrees, without destabilizing enzyme viability, and is more than sufficient for even a protracted saccharification rest if the mash tun is insulated.

## MONITOR STARCH CONVERSION

Starch conversion should be monitored every time the mash is uncovered. Before stirring up the mash, run a little of the mash liquid onto a white porcelain saucer kept especially for this purpose. Onto the extract squeeze several drops of iodine. Immediately check the color at the interface of the two substances. Native starch and amylose form an intense blue-black reaction with iodine. Amylopectin gives a red color, erythrodextrin and amylodextrin a faint violet to reddish color. Achrodextrins, oligosaccharides, and sugars give no color reaction from the yellow of the iodine.

Read the color only at the interface of the liquids. Where the iodine reaches husk particles in the sample, it will always turn blue-black. It is of no concern. However, starch particles that give a color reaction indicate that amylolysis should be continued.

Only when iodine testing gives no color change should the final decoction be made.

Unlike the first two decoctions, the thinnest part of the mash is boiled for the lauter decoction to destroy vestigal enzymes, but also because at this point we don't want to risk releasing any unconverted starch into solution. Any ungelatinized starch must remain with the thickest part of the mash, where residual alpha-amylase is strongest and can continue a subdued polysaccharide reduction while the lauter decoction is being processed. At least one third of the mash volume should be taken for the decoction, and up to one half of it where it has been made thinner during saccharification. Again, it is advisable to judge liberally the amount to be boiled, so that the 167 to 170 degree F strike temperature of the final mash rest is sure to be achieved.

Assuming that 1 gallon of boiling water has been infused into the mash to maintain saccharification temperature, 50 percent, or about 7 quarts, of the mash liquid would be removed for the decoction. Bring it to a boil and hold it for 15 minutes. Return to the main mash so that strike temperature is achieved precisely and uniformly. Coming to rest at much above 170 degrees F risks dispersion of residual malt gum, which no surviving enzymes are capable of breaking down. Causing starch to be carried into the runoff can only encourage oxidation in the wort. On the other hand, at below 167 degrees F sugars will not run freely, and the expansion of starch particles is insufficient for them to remain suspended while the later tun filter bed forms.

## ROUSE THE MASH

The strike temperature having been established, vigorously rouse the mash for 10 or 15 minutes, so as much insoluble matter as possible will be forced into temporary suspension. This is essential if the mash is to form an efficient filter bed.

It is advisable to transfer the mash to a separate lauter tun for filtering so that the space between the bottom of the mash and the bottom of the tun will not be filled with sediment, and because our rectangular cooler does not permit the even percolation of extract from the mash. Sparging is more successful where the mash is contained in a cylindrical tun.

That no manufacturer offers a lauter tun for grain brewing presents the greatest obstacle facing the would-be grain brewer. An insulated, cylindrical unit complete with a tight-fitted, slotted false bottom, a spigot, and sparging and flushing apparatus could be inexpensively produced and is direly needed, but none is available. The grain brewer must make do with whatever can be pieced together. Assembling a satisfactory approximation is imperative. The difference between being equipped to manage sparging properly and resorting to sparging dry grains or repeatedly flooding and then draining the mash is the difference between our 70 percent extraction and the 50 percent efficiency experienced by many grain brewers.

The best possible arrangement may be a 5-gallon polypropylene pail fitted with a spigot at its base (see illustration on page 283). The false bottom can be approximated by a perforated insert, such as a vegetable steamer. Hang a large, fine-mesh filter bag inside the pail to contain the mash. A rigid plastic tube connected to the sparge-water container by flexible tubing is used for flushing or underletting the mash. Disconnected from it, the flexible tube is used for sparging.

There are two drawbacks to this arrangement. The bag does just that—it bags at the bottom—and the pail does not hold heat well. A stoneware crock would perform better, but crocks with spigots are uncommon. If you are using a plastic pail, line a box with blocks cut from foam-insulated board (available from any home improvement outlet) and set the pail into it to reduce heat loss.

Making do with what is available, fill the bottom of the lauter tun to ½ inch above the false bottom with boiling water before pouring in the well-roused mash. Whip the mash with a final vigorous stirring, cover it, and allow it to settle undisturbed.

The object of sparging is to leach the mash extract from the particulate matter by diffusing it into water percolated down through it, and to strain suspended starch and protein gum from the extract as it runs through the husk fragments settled at the bottom of the filter bed.

## DENSE FILTER BED

The filter bed will be fairly well settled after 20 to 30 minutes' rest. It is denser than one made by an infusion mash, and filters the runoff more effectively. The filter bed should be measured between 12 and 16 inches thick. In the clear, black liquid displaced above it, a gelatinous cloud of protein trub will float. This must be settled by slowly drawing excess liquid from the tun. Open the spigot at the bottom of the lauter tun until a steady trickle just forms, collecting the runoff in a saucepan.

In the meantime, begin heating water for sparging. We require 5 gallons, less the amount of water used to maintain saccharification temperature. If this was 1 gallon, 4 gallons will be needed for sparging. If the water for mashing in required treatment, so will the water for sparging, so that as the runoff is diluted, the pH does not rise above 6.0.

Expect it to take 30 minutes to settle the filter bed. When the liquid displaced above it lies only ⅛ to ½ inch deep, close the lauter tun spigot. Smooth the settled protein sludge with the back of a spoon, leveling it and filling in any cracks in the surface. The sediment is lighter and more porous than the gummy sludge precipitated by an infusion mash, which is principally why the denser decoction filter bed does not present a correspondingly increased risk of set mash.

Thrust a thermometer into the filter bed to monitor mash temperature during sparging. Fill the sparge-water reservoir and connect the inlet tube. Manipulating both the reservoir and lauter tun spigots at once, introduce a surge of sparging water through the space below the false bottom, to flush away mash particles that might cloud the runoff or even obstruct it. The inlet and runoff rates must be evenly matched, so the liquid level above the mash is not changed. Close both spigots as soon as the runoff clears.

## HEAT ALL RUNOFF

Heat all of the cloudy runoff to 170 to 175 degrees F and return it to the lauter tun. Open the spigot until a steady trickle of extract just emerges.

When the liquid level drops again to just above the surface of the filter bed, attach a piece of flexible tubing to the lauter tun spigot, form it into a gooseneck to control the depth of the liquid, and open the sparge-water tap. Ideally its temperature should be between 170 and 175 degrees F, but where the temperature in the lauter tun has fallen below 167 degrees F, hotter water might be used. Expect the filter bed temperature to drop. It is nearly impossible to maintain lauter mash temperature at 167 to 170 degrees F. Nonetheless, every effort should be made to do so.

As sparging proceeds, the runoff rate can be increased beyond the trickle initially established, but keep in mind that the more slowly the extract runs off, the better the extract efficiency will be. Expect it to take between 1 and 2 hours to collect all the runoff. In the meantime, prepare and begin to boil this sweet wort. Cease sparging when runoff pales to 1.010 or so.

Don't be put off by the complexity of the chemistry involved in the decoction mash. Remember that for centuries brewers using simple equipment have successfully decoction-mashed without even a glimmer of understanding of why it worked so well. The mash is more time-consuming but not more difficult than any other kind, and the results are well worth the labor.

### 1.049 Lager

- 8¼ lbs. pale malt
- 7¼ gal. brewing water
- 2½ oz. Cascade hops
- 1¼ oz. Saaz or Tettnanger hops
- lager yeast
- gelatin (optional)

1. Dough in the crushed malt with 6 quarts cold water.
2. After 15 minutes, mash in to 100 degrees F with 3 quarts boiling water.
3. Withdraw densest 3½ quarts of mash to kettle after 15 minutes and heat to 155 degrees F in 10 to 15 minutes. Hold for 10 minutes.
4. Heat to 167 degrees F in 15 minutes and then to boiling.
5. Boil for 20 minutes.
6. Return decoction to mash tun in 5 to 10 minutes; strike temperature stabilizes at 152 to 155 degrees F.
7. Add boiling water to maintain temperature until iodine starch test is negative (up to 1 hour).
8. Remove thinnest 5 to 7 quarts of mash. Heat to boiling; boil for 15 minutes.
9. Return to mash tun; lauter-rest strike temperature is 170 degrees F.
10. Rouse for 15 minutes.
11. Transfer to lauter tun, rest there until settled (20 to 30 minutes).

12. Sparge with 5 gallons of 175-degree F water.
13. Add Cascade hops to runoff and heat to boiling.
14. Boil past hot break (1½ hours).
15. Add finishing hops (Saaz or Tettnanger), boil 15 minutes more, remove from heat, cover, and allow to settle.
16. Filter wort through hops into sterile carboy; collect 5 gallons of wort.
17. Collect excess (2 quarts) in sterilized bottles; cap and refrigerate.
18. Cool wort to 50 degrees F, rack into primary fermenter, and pitch with yeast starter.
19. Transfer to carboy after 5 to 8 days (SG 1.022) and cool to below 39 degrees F.
20. After two weeks (SG 1.012), transfer to second carboy for lagering.
21. Store at 33 to 36 degrees F for 2 months. Specific gravity should be 1.002 to 1.004.
22. Fine with vegetable gelatin to clear, if necessary.
23. Pitch 2 quarts of wort for bottle priming.
24. Siphon into twenty-two 1-quart bottles.
25. After 3 days at 50 degrees F, gradually cool to 35 to 40 degrees F.
26. Allow four weeks for bottle conditioning before consuming.

# THE DETRIMENTS OF HOT-SIDE AERATION

### by George Fix

*This article originally appeared in* Zymurgy, Winter 1992 (vol. 15, no. 5)

The next time you visit Colorado and have time to tour the Coors Brewery in Golden, ask if you can see the pilot brewing system. The 50-barrel system has a stainless steel tube attached to the inlet near the top and extending to the bottom. The tube is not part of the original equipment; it was added to dramatically alter the hot wort processing. Before modification, the wort came splashing into the kettle; now they get a smooth fill that significantly reduces oxidation of the hot wort. Before pitching the yeast, it is necessary to oxygenate wort to make sure the yeast goes through the respiratory cycle. In the new Coors system with the kettle modification, wort oxygenation is done on the cold side, i.e., after wort chilling.

What is the relevance of procedures used by megabrewers for small-scale brewers? And what harm is done by hot-side oxidation either in wort transfer or in other parts of wort production? First, because oxidation is strongly affected by surface-area-to-volume ratios, something that increases with decreasing brew size, the effects are more relevant to small-scale brewing than to larger systems. Second, the materials that get oxidized on the hot side are derived from malt. Thus the bigger the beer, the more relevant the effects. Third, oxidation rates increase exponentially with temperature, so the oxygen picked up on the hot side will be quickly bound up with malt constituents and not be available to yeast.

Finally, the materials oxidized on the hot side will be passed on to the finished beer, where they will play a role in staling. Their effects are somewhat complicated and different from the formation of papery-cardboard tones created by head-space air in beer bottles. Some brewers may find this a highly relevant issue, while others may find the effects are marginal. In small-scale commercial brewing, hot-side aeration is a "hot" subject. Just about all fabricators of 5- to 20-barrel brewhouses are taking great pains to engineer "mild mashing systems" where hot-side aeration is kept to an absolute minimum. In these contexts the hot-side aeration issues have proven to be very relevant.

The abbreviation HSA will be used for the term *hot-side aeration* throughout. For those who disapprove, I can only offer condolences. After many years of fighting such things, I have given in to the jaded and cynical view that it was a lost cause!

It should be noted that HSA is one of several issues associated with beer stability.

## A SURVEY OF THE QUALITATIVE THEORY

The wort constituents most relevant to the HSA issue are a class of pigments called *melanoidins*. These compounds are formed by amino acid–carbohydrate reactions induced by heat. This is a special case of a family of reaction systems that are generally called browning or Maillard reactions.

The place where Maillard reactions take place first is during malting. The production of color malts (amber to dark) would not be possible without these reactions. This is the main reason color malts are rich in melanoidins. Even though the melanoidin content of pale malts is lower, they will be present at sufficient levels. Therefore the transformation systems described below—the heart of the hot-side issue—are relevant to worts produced exclusively from pale malts as well as for worts of amber and dark beers.

The other place where Maillard reactions are relevant is in the kettle boil. Here the process is remarkably efficient in the sense that very little of the wort carbohydrates and amino acids are used, yet there is typically a nontrivial increase in color. This in turn leads to a nontrivial increase in the wort melanoidin level. For example, it can happen that wort color measured in degrees Lovibond (SRM) doubles during the boil with only a 1 percent loss in extract. The actual extract loss can be measured by hydrometer, once the volume reduction in the boil is taken into account. The actual change in color can be measured using the color curve chart, Table 10-4.

Oxidation-reduction reactions (called *redox reactions*) occur throughout the brewing process. The redox reactions that occur during wort production are just as important as those that occur later. Oxidation is an electron transfer system. A compound oxidizes by giving up electrons, and conversely a compound is reduced when it takes on electrons. A compound can be oxidized if and only if another compound is reduced.

The heart of the HSA issue deals with the fate of the melanoidins in the mash, sparge, kettle boil, and wort cooling. They will all start out in their reduced state, and will remain that way unless the hot wort is abused. Oxidation of the hot wort will promote oxidation of the melanoidins. Temperature is very important because oxidation rates increase exponentially with temperature. For example, if one were to put 12 ounces of wort at 68 degrees F (20 degrees C) into a beer bottle and inject up to three milliliters of air, then several days would pass before oxidation reactions would be detectable even at low levels. Repeat this experiment with wort at 140 to 212 degrees F (60 to 100 degrees C), and reactions would be detectable in minutes if not seconds.

Melanoidins, either oxidized or reduced, will pass through the fermentation more or less unmodified. Reduced melanoidins are favorable and oxidized melanoidins are unfavorable in beer production. For example, in packaged beer the oxidized melanoidins can play the role of oxidizers by reacting with alcohols and producing staling aldehydes. The latter have an astringent character that

## Table 10-4. Intensity/Dilution Chart

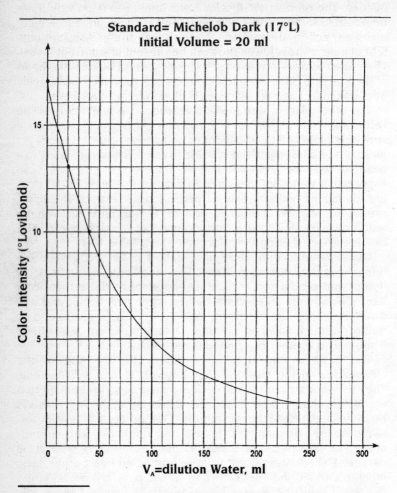

Chart reprinted with permission from *Oktoberfest, Vienna, Marzen*, Brewers Publications, 1992

sometimes takes on a metallic tone. De Clerck once characterized this process as "oxidation without molecular oxygen" because it can occur even if there is little, or no, air in the bottle's head space.

Conversely, melanoidins in the reduced state can act as flavor protectors in packaged beer, which is why they are called *reductones*. They do this by reacting with oxygen dissolved in beer from head-space air, which prevents the oxidation of alcohols and other beer constituents. This in turn delays the emergence of stale flavors. Amber and dark beers are rich in melanoidins, a point that explains why some of these beers have such remarkable flavor stability. It also explains why some amber and dark beers literally fall apart when packaged and subjected to even mild abuse. The key to this is proper wort-production procedures, and in particular the avoidance of HSA.

Phenols, sometimes called *tannins*, are another wort constituent of interest. For the most part they are extracted from malt husks during the sparge. They are somewhat similar to melanoidins in the sense that the phenols in their reduced form act as reductones and flavor protectors. In fact, the presence of selected phenolic compounds in their reduced state is thought to be responsible for the freshness of flavor in many foods, including beer. Oxidized phenols, on the other hand, bring their own flavoring to beer. The German word *herbstoffe* (which roughly translates to "grain astringent") is often used to describe their effect. Many regard these flavors as being as unpleasant as the staling aldehydes that arise from the oxidation of alcohols.

There are highly complex redox, oxidation-reduction reactions involving hops in the kettle boil. These reactions are poorly understood, and some might be beneficial to beer flavoring. There is strong evidence that these are completely unrelated to the issues discussed in this article. Indeed, because boiling expels oxygen, the kettle boil may be a benign period for HSA.

Once beer is packaged, the redox reactions will start, and the higher the storage temperature, the faster the effects will become apparent. The net result is reminiscent of a scene from a Marx Brothers movie, wherein characters appear and disappear in a state of chaos. Head-space air reacts with alcohols, hop constituents, and fatty acids derived from wort to create staling products like trans-2-nonenal. Oxidized melanoidins and tannins will do the same, creating other types of staling compounds. Countering this is melanoidins in their reduced state and other reductones that initially block staling. They will ultimately join the fray once they get oxidized. As a consequence, the forces promoting staling will always win. Natural organic foods like beer will never have an infinite shelf life. However, by paying attention to appropriate details, a respectable period of flavor stability can be achieved. In the end this is all we can expect.

## SOME PRACTICAL TIPS ON REDUCING HOT-SIDE AERATION

Perhaps the most important thing to avoid is excess splashing of wort during hot-side transfer, but exactly where the hot side ends and the cold side begins is somewhat arbitrary. For purposes of discussion we will make it at 86 degrees F (30 degrees C). I first became interested in the splashing problem when I had difficulties with transfer of the sparge to the brewkettle. This was achieved at the time by means of an ill-conceived pump. At both the start and the end of transfer there was considerable frothing and foaming. Using procedures I will describe, I concluded the net effect of pumping was identical to what would have occurred if the mash were literally dumped into the kettle. There are pumps capable of a smooth laminar transfer, but this one was not. Moreover, serious flavor problems developed after the pump was installed. Removal of the pump and introduction of "mild" transfer techniques, in this case a gravity-based system, removed the defects. The results were dramatic, particularly in the stability of fresh beer flavor. Consequently, anyone wanting to do the obvious experiments can readily verify the relevance of these issues.

Many use a pump to pull wort through a heat exchanger or counterflow chiller. This is fine even with a "frothing" type of pump described above. It is important to keep the pump on the cold side of the transfer. Any oxygen dissolved there will hang around as an inert gas until the yeast use it in a highly constructive manner during their aerobic respiration.

The next area to watch is the way the mash is stirred. In fact, as far as oxidation issues are concerned, the ideal mash is a single-step infusion where there is no stirring. This is not to say that multiple-step infusion or decoction mashing are necessarily going to lead to unacceptable levels of HSA. It is, however, important to minimize abusive procedures including excessive stirring of the mash.

There is a point of historical interest here. In the 1950s many small and regional breweries purchased new brewhouse equipment as a part of the post–World War II economic expansion. The brewhouse technology at that time for regionals featured raking systems that literally hurled grains to and fro in the mash tun. This was a period when many of the regionals started to go under. The conventional wisdom at the time held that the national beers were more appealing to customers because they were less bitter. The implication was that the regionals were overhopped for popular taste. C. Prechtl, in a very perceptive article "Some Practical

Observations Concerning Grain Bitterness" (*Master Brewers Association of the Americas Technical Quarterly*, vol. 4, no. 1, 1964), made the point that brewhouse procedures and not hop levels were responsible for the unpleasant bitterness found in many regionals. He focused in on excess stirring (the raking systems) and excessive sparging, and from the perspective of modern brewing science, he got it exactly right in these areas. Unfortunately, this article was largely ignored, unfortunate not so much for Prechtl as for the regional breweries that are no longer with us.

Direct measurement has shown that sprinkler-type sparge water systems do not induce much HSA. However, beers that have been oversparged (when the volume of sparge water exceeds 1.5 times the amount of mash water) invariably have elevated phenols, and a large proportion are in the oxidized state. In fact, some of the world's smoothest beers have not been sparged at all.

The best advice is for brewers to use common sense about such matters—there is no need to become hyper about HSA. Perhaps the best posture is to think "gentle" and "mild" as one brews.

## MEASUREMENT OF HOT-SIDE AERATION

A brewer can evaluate the extent to which HSA has occurred in a particular brewing situation through a test brew and through direct measurement.

The test brew is relatively straightforward. Take your favorite formulation, use your standard procedures, but omit the sparge. It is desirable to keep the brew length (size) and starting gravities the same, so you will need to use more grains, roughly a factor of four thirds more. The following is a specific example I have used on several occasions. The brew length is 50 liters (13.3 gal.), but any other size could be used as well. For example, if brewing 25 liters (6.7 gal.), simply divide the appropriate numbers by two.

**Control Batch**

| | |
|---|---|
| Brew size | = 13.3 gallons (50 l.) |
| Grain bill | = 22 pounds (10 kg.) pale malt |
| Mash water | = 8.5 gallons (32 l.) |
| Sparge water | = 8.5 gallons (32 l.) |
| Volume at the start of boil | = 14.8 gallons (56 l.) |
| Starting gravity | = 1.048 (12 °Plato) |

### Experimental Batch

| | |
|---|---|
| Brew size | = 13.3 gallons (50 l.) |
| Grain bill | = 30 pounds (13.3 kg.) pale malt |
| Mash water | = 11.5 gallons (44 l.) |
| Water directly added to kettle | = 5 gallons (20 l.) |
| Volume at the start of boil | = 14.8 gallons (56 l.) |
| Starting gravity | = 1.048 (12 °Plato) |

Note the mash thickness is just about the same in both batches. In the experimental batch the extra water not used in the mash is added directly to the kettle.

The phenol level in the experimental batch will be 10 to 100 times lower than the control batch, so there will be significantly less material around in the former to be oxidized. This test does not touch on the oxidation of other reductones like melanoidins, but the assumption is that problems with phenols will always be accompanied with other types of HSA problems.

The experimental batch invariably will be a smoother brew, but ideally not by much. This is a subjective area in which brewers may disagree. However, if the differences are unacceptably large, then the next step is to find out exactly where HSA is occurring. For this, direct measurement is needed.

One may possibly be familiar with Zahm and Nagel air testers or dissolved oxygen (DO) meters. These are excellent for measuring the amount of head-space air in beer bottles, and/or the oxygen dissolved in beer during cold storage. Such instruments are, however, worthless for hot-side work because any oxygen present will disappear (binding up with wort constituents) before it can be measured. These instruments measure the potential for possible redox reactions. What is needed is a measurement of the amount of oxidation that has actually occurred.

A classical procedure well suited to small-scale brewing is the indicator time test (ITT). The theory is simple. One takes a solution in an appropriately oxidized state that will change colors when it undergoes redox reactions. When it is added to a highly reduced solution (for example, a 4 percent ethanol-deaerated water mixture), the decoloration will take place in seconds. On the other hand, when it is added to a solution in a higher oxidation state (for example, an ethanol–acetic acid–water mixture), the decoloration will take much longer. Beer wort is a good deal more complicated and will contain diverse constituents in a number of different oxidation states. Nevertheless, ITT measurements will give a bulk measure of the extent to which HSA has occurred. Moreover, by taking measurements at selected points in the wort production cycle, we can bracket locations where major increases occur.

The following is the detailed procedure. It is a modification of one recommended by De Clerck (see References). The reactant is a simple salt/dye called 2,6-dichlorophenolindophenol. Rather than do that one again, we'll simply call it DCI! It is available from Kodak and other sources.

## PROCEDURE

1. Buffer solution: Add an acid to distilled water until the pH of the wort to be tested is obtained.

2. Comparison solution: Dissolve 0.25 milliliters of DCI in 10 milliliters of distilled water. Increase the volume to 50 milliliters with distilled water. Add 10 milliliters of this solution to a test tube and place upright in a rack.

3. Wort solution: Dissolve 0.25 milliliters of DCI in 10 milliliters of room-temperature wort in a test tube and put in the rack next to the comparison solution. Start measuring time from the moment the DCI is added. Stop when the wort solution has changed from a reddish violet color to that of the comparison solution. (This amounts to 80 percent decoloration of the wort solution.)

The following are general criteria:

| Time for Decoloration | Comments |
|---|---|
| < 100 seconds | great! |
| 100 to 500 seconds | good |
| 500 to 1,000 seconds | poor |
| > 1,000 seconds | unacceptable |

It is a good idea to try this procedure first on simple mixtures to get a feel for things. An ethanol-deaerated water solution should decolor rapidly. A water-vinegar solution may not ever fully decolor.

With these two tests and an understanding of the detriments of HSA, you can modify your brewing procedures and improve your beer.

## REFERENCES

Fix, George, *Principles of Brewing Science*, Brewers Publications, 1989.
Fix, George and Laurie, *Vienna, Märzen, Oktoberfest*, Brewers Publications, 1992.
De Clerck, Jean, *Textbook of Brewing*, vol. 2, Chapman-Hall, 1957, pp. 380–81.
Master Brewers Association of the Americas, *The Practical Brewer*, 1977.

*A native Texan, George Fix lives with his wife, Laurie, in Arlington. He earned a doctorate at Harvard University and has been on the faculties of Harvard, Michigan, and Carnegie-Mellon. He is Chairman of the Mathematics Department at the University of Texas at Arlington and is the Senior Consultant for Brewers Research and Development Company. Fix has won sixty brewing awards, including two Best of Shows in AHA- and HWBTA-sanctioned competitions.*

# pH AND THE BREWING PROCESS

## BY ERIC WARNER

This article originally appeared in Zymurgy, Spring 1993 (vol. 16, no. 1)

The pH of brewing water, mash, wort, and beer is one of the most indicative parameters any brewer can use to monitor the progress of the brewing process, and to determine the quality of the raw materials. Measuring the pH throughout the life of a brew can lend insight into what may have gone wrong if the brew doesn't taste right. Best of all, measuring pH can be done very simply and at minimal expense.

Before going further, it would be wise to revisit the good ol' high school chemistry lecture. I know all of you were paying close attention to your teachers, but just as a refresher, pH is a measure of the acidity of a solution. Specifically, it is a measure of the hydrogen ion concentration of a solution. Pure water dissociates into hydrogen $(H^+)$ and hydroxyl $(OH^-)$ ions. pH is actually an expression of $\log_{10} 1/(H^+)$, where $(H^+)$ represents the concentration of hydrogen ions. Since the concentration of hydrogen ions in pure water is $10^{-7}$ mole/dm$^{-3}$, the pH of pure water is 7. Solutions that have a greater concentration of hydrogen ions than this are acidic, and have a pH that is less than 7. Conversely, solutions having more dilute concentration of hydrogen ions are basic and have a pH that is greater than 7. The range of the pH scale is 1 to 14.

Brewers are interested primarily in the pH range of 4 to 7, but the brew should have a very specific pH at each step of the brewing process. Because beer is about 90 percent water, it probably makes sense to look first at the ideal pH of brewing water. A normal municipal water supply should yield a water that has a pH of about 7. To the brewer, the pH of a given water is not nearly as critical as the spectrum of ions within that water. Certain ions have an effect on the acidity of the mash, in particular the hydrogen carbonates and the alkaline earth metals, calcium and magnesium. Malt reacts acidic in the mash, and these ions affect the overall acidity of the mash. The carbonates increase the pH of the mash and the alkaline earth metals decrease it, so the ratio of these ions is important.

This all ties in to what is called the *hardness* of the water. Water hardness is defined by the content of alkaline earth metals, calcium, magnesium, strontium, and barium. Because the latter two are rarely present in significant amounts, hardness mainly refers to the concentration of calcium and magnesium in a certain water supply. Carbonate hardness, or what was once referred to as *temporary hardness*, corresponds to the portion of alkaline earth metals that is equivalent to the hydrogen carbonates of water. Hardness is usually greater than carbonate hardness, and the difference is called *noncarbonate hardness*. What a brewer is ultimately looking for is a ratio of noncarbonate to carbonate hardness that will result in an optimal mash pH. If this ratio is not ideal, then modern brewers will seek to *increase* this ratio by either reducing the hydrogen carbonates, increasing the calcium ion concentration by using calcium salts, or by employing some combination of the two.

The brewing water isn't the only raw material that will affect the mash pH. The malt selection for a given brew can also greatly influence the mash pH. In general, darker malts will reduce the mash pH because the high level of melanoids developed during the higher-temperature kilning of a darker malt react very acidic. On the other hand, malts from heavily fertilized barley crops will increase the pH of the mash. The ongoing overfertilization of agricultural soils has the effect of increasing the pH of the soil, which in turn decreases the acidity of the barley and ultimately the malt.

So what is the ideal mash pH? Ideal is often difficult to achieve in small-scale brewing, but if the mash pH is between 5.4 and 5.5, you are in good shape. This range is ideal for the spectrum of enzymatic reactions that take place in the mash. The amylases have pH optima that are just above the ideal mash pH (alpha-amylase, 5.6 to 5.8; beta-amylase, 5.4 to 5.6), and the key peptidases have pH optima of around 5.0 to 5.2. It is rare for a mash to have a pH below 5.3, but it is conceivable for the mash pH to be well above 5.5. Mash pH higher than 5.5 will cause:

- poor saccharification
- increased mash viscosity
- reduced protein breakdown
- darker than desired color in paler beers
- harsh bitterness in the beer
- sluggish fermentation
- poor foam retention

The ideal wort pH, about 5.2, is ideal for protein coagulation during wort boiling. Alpha acids are most soluble at a higher pH of 5.9, but the tradeoff of having a wort pH at this level isn't worth the increase in the yield of bitter substances, even for commercial breweries. If the wort pH is closer to 5.5, this isn't a big problem for ales, as top-fermenting yeasts are able to overcome a higher-wort pH better than lager yeasts.

In young beer, the pH of an ale should drop to below 4.4 within a few days of pitching the yeast, and final pH of a top-fermented beer should be between 4.0 and 4.4. With lagers the decrease in pH is more gradual, but by the end of primary fermentation the pH should be between 4.4 and 4.6, and the pH of a finished lager beer is usually between 4.3 and 4.6. If the pH of an ale or a lager is below its respective pH norm, then there is a good possibility that the beer has been infected by some form of lactic acid bacteria. This is usually evident by the taste of the beer. Obviously, with Berliner Weissbier or lambics, the pH should be lower than 4.0 and can be as low as 3.0. If the pH of an ale or lager is higher than its respective norm, then there is a good chance that the beer is old or overaged. Particularly if the pH is being continuously monitored, an increase in beer pH is a sign that autolysis is beginning to take place and that the yeast are beginning to excrete amino acids that react basic. Of course, if the wort pH was above 5.5 to begin with, then it becomes less likely that the beer will be in its ideal pH range, particularly if the beer is a lager.

The simplest and least expensive way to measure pH is with litmus paper. I use paper that has a range of 4.0 to 7.0. With this I can determine the pH of my water, mash, wort, and beer within 0.1 to 0.2. For those who desire greater accuracy, pH meters can be purchased from lab equipment suppliers for $150 or more. The accuracy of such a unit is 0.01. Some of these units are very handy and easy to use, though most must be calibrated with a buffer solution prior to use.

Accurate pH measurement during the brewing process can keep your beer on the right track to avoid sluggish fermentation, harsh bitterness, poor foam retention, and other undesirable outcomes. Adjusting brewing water pH is easy enough and keeps your brew true to style.

*Eric Warner is President and Brewmaster at Tabernash Brewing Company, Denver, Colorado. He is a certified brewmaster, having obtained his brewing diploma from the Technical University of Munich at Weihenstephan. He has been published in* Zymurgy *and* The New Brewer *and is the author of the Classic Beer Style Series book from Brewers Publications,* German Wheat Beer.

# eleven

# Beer Evaluation

## THE SENSORY ASPECTS OF ZYMOLOGICAL EVALUATION

BY DAVID W. EBY, PH.D.

*This article originally appeared in* Zymurgy, *Winter 1992 (vol. 15, no. 5)*

This spring the American Homebrewers Association hosted one of the largest amateur beer-brewing competitions in the world. The nearly 2,400 entries came from all over the United States and from Canada, the Virgin Islands, Australia, Japan, and Sweden. Using this annual competition as a gauge, the interest in having homebrew evaluated has increased dramatically over the last several years.

Commercially, the evaluation of beer has become an important advertising boon. The Boston Beer Company has built an advertising program on the claim that their Samuel Adams Lager was voted "the winner of the Great American Beer Festival for three years in a row." This claim is based on the fact that Samuel Adams was voted best beer in a consumer-preference poll at the 1989, 1990, and 1991 Great American Beer Festivals (a national competition for commercial breweries). With so much interest, emphasis, and capital placed on the outcome of zymological competitions, it is important to understand the task of the evaluator.

Many factors are involved in being a good judge, including training, experience, and understanding of the brewing process. For these reasons the AHA and Home Wine and Beer Trade Association organized the Beer Judge Certification Program (BJCP) to provide training, set standards, give experience in judging, and evaluate the judges themselves for competence and experience. I think that equal importance should be placed on understanding of the psychological aspects of the evaluation process. This point was made by Charlie Papazian in "Evaluating Beer," *Zymurgy*, Winter 1990 (vol. 13, no. 5).

When tasting a beer, a person is having a psychological experience that is primarily perceptual in nature, but other psychological factors (such as mood) can affect the experience. (Sensations like taste and sight are considered to be psychological, just like emotions and thoughts.) This article intends to educate homebrewers about how their sensory systems are used when tasting beers (or other things) and will analyze the factors that can affect a sensory judgment. It is my hope that this information will enhance the quality, consistency, and efficiency of judging.

We acquire information about the world through our sensory systems in a process known as *perception*. A common misconception about perception is that we have only five senses: seeing (vision), hearing (audition), touching (tactile), tasting (gustation), and smelling (olfaction). Sensory psychologists add several more, including *flavor* perception, a combination of taste, touch, smell, and, perhaps, vision. When evaluating a beer, a judge uses all of these senses to gauge various characteristics, but certain senses have a strong influence while others contribute minimally.

Another common misconception about perception is that we perceive exactly what is out there in the world (or in a beer), but this is not always the case. For example, a full moon appears bigger on the horizon than when it is overhead. This is a perceptual illusion because the size of the moon is not changing, only your *perception* of its size. In terms of beer, perception is affected by factors not related to the beer itself. The same beer can be perceived differently depending on many factors, several of which I will discuss below.

Following is a description of the senses and their relation to zymological evaluation, including a discussion of some of the factors that can affect a perceptual judgment, and hence the score a judge might give a beer.

## SEEING

While visual perception is arguably the most important sense in everyday perception, it is of lesser importance in beer evaluation. Vision is used to determine the fill level and amount of sediment in the bottle and to assess the head density and thickness, clarity, and color of the beer. The scoring system used at AHA-sanctioned events allocates 8 points (of a 50-point scale) to the beer's appearance. Thus 16 percent of the possible points are judged using vision.

A person with normal vision should have little difficulty in evaluating a beer's appearance. No psychological factors are known to affect the accuracy or consistency of appearance judgments, with one notable exception. The perceived color of a beer can be influenced by a number of factors.

Vision operates when the eye is stimulated with light of certain wavelengths. Combinations of different wavelengths roughly correspond to different perceived colors. When light composed of many wavelengths (such as sunlight) passes through a bottle or mug of beer, most of the wavelengths are filtered out—only a few reach the eye. A person looking at the beer from the opposite side of the light source will see only those wavelengths passing through and will perceive the color that corresponds to the particular combination of unfiltered wavelengths. Different types of beers and colored glass filter out different combinations of wavelengths and thus have different colors. (Stouts filter out all wavelengths and appear black or without color.)

Moreover, the perceived color, to some degree, is a function of the light source passing through the beer. Consider what happens if you start with a light source that contains some wavelengths (such as colored light, the light from a standard incandescent bulb, or candlelight). The combination of wavelengths that passes through the beer will be different from what is seen when the light source contains all wavelengths. To obtain the most consistent perception of color, judge a beer using a light source that contains all (or a large proportion) of the wavelengths that the eye can sense. Good light sources are sunlight, fluorescent lights, and high-pressure xenon lamps.

Another relevant aspect of color perception is something known as the *contrast effect*. Visual perception of an object varies as a function of what is located near the object. Figure 1 shows the contrast effect. All four beer bottles are the same shade of gray, while the surrounding squares are drawn with different gray shades from light to dark. The brightness of the bottles appears

# The Sensory Aspects of Zymological Evaluation

**Sensory Figure 1**
JOHN MARTIN

different depending on the shade of the surrounding patch. The beer appears darker when judged against a light background than when judged against a dark background. In actual judging, the contrast effect means the background a beer is judged against can alter perception of the color. For example, a light red background might make an amber beer appear dark red, or a green background might give an amber beer a nonred tint. To obtain consistent judgments of a beer's color, it is best to use a uniform background, preferably a piece of white paper.

## HEARING

Next to vision, audition probably is the most important sense for humans, but it plays almost no role in zymological evaluation. The only information you can obtain through audition is a general impression of the carbonation level when the bottle is first opened. If no hiss is heard (an event that has saddened the hearts of many beginning homebrewers), you can conclude the beer is flat. Louder and higher-pitched hisses indicate more carbonation. But this information is questionable in zymological evaluation because the carbonation level can be detected in other ways. The one way that auditory information can lead to inconsistent judging could be when one judge opens the bottle while the others are not listening.

## TOUCH

The sense of touch mainly involves the skin, including the lips, mouth, and tongue. Tactile perception is surprisingly important in zymological evaluation, providing two primary types of information: temperature and pressure (or texture). Because the viscosity (texture) and the rate of release of carbonation are affected by the temperature, it is important to chill beers properly for the style. The AHA scoring system allocates 5 points for body and 19 points for flavor, so nearly half of the total points are for the sense of touch.

This sense works by applying pressure to the skin that stimulates nerve cells located there. Because certain areas of skin have a high concentration of nerve cells, some parts of the skin are better at detecting pressure than others. Fortunately for zymological evaluators, the lips and tongue are among the most sensitive parts. The tongue and jaw push the liquid against the roof of the mouth and teeth, creating pressure that stimulates the touch nerve cells. Full-bodied beers create more pressure than do light- or medium-bodied beers. You may have heard someone make the comment, "Guinness Stout is so thick you have to chew it." Such a comment implies it is possible to discriminate among the bodies of beers solely on the basis of touch.

This ability can be affected because the sense of touch exhibits something known as rapid *adaptation*. Prolonged pressure on a certain part of the skin will make that part unable to signal it is being pressed on. For example, if you were to rest your arm on a tabletop, you would initially be aware of the pressure, but after a couple of seconds you would no longer feel the tabletop unless you move your arm. This adaptation occurs in all of the body's skin, including the mouth, lips, and tongue. Therefore, zymological evaluators should make sure they judge the body of the beer soon after they sip the beer. Otherwise their ability to make this judgment will rapidly decrease unless they take another sip.

In addition, the inside of the mouth is about 98.6 degrees F. Physicists tell us the viscosity or body of the beer and the rate of

# The Sensory Aspects of Zymological Evaluation

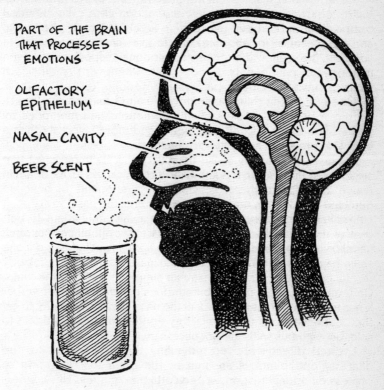

**Sensory Figure 2**

carbon dioxide release are directly related to the liquid's temperature. Increases in temperature will lighten the body and increase the rate of carbonation release in the mouth, so the longer the beer remains in the mouth, the higher its temperature will become and the more difficult an accurate judgment of body will be.

## SMELL

As most of us who enjoy a fine beer know, an important part of the enjoyment is the smell or aroma. In fact, the AHA scoring system allocates 10 points or 20 percent to aromatics. Because a beer's aroma is an important component of its flavor (19 points), the olfactory sense is involved in about 60 percent of the total score.

The olfactory sense provides information about the chemical composition of the beer, the relative levels of certain chemicals, and the presence (or absence) of additives and contaminants. In addition, the scent will often recall emotional responses and even memories of past events. For example, whenever I smell peach lambic, I invariably think back to my wedding reception when, instead of drinking champagne, my wife and I toasted with peach lambic. A beer that evokes positive emotions and memories by virtue of its distinct smell may be judged more favorably than a beer that provokes negative recollections.

Because the olfactory sense detects chemicals (molecules) that are diffused into the air, it is known as a chemical sense. As shown in Figure 2, the area where the chemical components of beer are detected is hidden in the top, back portion of the nasal cavity at an area called the *olfactory epithelium*. During a sniff, only about 10 percent of the inhaled air reaches the olfactory epithelium. Research has shown that the judged intensity of a scent is not increased if one sniffs harder to get more air into the nasal cavity. This probably results from the fact that the extra air simply moves into the lungs. On the other hand, a normal breath does not create enough turbulence to get any air into the back of the nasal cavity. Therefore, to get the beer's chemicals to the olfactory epithelium, it is important to sniff, but vigorous sniffs are not necessary.

Located directly above, and many think connected to, the olfactory epithelium is an area of the brain that is known to process emotional responses and certain memories. This connection between the area for smell and the brain probably accounts for the emotional responses and memory flashes associated with certain distinct smells.

The olfactory epithelium contains a layer of cells that are sensitive to various molecules. Surprisingly, these cells are covered with a layer of mucus (the substance that normally coats the inside of your nose and increases in quantity when you get a cold). Chemicals must pass through this mucus to get to the cells of the olfactory epithelium and be detected.

Four main factors can affect the ability to efficiently perceive the aroma of beer. First, changes in the thickness of the mucus layer will change the amount of molecules that make it through the mucus to the olfactory epithelium. As you have probably noticed, having a cold can reduce your sense of smell because clogged nasal passages do not allow air to enter the nasal cavity, and also because the mucus layer over the olfactory epithelium is thickened. Smoking, eating spicy foods (horseradish), using nasal sprays (of the type used for colds), and taking certain drugs (antibiotics, cocaine) can affect the thickness of this layer. Judges who believe their olfactory ability is reduced for whatever reason should voluntarily refrain from judging a competition until their sense of smell returns. Furthermore, to prevent diminished olfactory ability, it is best to prevent cigarette smoke from wafting around the judges during a competition.

A second factor influencing the sense of smell is the state of the cells in the olfactory epithelium. Certain fumes (such as those from paint and ammonia), smoke, and nasally ingested drugs can damage these cells so they cannot respond to scent chemicals. Following such damage, it can take up to three weeks for the cells to be replaced (in some cases they are never replaced), knocking out the sense of smell in the meantime.

A third factor is simply individual variations in olfactory ability. Some people have a superb sense of smell and are able to detect a wide range of chemicals in very small quantities. In general, this ability is best for women and decreases with age. However, many older people have excellent olfactory abilities.

The fourth factor influencing smell perception is that olfaction shows rapid adaptation, similar to what we discussed for touch perception. This means that an initial perception of a strong odor will quickly decrease in perceived intensity until it is no longer detectable. You can experience this by conducting the following demonstration. Get a strong odorant like nail polish, ammonia, or an onion and place it next to you while you read the rest of this article. In about 10 minutes you will probably notice the smell has dramatically decreased in intensity, if you can even smell it anymore. As with touch, beer evaluators should make their judgment of aroma as soon as possible after the first sniff. Otherwise, their sensitivity to the chemicals in the beer will rapidly diminish, as will the ability to effectively evaluate the aroma.

## TASTE

Gustatory, or taste, perception probably is the most heralded sense in the culinary arts (such as beer-making). It is this sense that most people believe forms the foundation for zymological evaluation. As we already mentioned, the AHA scoring system allocated 19 points to the beer's *flavor*. However, taste and flavor perception are not the same thing. A beer's flavor is greatly influ-

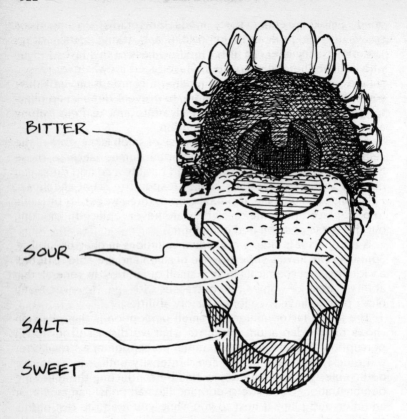

**Sensory Figure 3**

enced by its taste, but is also affected by its smell, feel, and probably even its appearance. Therefore, based on the AHA scoring system, the information obtained from the gustatory sense is involved in somewhere around 30 percent of the total score.

Gustatory perception is similar to olfaction in that it is a chemical sense. The gustatory sense, however, detects chemicals dissolved in a solution rather than diffused in the air. It allows a person to determine the basic chemical composition of substances in the mouth.

The taste sensations can be divided into four categories: sweet, salty, sour, and bitter. Many sensory psychologists believe all tastes are a combination of these four basic units. The organ of taste is, of course, the tongue. Located on the tongue are a bunch of cells called *taste buds*. (The name is derived from their rosebud appearance rather than their ability to detect a certain domestic beer!) These cells seem to be arranged so that groups of taste buds signal a certain basic taste sensation.

Sensory psychologists have mapped the locations of these groups on the tongue (Figure 3). The area that senses sweet is on the tip. To perceive the sweet quality of a substance, you must place that substance on the tip of the tongue. If something is placed only on the tip of the tongue, you will not be able to taste sour or bitter characteristics. You can test this with a simple demonstration. Carefully place a drop of lemon juice on the tip of your tongue. While your tongue is sticking out of your mouth, close your eyes and determine if you can taste the sourness. (This part must be done with the tongue out of your mouth. If you close your mouth, the juice will quickly coat your tongue. I recommend doing this demonstration in the privacy of your own home!) Now pull the tongue back in and close your mouth, allowing the lemon juice to move to the sides of the tongue where you will likely taste the sour characteristics.

The taste-bud map has several consequences for zymological evaluation. If you have never seen beer being evaluated (or have never judged a competition yourself), you may be unaware that beer judges swallow their sips of beer. This would be a major breach of etiquette during a wine-tasting competition. However, there is a good reason why beer judges swallow in order to evaluate a beer's taste. Unlike wine, one of the most important taste characteristics of a beer is bitterness. As shown in Figure 3, the taste buds that signal bitterness are located far back and partly down the throat. To detect the bittering hops, the beer must pass over these taste buds, and by the time these taste buds are stimulated, the beer is well on its way down your throat. This is why the aftertaste often consists of the bitter characteristics. I jokingly tell perception students that because a beer's taste is a combination of all four of the basic taste categories and a wine's taste is only a combination of three, the argument can be made that a beer drinker's palate is more sophisticated than a wine drinker's!

As in gustatory perception, the ability to sense different tastes can be influenced by several factors. The two most important are damage to the taste buds and interference of tasting abilities caused by the presence of other chemicals in the mouth. The taste buds have a life span of about 10 days and are continually dying and being replaced. Fortunately some taste buds are being replaced while others are dying, so there are always live taste buds ready to detect chemicals. However, a large number of taste buds in a localized area of the tongue can be simultaneously damaged, causing a loss of tasting ability in that area until they have been replaced. This type of damage most frequently occurs when you eat food that is too hot. The heat may wipe out a whole area of taste buds, requiring several days to be replaced and diminishing a person's ability to perceive bitterness, sweetness, saltiness, or sourness (depending on the injured area). Other events that can prematurely destroy taste buds (and temporarily alter tasting abilities) are smoking, eating spicy or acidic food (fresh pineapple), and using certain chemicals (some mouthwashes).

The residual chemicals that remain in the mouth after a beer is sampled can also affect taste perception. To prevent this from happening, many judges eat crackers or some other food to "clear their palate." However, this introduces different chemicals into the mouth that could affect later taste perception (and judging scores), especially if the food has a high salt or sugar content. To minimize this problem, wine evaluators eat unleavened and unsalted crackers. Another way to remove many of the residual chemicals in the mouth is to rinse with a neutral liquid, such as water at room temperature. Some chemicals, however, cannot be easily removed. You may have noticed this if you have brushed your teeth right before going down to your favorite brewpub.

## FLAVOR

I have saved flavor perception for last because it is the product of combining the information gathered from smell, taste, and touch (and possibly vision). I distinguish flavor perception from other perceptions because, when a substance is placed in the mouth, it has a smell, texture, and taste. Alterations in any of these sensations will affect the flavor of the substance. It is perhaps easier to understand what I mean by trying another demonstration that, again, you will probably want to try in the privacy of your home. Place a small piece of apple and a piece of potato of equal size on a plate in front of you. Close your eyes and spin the plate several times so that you don't know which piece is which. Pinch your nostrils shut and place one of the pieces in your mouth. Try to guess which piece you are chewing. Most likely you will have a difficult time making this judgment. By pinching your nostrils you have removed the olfactory information that contributes partly to the flavor of foods.

The AHA scoring system has a category titled "flavor," worth a total of 19 points. In the AHA system, the term *flavor* is used in a different way than it is used here. The AHA system includes gustatory (taste) and flavor information in this category, with no designation as to the relative contributions of each type of information. According to my definition, flavor is the perception that results from the combined information from the other senses. Thus flavor perception merits a separate judgment category right beside that for taste perception. Regardless of this difference, because the perceived flavor is affected by the perceived smell, texture, taste, and appearance of the beer, flavor perception may be involved in as much as 40 of the 50 points in the AHA judging scale.

Because the flavor of beer is a combination of all the sensory information, any of the factors that influence a certain sense (such as smell) also will influence the perceived flavor. For example, a judge who has a temporarily deficient olfactory sense could exhibit inconsistencies not only in the aroma category (10 points), but also in the flavor category (19 points). This could result in inefficient judging for 29 of the 50 points used in evaluating a beer.

## CONCLUSION

We have reviewed the basic perceptual systems used by a zymological evaluator when judging beer and some of the factors that can affect the performance of these systems. You may have noted that in this entire discussion we did not talk about any factors that change the *physical* characteristics of the beer; the discussion was limited to the factors that affect the *perceived* components of beer. Thus you should be aware that the perceived characteristics of any given beer can vary greatly depending upon the factors discussed in this article, *even when the physical characteristics of the beer have not changed*.

The drinking or evaluation of beer is a sensory process. By having a thorough understanding of how each sense acquires information, zymological evaluators and all people who enjoy quality beer should be able to increase their consistency and accuracy of judging competitions as well as increase their enjoyment of drinking beer.

*David W. Eby received his masters and doctorate from the University of California, Santa Barbara, in experimental psychology. He has a postdoctoral fellowship at the University of California, Irvine, in the Cognitive Sciences*

Department researching visual perception. Since beginning to homebrew in 1981, Eby has entered numerous AHA competitions. He enjoys science fiction, sports, and art.

For information on the BJCP, contact the AHA or see "How Can the Beer Judge Certification Program Benefit You?" Zymurgy, Fall 1991 (vol. 14, no. 4).

This article also appears in Evaluating Beer (Brewers Publications, 1993).

# twelve

# World of Worts

## THISTLE DO IT EXPORT PILSENER
### by Charlie Papazian

*This article originally appeared in* Zymurgy, *Winter 1994 (vol. 17, no. 5)*

Welllll, it started out, at least in my mind, as a classic German-style Pils, but like many good recipes, things didn't exactly turn out the way I expected them to. I admit I did have the option to add more water and get my gravity back into the range of what I personally consider a German-style Pilsener to be, but after two homebrews consumed during the process, I decided to leave well enough alone and let 'er rip. I was shooting for about 1.043 and ended with 1.050. What to do? Easy—relax, have another homebrew, and call it Export Pilsener right then and there. Having fermented and cold-lagered this Pilsener for a couple of months, I have no doubt that Thistle Do It. A wonderfully hoppy, clean, crisp yet malty extrastrength Pilsener-style lager, enhanced by cool fermentation and slow cold lagering—and noticeably excellent head retention.

This is a mash-extract recipe; however, malt-extract brewers can replace the grain malts with a tad more than 3 pounds of dried amber malt extract.

So let's cut the shuck and jive and get on with the recipe.

Steve Lawing

## Mash-extract recipe for 5 gallons (19 l.)

*For the mash*
- 3 lbs. (1.36 kg.) two-row Pils or pale malt
- ½ lb. (0.23 kg.) Munich malt
- 7⁄10 lb. (0.32 kg.) light German crystal malt or dextrin malt

*Add to the mash runoff*
- 3½ lbs. (1.58 kg.) light dried malt extract

*And boil with hops*
- 6 Homebrew Bitterness Units:
  I used ¼ oz. (7 g.) German Northern Brewers whole hops, 9% alpha acid; ½ oz. (14 g.) German Hersbrucker whole hops, 3% alpha acid; and ½ oz. (14 g.) Czech Saaz whole hops, 3.7% alpha acid
- 3 Homebrew Bitterness Units of flavor hops: I used ¾ oz. (21 g.) Czech Saaz whole hops, 3.7% alpha acid, for 30 minutes
- 3 Homebrew Bitterness Units of late flavor hops: I used 1 oz. (28 g.) German Hersbrucker whole hops, 3% alpha acid, for 15 minutes
- ¼ tsp. Irish moss (15 min.)
- 1 oz. American Tettnanger hops for aroma
- ½ oz. German Hersbrucker hops for aroma
- ¼ oz. Czech Saaz hops for aroma

Pilsener-style lager yeast is recommended
¾ c. (177.4 ml.) corn sugar (to prime)

- Original gravity: 1.048 to 1.052 (12 to 13 °B)
- Final gravity: 1.010 to 1.014 (2.5 to 3.5 °B)
- IBUs: about 35

Use a step infusion mash. Add 4 quarts (3.8 l.) of 135-degree F (57-degree C) water to the crushed grain, stir, stabilize, and hold the temperature at 130 degrees F (54 degrees C) for 30 minutes. Add 2 quarts (1.9 l.) of boiling water, stabilize temperature at about 148 to 152 degrees F (64 to 67 degrees C), and hold for about 45 minutes. Temperature may be allowed to drop from 152 to 148 degrees F (67 to 64 degrees C) with no worrying. Then raise temperature to 160 degrees F (71 degrees C) and hold for 5 to 10 minutes to complete conversion.

After conversion, raise temperature to 167 degrees F (75 degrees C), lauter, and sparge with 2 gallons (7.6 l.) of 170-degree F (77-

degree C) water. Collect about 2½ to 3 gallons (9.5 to 11.4 l.) of runoff and add the malt extract and bittering hops then bring to a full boil.

The total boil time will be about 70 minutes. Strain wort into a sanitized fermenter to which you've added 2 gallons of cool water. It helps to prechill the water to 33 degrees F (1 degree C) before adding to the fermenter rather than simply adding warmer tap water.

IBU bitterness of about 35 IBUs was calculated for this recipe by making the following assumptions: (1) Whole hops were used; (2) the wort boil was a concentrated boil with about 2 pounds (0.9 kg.) of extract per gallon (3.8 l.) of liquid boiled; (3) 26 to 27 percent utilization was assumed for 70 minutes of boiling, 13 to 14 percent utilization was assumed for 30 minutes of boiling, and 7 percent utilization was assumed for 15 minutes of boiling. Beginners and intermediate brewers should relax, don't worry, and have a homebrew.

Primary ferment at temperatures between 50 and 65 degrees F (10 and 18 degrees C) with lager yeast. Rack your brew after primary fermentation into a secondary fermenter and lager at 38 to 45 degrees F (3 to 7 degrees C) for four to six weeks. Prime with sugar and bottle when fermentation is complete.

Yep, with your first taste of this brew I'm sure you will agree that Thistle Do It is the beer that arouses your appetite.

# HERE TO HEAVEN OKTOBERFESTWINE ALE

## by Charlie Papazian

*This column originally appeared in Zymurgy, Winter 1993 (vol. 16, no. 5)*

Wha . . . ?

Oktoberfestwine Ale?

No, I haven't gone bonkers. It's just that I have found the sublime, and it appears to be a direct connection from here to heaven. I really appreciate the rich, malty, tawny character of German-style Oktoberfests, while at the same time knowing a barley wine ale made in the style of a superstrength Oktoberfest would suit my fancy. Rich, malty, amber-hued, with a distinct character of German hops and a hint of American Cascade for a German-

British-American hybrid. Sure, there are a lot of hops in this recipe, but because of the richness and abundance of malt, the overall bitterness is subdued.

This sipping brew has 9½ percent alcohol by volume. As with every single homebrew you've made, this, too, should be brought out for special occasions. It ages well. Bottling to within ½ inch (1.2 cm.) of the crown cap will minimize the effects of oxidation. Oxygen-absorbing Pure Seal Caps are also recommended to help extend freshness.

So let's cut the shuck and jive and get on with the recipe.

## Mash-extract recipe for 6½ gallons (25 l.) because 5 gallons (19 l.) isn't enough.

*For the mash*
- 3 lbs. (1.4 kg.) crushed pale malt
- 1 lb. (0.45 kg.) crushed Munich malt
- 1 lb. (0.45 kg.) crushed Vienna malt
- 1 lb. (0.45 kg.) crushed crystal malt
- ½ lb. (0.23 kg.) dextrin malt (light crystal or CaraPils)

*Add to the mash runoff*
- 13 lbs. (5.9 kg.) light dried malt extract

*And boil with hops*
- 36 Homebrew Bitterness Units: 4 oz. (114 g.) 9% alpha-acid-rated German Northern Brewer whole hops for bittering
- 6 Homebrew Bitterness Units: 1½ oz. (43 g.) 4% alpha-acid-rated German Hallertauer whole hops for flavor
- 1 Homebrew Bitterness Unit: 2 gal. (7.6 l.) prechilled water for primary (see note in directions)
- ⅞ c. (207 ml.) corn sugar for bottling
  ale yeast (American Ale Wyeast No. 1056 is recommended) with a healthy and vigorous starter

- Original gravity: 1.098 to 1.102 (24.5 to 25.5 °Balling)
- Final gravity: 1.028 to 1.034 (7 to 8.5 °Balling)

Use a step infusion mash to mash the grains. Begin by adding 6½ quarts (6.2 l.) of 130-degree F (54.5-degree C) water to the crushed grain, stir, stabilize, and hold the temperature at 122 degrees F (50 degrees C) for 30 minutes. Add 3½ quarts (3.3 l.) of boiling water, stabilize the temperature at about 148 to 152 degrees F (64 to 67 degrees C), and hold for about 60 minutes. Temperature may be allowed to drop from 152 to 148 degrees F (67 to 64 degrees C) with no worrying.

After conversion, raise temperature to 167 degrees F (75 degrees C), lauter, and sparge with 3 gallons (11.4 l.) of 170-degree F (77-degree C) water. Collect about 4 gallons (15.2 l.) of runoff, add the malt extract and bittering hops, then bring to a full boil.

Boil for about 90 minutes. When 20 minutes remain, add 6 HBUs of flavor hops. After a total wort boil of 90 minutes, turn off the heat and add 1 HBU of aroma hops and let steep 2 to 3 minutes before straining and sparging into a sanitized fermenter to which you've added 2 gallons of water. It helps to prechill the water to 33 degrees F (1 degree C) before adding it to the fermenter rather than simply adding warmer tap water.

Note to advanced brewers: Because of the high density of the boiled wort, it is very difficult to calculate the exact bitterness of this recipe in terms of International Bitterness Units. A rough approximation would be from 55 to 70 International Bitterness Units.

Primary ferment with ale yeast. Rack your brew after primary fermentation into a secondary fermenter and ferment to completion.

Prime with sugar and bottle when fermentation is complete.

Let age at least three or four months in the bottle before sampling. Heaven can wait, but when this brew is ready, you'll feel the connection from here to heaven. I guarantee it.

# SLUMGULLION AMBER ALE

## by Charlie Papazian

*This article originally appeared in Zymurgy, Summer 1994 (vol. 17, no. 2)*

Let's see, the last five columns included a German dunkel, Kölsch, festwine, Oktoberfest, and a stout. I do believe it's time to revisit a good ol' middle-of-the-road-everyone-will-like-this kind of a beer. Slumgullion Amber Ale is just that. The name was inspired by Slumgullion Pass between Creede and Lake City, Colorado. Slumgullion also refers to a pot of stew that contains all kinds of things.

Slumgullion is not so sharp and bitter as a British pale ale, yet has plenty of smooth hop character to inspire comment and praise such as "This is good beer. I like this. It's not too bitter, not too malty, not too dark, not too light, yet is thirst-quenching. Yeah. Good job. I like this beer."

The artful touch of specialty malts adds delicate complexity. The dry hops don't wallop you, but do offer a nice finish. Slumgullion is everyone's type of beer and is similar in character to many specialty amber lagers and ales on the beer shelves these days.

Note the high-temperature protein rest. This promotes the type of protein development and degradation that enhances foam quality and stability. When brewing an all-malt beer, it is not so critical to develop nutrient-type proteins at lower-temperature protein rests. All-malt beer formulations almost always have plenty of protein nutrients regardless of protein rests.

So let's cut the shuck and jive and get on with the recipe.

## Mash-extract recipe for 6½ gallons (25 l.) because 5 gallons (19 l.) isn't enough:

*For the mash*
- 4 lbs. (1.8 kg.) crushed pale malt
- 1 lb. (0.45 kg.) crushed wheat malt
- 1 lb. (0.45 kg.) crushed Vienna malt
- 1 lb. (0.45 kg.) crushed 40 °L crystal malt
- ½ c. (118 ml.) crushed chocolate malt

*Add to the mash runoff*
- 2¾ lbs. (1.24 kg.) light dried malt extract

*And boil with hops*
- 6.6 Homebrew Bitterness Units: I used ¼ oz. (7 g.) 10% alpha acid Centennial hops plus ¾ oz. (21 g.) 5.5% alpha acid Willamette hops

4 Homebrew Bitterness Units of flavor hops: I used 1 oz. (28 g.) 4% alpha acid American Tettnanger hops
¼ tsp. Irish moss
1 oz. (28 g.) Cascade hops (dry)
⅞ c. (207 ml.) corn sugar for bottling
ale yeast (your favorite brand)

- Original specific gravity: 1.043 to 1.047 (11 to 12 °B)
- Final specific gravity: 1.008 to 1.012 (2 to 3 °B)
- IBUs: about 25 or 26

A step infusion mash is employed to mash the grains. Add 7 quarts (6.7 l.) of 135-degree F (57-degree C) water to the crushed grain, stir, stabilize, and hold the temperature at 130 degrees F (54 degrees C) for 30 minutes. Add 3½ quarts (3.3 l.) of boiling water and stabilize the temperature at about 148 to 152 degrees F (64 to 67 degrees C) and hold for about 45 minutes. Temperature may be allowed to drop from 152 to 148 degrees F (67 to 64 degrees C) with no worrying. Then raise the temperature to 160 degrees F (71 degrees C) and hold for 10 to 15 minutes to complete conversion.

After conversion, raise the temperature to 167 degrees F (75 degrees C), lauter, and sparge with 3 gallons (11.4 l.) of 170-degree F (77-degree C) water. Collect about 4 gallons (15.2 l.) of runoff and add the malt extract and bittering hops and bring to a full boil.

The total boil time will be about 90 minutes. Expect to boil off about 1 gallon of wort. When 20 minutes remain, add 4 Homebrew Bitterness Units of flavor hops and Irish moss. After a total wort boil of 90 minutes, turn off the heat, strain, and sparge into a sanitized fermenter to which you've added 2 gallons of water. It helps to prechill (33 degrees F or 1 degree C) the water added to the fermenter rather than simply adding warmer tap water.

Bitterness of 25 to 26 IBUs was calculated for this recipe by making the following assumptions: (1) Whole hops were used; (2) the wort boil was a concentrated boil with about 2 pounds (0.9 kg.) of extract per gallon (3.8 l.) of liquid boiled; (3) 24 to 28 percent utilization was assumed for 90 minutes of boiling and 14 percent utilization was assumed for 20 minutes of boiling. Beginners and intermediate brewers should relax, don't worry, and have a homebrew.

Ferment with ale yeast. Rack your brew after primary fermentation into a secondary fermenter and add 1 ounce (28 g.) of whole or pelletized Cascade hops and let ferment to completion, or "cellar" for two more weeks. Most of the dry hops you've added will either end up floating on the surface of the beer or sink to the bottom. Take care when siphoning during bottling to avoid hops.

Prime with corn sugar and bottle when fermentation is complete.

Age until clear and carbonated and enjoy the fresh maltiness and lively but not overpowering hop character. Chilled on a hot summer day, this is a real winner.

# thirteen

# Dear Professor Surfeit

## DEAR PROFESSOR

*This column originally appeared in Zymurgy, Summer 1992 (vol. 15, no. 2)*

### PRECIPITROISKA

Dear Professor,

Is this a job for the CIA or the AOB? Despite the fall of the Berlin Wall, I have reason to believe that we are in for a new kind of Cold War.

Late last December, as most of the nation was in the grips of a Siberian deep freeze, I spied a strange occurrence. A variety of recently purchased bottles of beer were carefully placed in my garage to make room in the refrigerator for the holiday meals. As the temperature plummeted, the beers were subjected to an unduly harsh cold. Miraculously, none of the bottles cracked or exploded, but two of them showed signs of liquid leaking past the caps due to the Eastern bloc nations. Moscova Beer from the Moscow Brewery in the Soviet Union and Red Star Beer from the Berliner Brewery in East Berlin stood alone among the twelve or so beers left in the garage that day. I did notice upon purchasing these particular products that they were hazy with particulate matter in suspension.

**The Professor**
JOHN MARTIN

Have I unwittingly stumbled on a communist plot or would a centrifuge eliminate this subterfuge?
AGENT 170, A.K.A. MARTY NACHEL
Frankfort, Illinois

Dear Agent 170,
Sounds like you've been the victim of a spy-counterspy fake drop. That is not Siberian snow in those bottles, but precipitate from old beer. Now that the wall is down, I wonder if they are cleaning out their cellars and exporting that "age-old" import taste to us capitalist Americans. Actually, because of the pressures generated by freezing, even our own homebrew could spring a leaker.

Strange thing about freezing weather and beer. I've had beers and meads exposed to minus 30 degrees F that survived this exposure over the winter. If the beer is not disturbed, crystallization may not occur. What you have is a supercooled liquid that could freeze almost instantaneously if jarred.

The centrifuge won't take out stuff that is in solution to begin with. Sorry, but you got some aged beer.
Your Comrade,
THE PROFESSOR, HB.D.

## LOVE A BOND

Dear Professor,

I enjoy reading Zymurgy very much and have a suggestion: A lot of articles talk about the beer color expressed in "degrees Lovibond." I don't know what one degree Lovibond looks like. So could you please use the front cover of Zymurgy to print a Lovibond color chart?

Keep on the good work.

    Jacques Bourdouxhe
    SAINT-LAURENT, CANADA

Dear Jacques,

*We thought of doing just that years ago, but then when we consulted with experts, they advised us that we would be misleading people to think they could use a printed color to compare to beer color.*

*You see (I see too), beer color is transparent and depends on the volume of beer you are looking through. Have you ever siphoned a stout? It looks light brown, doesn't it? Put it in a glass and all of a sudden it's quite a bit darker.*

*The best way to get a feel for beer color is to notice the colors of beers with known Lovibond colors (now called SRM rather than Lovibond). Michelob dark is a 17; Michelob light is about a 2.5 to 3, for example.*

*There's an excellent article in Zymurgy, Fall 1988 (vol. 11, no. 3), by George Fix on beer color and how to calculate it. Based on his data, clubs or individuals could easily make sample beers representing different color ranges.*

    *Oars de wavre,*
    THE PROFESSOR, HB.D.

*This column originally appeared in* Zymurgy, Spring 1992 *(vol. 15, no. 1)*

## DAMN THE INSTRUCTIONS— FULL SPEED AHEAD!

Dear Professor,

I have been making beer for about a year now. I just received my first copy of Zymurgy and it didn't take very long to realize that I could improve on the recipe that comes with a can of malt.

I have quite a large supply of malt extract on hand, and I wonder if you could send me a better recipe than the sugar one that comes with the malt, while I continue to study your magazine and learn more about homebrewing.
Thank you,
>BILL TYSON
>NEATON, TEXAS

Dear Bill,
*Well, it won't take you very long to discover that improving on the sugar recipes supplied with many malt extracts is a pretty easy thing to do. It doesn't take a whole lot of reformulation!*

*For starters, just substitute extract for sugar, pound for pound. That is the simplest thing you can do. If the extract you are using is hop-flavored and the straight substitution turns out too bitter for you, then buy the plain unhopped version of that malt extract and substitute that.*

*Don't worry too much about overbitter beer when you first try the recipe because, although you are adding more bitterness by substituting hopped extract for sugar, you are also adding more body and malt.*
*Sweetly,*
>THE PROFESSOR, HB.D.

## LOTS OF GOOD QUESTIONS

Dear Professor,
I've been a homebrewer for thirteen years. Don't conclude that by now I'm a master brewer; I've just last year lost the habit of adding 3 pounds of dextrose per 6-gallon batch of beer to get the gravity up. Only in the last six batches or so have I actually recorded what I added and what I did. You see, I'm an analytical chemist by profession and don't believe in recipes. I just do it (or try to), you know. Let me shoot several questions your way, please.

1. Ever since my first homebrew, I have refused to boil my wort and hops together for the requisite duration. Why? Because I can't relax knowing that a semiviscous, potentially sticky mass is just waiting to leap from my brewpot when I'm least attentive. Instead, I treat my water and boil my hops in 5 liters or so for the bittering/flavor length, then either steep my aroma hops or dry-

hop. In the brewpot I mash and add malt extract, specialty grains, and anything else. Then I bring the wort up to a boil and just boil long enough with added Irish moss to get the hot break and feel relaxed about sterilizing the wort. I then combine everything and don't worry. Am I really missing anything by going this route?

**2.** Whenever I mash, I always throw in 2 teaspoons or so of amylase enzyme to try and absolutely ensure complete conversion. I never use nonmalt adjuncts. Addition of amylase can only help, right?

**3.** Fred Eckhardt says in *The Essentials of Beer Style* that beer just has got to be consumed as quickly as possible for it to be at its peak of flavor, aroma, and bouquet. Now, a lot of the "Winner's Circle" brews I see (that aren't long-maturation beers to boot) are six to eighteen months old when judged. Who's zoomin' whom here?

**4.** On page 45 of Dr. Foster's *Pale Ale*, he states that it doesn't make any difference if one uses cane or corn sugar as sugar adjuncts. Now, I believed that sucrose, being a disaccharide, was unfermentable by yeast until they generated sucrase/invertase enzyme to cleave the oxygen linkage, thus breaking the molecule. This enzyme hanging out in solution yielded the cidery, winy taste. Straighten us out, please.

**5.** Just one more: At the risk of being nosy, what specifically happened to Dave Line so that he couldn't relax and have any more homebrews?

    Have I asked too many questions?
        CHRISOTPHER CAPE
        CLARK, NEW JERSEY

Dear Christopher,
    *I do believe you are missing a boat.*

*1. One of the primary purposes of boiling hops with malt is to provoke chemical reactions that improve the flavor, appearance, fermentation, and overall quality of your beer. You may be happy with the quality, but quite frankly, I'd eliminate your elaborate combinations and isolated boils and combine it with an* equal amount of attention so you don't boil over. Six of one is a half dozen of another.

**2.** Wrong. Actually, the teaspoons of amylase are activated at temperatures not the same as mash temperatures. Also, if it were active, you would lose some control over the dextrin (for fuller body) and fermentable balance. However, if you don't mind your beers on the dry and thin side, amylase used at the proper temperatures can only produce more fermentables.

**3.** That ole Fred, yessiree he can be a schmoozer, all right, but a nice one at that. It's true what he said, partly. If you buy commercially bottled beer, or any beer that has had the yeast filtered out, then the beer is best just before it leaves the warehouse doors of the brewery. Drink it at the freshest date you can find it. Now, homebrews are quite different. They have yeast in the bottle. Yeast can protect the shelf life of any beer. As a matter of fact, some major Japanese breweries, I hear, actually dose their light lagers with a minute quantity of yeast just before bottling to help maintain the quality. And keep in mind those eighteen-month beers are usually the strong, full-bodied, or hoppy brews that are complex and mature in wonderful ways . . . up to a point. You be the judge.

**4.** I think you are straightened out. There are many opinions about cane sugar versus corn sugar. I think you've got it right—partly. It's not the enzyme hanging out in solution, it is the by-products of the extra metabolism required to hocus-pocus the sugar into being fermentable by the yeasts. I believe there is a difference. Others don't. Let's have a glass of homebrew someday and have a friendly argument and then get on with how we can make the world a safer place with homebrew.

**5.** On a sad note, Dave Line died of Hodgkin's disease in the prime of his life. He was one of the pioneers in the world of British homebrewing. It would have been a real treat to have met the man. We all owe him a toast toward the stars.

If you didn't ask questions, I wouldn't have a job,
   THE PROFESSOR, HB.D.

*This column originally appeared in Zymurgy, Winter 1991 (vol. 14, no. 5)*

## WALLEYE'LL BE A BLUE NOSE GOPHER

Dear Professor,

All-grain brewing takes up the better part of Saturday or Sunday, which cuts into precious fishing time. I'd much rather brew midweek and leave weekends open for the wild 'n' woolly outdoors. However, even if I'd crush grains and weigh out all ingredients one evening, the mash/sparge/boil/chill/pitch would leave me sleepless the next evening.

One possible solution would be to mash the first evening, then boil the next. I'd appreciate your input on leaving the wort in a 10-gallon covered stainless brewpot overnight, ready for boiling the next. My concerns are contamination and the overall effects of this delayed method. Will my 60- to 90-minute boil kill all bacteria that have surely started "eating" my wort? And what flavor changes can I expect, if any? Please don't suggest extract brewing or speeding up my process; I'd rather go to work following a sleepless night or keep my fishing line dry on a weekend.

Hope to see you at a homebrew fest.

    GARY "MR. WALLEYE" HOFF
    GRESHAM, OREGON


Dear Mr. Walleye,

*Now, just a dad-garn minute there, pardner. You a-tellin' me ya'll got fishun' ta do? Or what? Tell me, do you want to spend more time brewin' or fishun'? I can't finger you out there, sonny.*

*Let me give ya the lowdown on how to relax a right bit more. I use that there Zapap lauter-tun method and do all-grain brews now and then, and I'm start to finish in less than 5 hours. Now, I hear tell a lot of folks are up in the neighborhood of 6 to 10 hours. Well, if ya love it that much, don't read any further.*

*Here's the scoop. Once ya get up to mashin' temperatures of 155 to 160 degrees F (68 to 71 degrees C), you're a-gonna get conversion within 20 minutes. You want to squeeze out that extra 5 or 8 percent, then go another 40 to 60 minutes; go ahead, be my guest. Or ya can throw an extra 50 cents worth of pale malt in the mash and cut your time down.*

*Now, about sparging. You can do it in less than 45 minutes for sure. Take the first runnings off and get it on the stove to start it towards boiling, save some time. So you lose 3 or 5 percent because your runoff is a bit on the fast side, or* you don't get every last molecule of sugar (and all those nasty polyphenols; husks, ya know). So throw in 30 cents more worth of grain and take the first runnings and don't sparge so long.

Get yourself a propane burner stove and get'er to a rollin' boil fast, fast, fast.

Now, listen, chillin, and you shall hear the midnight brew of Paul and his beer. I just made that up, but it don't mean a thing. But, chillin, your brew should take 15 to 20 minutes. If it's a-takin' more time, then do some investigating.

Remember, ya ain't a penny-pinchin' accountant/brewer at a megabrewery trying to get every last drop of decent extract. You're a fisherman/brewer, ain'tcha?

Leaving wort sittin' around overnight is gonna cause some down and funky brews, tasting sour and weird vegetablelike if nothing more.

But you did want to shorten your brew time, didn't ya?

Catcha fishin'

THE PROFESSOR, HB.D.

This column appeared in Zymurgy, Fall 1991 (vol. 14, no. 3)

## PUCKERED AND BITTER

Dear Professor,

I'm all puckered up and I don't like it.

It's been about a year and a half since I started brewing and reading your magazine (both with great enjoyment).

With each new batch it seems like I'm getting closer to that perfect brew. I get a nice malty aroma, a clean taste, just a hint of hops, and then whammo, this astringent aftertaste. It's driving me nuts.

Mostly I do extract ale brews with various amounts of grain adjuncts and hop pellets. After two to three days, I rack to a secondary and then bottle within two weeks.

I'm trying to relax, but I'm so puckered up it's hard to drink a homebrew. I'm desperate, so no shuck and jive. What can it be?

A PROUD BACKDOOR BREWER,
RICK PAULY
CHARLOTTESVILLE, VIRGINIA

Dear Rick,

I'm glad you sent a bottle of beer for me to taste. What I surmise is that the astringent pucker you refer to seems to be nothing more than a bad case of overdosing with hops. Mostly. Your beer is very bitter.

Yes, there is some astringency. When steeping your grains, don't steep above 150 degrees F (65.5 degrees C). That will help.

But it's mostly an intense lingering bitterness you have in your beer. If your water is very hard or high in carbonate or bicarbonate, this could also contribute to the bitterness and astringency significantly.

Sincerely,
THE PROFESSOR, HB.D.

*This column originally appeared in Zymurgy, Summer 1991 (vol. 14, no. 2)*

## TAKE A BATH, BUDDY

Dear Professor,

Hope you can help me with a problem I have each summer during the hottest months.

On the way home from work I spend most of the time thinking about my frosty homebrew in the fridge and/or a refreshing shower.

As I arrive home, I'm burdened with the weight of indecision. Should I quaff a quick one or hit the showers?

Sometimes the decision takes an excruciatingly long time. Once it took 30 seconds. I know you can help.

Trying to avoid the stress,
MARTIN JORDAN
Toronto, Ontario, Canada


Dear Martin, ole buddy,

Hey, listen, just take a bath. Relax. Don't worry and have a homebrew. You deserve it. But if you really must take a shower, decant that bottle of homebrew into another clean bottle and long-neck it to your thirsty lips.

No stress in Canada please,
THE PROFESSOR, HB.D.

## FERMENTATION FOREVER?

Dear Professor,

I am a new brewer and have encountered a problem with a large percentage of the beers I've brewed becoming overcarbonated as they age until they reach the point of gushing. Some of the beers have shown signs of autolysis—bubbles rising in the air lock after days of inactivity—and some have not, so I cannot relax when brewing. I have lost many dollars in wasted materials.

I have read the *Complete Joy of Home Brewing* over and over searching for the answer, but after paying particular attention to sanitation and trying to control temperature, my problem continues. Could either of the following be the cause? They were not addressed in the book.

1. Water—my water comes from a private well. I have had it tested for contamination, and the results show it is as good as city water. Remember, I must rinse with this water after sanitizing, right?

2. Temperature—I do not have a basement. The lowest temperature I can maintain is 75 degrees F (24 degrees C) in the summer, 68 degrees F (20 degrees C) in winter. Storage temperature is about 70 to 75 degrees F (21 to 24 degrees C) in summer, 60 to 65 degrees F (15.5 to 18.5 degrees C) in winter.

Please send anything that may help me relax.
Thanks,
BERNARD DOTTS
Charlotte, North Carolina


Dear Bernard,

*Oh, how you play that same old song. But it ain't sweet music, is it? Yep, I've heard of this prolonged fermentation a lot. In fact, I automatically ask (in your case I looked at your letterhead), "Where ya from?" Not that I'm looking to change the subject or anything like that, but it seems that long fermentations are more prone to the warmer climates and seasons.*

*The way I figure it, long fermentations are often a result of some kind of wild yeast that just likes to break down those unfermentables and ferment them—slowly. Where'd you get them? Maybe from the air. Maybe from your yeast source. Some dried yeasts will have wild yeast counts large enough to make a difference.*

*Bernard, try changing your yeast brand. Then try looking at your sanitization techniques.*

By the way, autolysis isn't known for producing a lot of bubbles. You just got some wild yeast in there. You say you rinse with tap water. Hot tap water is best.

I bet it's the yeast you started with,
THE PROFESSOR, HB.D.

🍺

This column originally appeared in Zymurgy, Spring 1991 (vol. 14, no. 1)

## AN OUNCE IS NOT AN OUNCE

Dear Professor,

I haven't bugged you in years, but I guess it was inevitable that I would again. This question is actually after the fact, but hoppily it came out well.

When using fresh undried hops, what would be your weight equivalents? Someone gave me some of undetermined bitterness. I thought I'd seen some information on how to use undried hops in a past issue of Zymurgy, so of course, I procrastinated looking up the information until I was mashing. Being unable to find the information, I decided to wing it and I threw the whole 5.5 ounces of undried hops in the batch (a moderately heavy, 1.054 OG porter), and an ounce of Chinook to boot. I'd rather err on the hoppy side than have an insipid beer, but for future reference and to decrease the unknowns, how many ounces of green hops does it take to get an ounce of dry hops? Should you use them green (undried)? They seem to have imparted a wonderful hop taste to the brew (and the dreams would make Carlos Castaneda jealous).

Just tryin' to maintain
a hoppy medium,
DON SMITH
Caribou, Maine

Dear Don,

Hey, old buddy, long time no hear. Yeah, we go way back. I was just perusing a back issue this morning and was wondering whatever happened to ole Don Smith.

Anyway, freshly picked hops contain about 80 percent moisture. After they are dried they're down to about 8 percent. Now, let's round off 8 percent to 10 percent and see that if you had 10 ounces of freshly picked hops, then 8 ounces would be water and 2 ounces would be just hop. So absolutely dry hops would weigh in at 2 ounces. But remember, they have 10 percent moisture. So if you figure 2.2 ounces at "dry" weight, then 2 ounces are just dry hop and 0.2 ounces are water. Looks like the ratio is about 5 to 1 to me.

Sounds like your after-the-fact beer will be just fine.

    Not ten-four over
    and never out, good buddy,
THE PROFESSOR, HB.D.

This column originally appeared in Zymurgy, Winter 1990 (vol. 13, no. 5)

## FRESH HOPS GET A GOOD HEAD!

Dear Professor,

Let me start off by saying that I enjoy your column immensely and always look forward to reading it as much as any section of the magazine. I also hope you will keep Michael Jackson's ruminations as a section.

I get the feeling from your column that many homebrewers get frustrated with the amount of head they get with their brew. I know in my own case I nearly gave it up because of problems with head formation. I mean, come on, one of the main reasons I want to make beer is so I can get one poured that has a head! I can't count the number of times I have waltzed into a place that supposedly is proud of its beer, on tap or otherwise, only to be poured a product with either no head or one that disappears in seconds. What a sickening sight, after a big buildup of hope and anticipation, to watch that beautiful top of white foam start the unmistakable fizzing action around the edges that can only mean, again, we are cheated out of the head, cheated out of the lacing effect as the glass is drained, cheated out of the creamy lid that should linger even as most of the head diminishes while the beer is drunk. All this is not to say the beer is not carbonated, oh no, maybe too carbonated, and probably too cold. I feel like giving a variation on the "Where's the beef?" gag and yelling, "Where's the head?"

So anyway, when brewing one's own, the last thing you want to have happen is more of the same problems, even though you are willing to go with the best ingredients. So one reads the available literature for help; if I may condense, the usual suggestions are to use all malt, avoid contamination by microorganisms, avoid hard water, avoid oxidation, and use clean glasses. I am sure all these things are true, but when you try to do these things and get bad results, finding no other suggestions in print, the result is hyper paranoia that somehow you aren't clean enough, somehow detergent has crept in, etc. I was driving myself nuts, trying to get my glasses clean and trying to improve sanitation. The truth is, the glass needs to be clean and the sanitation needs to be good, but if a reasonable effort is being made, that *is* enough. I can get a good head now, because I scour your column for info every time it comes out. How many people realize you made a direct hit in the Fall 1989 issue? Almost as an aside, you mention using fresh hops to help head retention. I finally decided to start experimenting with this idea, and man, what a difference!

The first thing many of us start to get away from is the use of fresh hops, because they are the hardest thing to get in good condition. The books and articles may suggest fresh is better, but the feeling seems to be pellets are nearly as good and are more convenient. I can tell you that if you want to use pellets, you will also want to add heading agent and lots of it. Other remedies such as using grain will help, but you will still get the fizzing and disappearing act, just a little more delayed. My experiments use malt extract, either alone or with a grain additive such as crystal malt, and also with a mash/extract combination. I have never tried an all-grain mash with no extract and can't comment on that. Basically, I would brew a batch with hop pellets both in the boil and in the finish, then brew a batch with everything the same except to use fresh hops. I'm sure you can tell that I feel that the fresh hops made all the difference. I mean night and day. I even tried using an excess of fresh hops and got a batch with so much foam it was a nuisance. Try your own experiments if you like, but pass it on, fresh makes the difference.

One of the other things that I have been able to dispel is the notion that dishwashing liquid has an effect on the head if even a trace is left in the glass or whatever. Now, here's an experiment anyone can try: Grab two clean glasses and a beer known to produce a good head. Put a drop or so of liquid detergent in one of the glasses and rinse it out only partially so that you leave some soapy water in the glass—just a trace if you like, or you can leave a lot. I always buy off-brand, but I never have been able to see an effect on the head. I get the feeling this came from England, where maybe dishwashing liquids are made differently, maybe from animal fats or something, but over here what we have can be used pretty fearlessly. Either that or the brand matters, which I doubt.

On another subject, what is the best way to sanitize ceramic-top bottles? I prefer these, but have quit using them because of some contamination problems.

I'm sure there was something else I wanted to ask or comment on, but I think I will spare you. I hope some of this is interesting or useful. Thanks for reading.
    Sincerely,
    CARL WILLIAMS
    Riverdale, Maryland

Dear Carl,
    I'll pass the good words to Mike.
    And you, Carl, go to the head of the class this month!
    Sanitizing ceramic-top bottles? Do it the same way you'd do regular bottles if you immerse the bottles in a weak chlorine solution. Remove the gaskets and soak them, too. Replace the gaskets if they are cracking.
    As a rule of thumb, I always bottle about eight or nine or ten ceramic-top bottles, and because I know they are the most likely to become contaminated, I drink those first or at least monitor them, and if they started to go (which, by the way, has been never), then I'd consume them in a flash.
    Thanks for highlighting the fresh hops and head,
    THE PROFESSOR, HB.D.

## STRANGER THAN SCIENCE

Dear Professor,
    A comment followed by a question . . .
    Comment: I do not understand why people take all the trouble and make all the mess and worry about sanitizing siphoning tubes when there are perfectly dandy kettles available which are fitted with a perfectly dandy feature called a *tap* (or faucet), both for fermentation and for priming.
    Question: John Bull Master Class instructions call for two extra steps in single-stage fermentation. After the boil and before bringing the wort up to full volume and pitching the yeast, we are told to allow up to 2 hours settling, then decant, leaving behind the trub (frog shit). Then add water to the volume, pitch the yeast, seal, and ferment.
    Next, we are told to add a minuscule dab of sugar (1 ounce), and allow 2 more days for "bulk conditioning," with frequent "rousings," when fermentation is complete.
    I do the above when using these kits, and I get very fine beers, but I ask—is there any point in either or both of these extra operations? If so, would beers made with other malts profit by similar treatment?

Why do manufacturers continue to suggest adding raw sugar to the bottles instead of bulk priming?
Sincerely and thirstily,
ROBERT E. KILBRIDE
Chicago, Illinois


Dear Robert,
Hey, man, be careful of those tap/faucets. Yes, you can use them, but attach a hose nonetheless. You don't want to "pour" your brew into bottles or secondaries—that will aerate it. That's a no-no resulting in oxidation and quickly deteriorating quality.

The settling of the trub and then racking off is a wise, age-old practice, indeed employed by many commercial breweries. It makes for a cleaner taste in the beer and less of a problem with chill haze. But be careful; after 2 hours, your wort may have cooled to less than 160 degrees F (71 degrees C) or thereabouts and you encounter a risk of contamination when transferring.

Now, about that 1 ounce of sugar for bulk conditioning. I'm glad you enclosed the package instructions because your letter doesn't tell the whole story. What they are suggesting is again a justified procedure for helping remove some of the more volatile off flavors in beer. Many breweries do this. What is happening is that the small amount of sugar is commencing fermentation and $CO_2$ gas is evolved. $CO_2$ is an excellent "scrubber" in that it "washes" out a lot of off flavors; the off flavors "dissolve" from the liquid into the $CO_2$ gas and are vented. Note that the instructions state: Rouse thoroughly at intervals to give vigorous evolution of gas to remove "green" flavors and odors. This process works well under sanitary conditions.

What I question is—do you risk losing more than you gain? Yes, you may scrub out "green" flavors and aromas, but then by rousing you may be introducing contaminating bacteria that will add off flavors and aromas. If you can be careful, do it.

Another method for those fermenting in 5-gallon stainless steel soda canisters is to devise a system to pass purified $CO_2$ through the beer and vent it out. This will effectively achieve that same thing as lagering for a period of time. Try it. You'll probably like it.

And adding raw sugar to each bottle for priming, I think, is archaic. But for those of you who do and can make great beer, go for it.
Anachronistically,
THE PROFESSOR, HB.D.

*This column originally appeared in* Zymurgy, Fall 1990 (*vol. 13, no. 3*)

## THE BLUES ON BLUEBERRY HILL?

Dear Professor,
 I have been brewing beer for almost two years. I have refined my technique to a point where the beers I brew are very clean, not to mention well liked by friends. The problem that came up recently has to do with clarity.
 Because I have several blueberry bushes in my backyard, I have made a few batches of blueberry beer. I use the same techniques that I use with my other beers. The flavor is wonderful and has met with rave reviews, but it is cloudy. Is this to be expected from fruit beers or am I missing a step I don't know about? I have never had any fruit beer except for my own, so I have no standard to shoot for. Perhaps the cloudiness is due to the pectin of the fruit in suspension. Please shed some light on this problem.
 Thank you,
 TED SAKEHAUG
 Laurinburg, North Carolina


Dear Ted,
 *Yep, you're right, especially if you've boiled or near-boiled the fruit. The pectin haze is normal. Sometimes it will clear with age, but your best bet is not to boil the fruit. Adding pectin enzyme to the fermentation also may aid clarification. And then there's always filtration for those who do that kind of thing.*
 *By the way, I love blueberry beer and pie!*
 *With a purple tongue,*
 THE PROFESSOR, HB.D.


*This column originally appeared in* Zymurgy, Summer 1990 (*vol. 13, no. 2*)

## THREE CHEERS BEER

Dear Professor,
 I occasionally take a break from brewing beer to make a batch of root beer for my low-alcohol friends. The recipe I use calls for

mixing cane sugar, root beer extract, and champagne yeast, then immediately bottling. Please answer a few questions for me.

Why don't the yeast beasties eat up all the sugar and cause the bottles to explode? Can I substitute corn sugar and ale yeast?

Do you know of any literature that describes the process for extracting the root beer flavoring from native plants?

>Thanks,
>BILLY LECALVEZ
>Aiken, South Carolina

Dear Billy,
*From my experience, the yeast does cause the bottles to explode and I would never recommend this procedure—it is far too dangerous. I suspect that when the yeast stops producing gas short of an explosion, it is because the pressure is so great in the bottle it inhibits yeast activity. Because of the danger of even airborne yeast causing bottles to explode, I don't recommend making nonalcoholic root beer. So I can't suggest whether corn sugar and ale yeast would be okay.*

*For naturally flavoring a brew, try sassafras root, sarsaparilla, wintergreen, licorice root, teaberry, deerberry, checkerberry, boxberry, spiceberry, clove, cinnamon, anise, and/or vanilla.*

>*Hip, hip hooray,*
>THE PROFESSOR, HB.D.

## LONG-FERMENTED CORRESPONDENCE

Dear Professor,
What an onionhead I am! I've only been poring over *The Complete Joy of Homebrewing* for a year and a half, but I finally figured out where to send this long-fermenting letter! If you're the editor of *Zymurgy*, which I discovered in the fine print, why not send it in care of the editorship? Hey! I am no dummy! If I'm rambling, it's because I'm enjoying a few Kinky Slinker Ales—an evolutionary delight of Winky Dink Maerzen.

I would like to ask a few questions. First of all, why does my brew get sediment on one side of the bottled product in a very fine film? At what temperature, if I do try lagering, should I pitch the yeast? Why do my beers, fermented at temperatures of 58 to 67 degrees F (14.5

to 19.5 degrees C), take so much longer to ferment than many recipes say they should? No complaints, mind you, I've made some excellent beer! Why does my Winter's Patience Summer's Payoff Ginger Mead, a variation of Barkshack, smell so bad when I bottle it? I almost threw the first batch out, but ended up drinking it all before it was five months old! Phew! It stinks when I rack it, but mellows nicely.

Cheers, relaxing and better have one,
DAVE TONETTI
Guelph, Ontario, Canada


Dear Dave,
Yeah, I've had the same phenomenon happen to me on occasion. The film seems to be a peculiarity of certain yeasts. I've noticed that all of the bottles will have this yeast film on one particular side. I forget which, whether it's the wall-facing or non-wall-facing side of the bottle, but this indicates that it has something to do with uneven temperatures in the room or geomagnetic forces. I'm stumped. Does anyone else have any hypotheses?

Most of your questions have been answered in Zymurgy, but for your benefit I will quickly answer. If you are using a true lager yeast from a liquid culture, the general norm would be to pitch the yeast at about 50 degrees F (10 degress C), but this will vary with each strain of lager yeast. Forget the dried lager yeast and pitching at low temperatures; you won't get much action.

Longer fermentations? Maybe your yeast isn't as fresh and viable as the ones I used. Try adding double the amount of yeast. Also cooler temperatures do inhibit ale fermentation, but don't worry. No problem.

Yeah, ginger root in dry meads really gives a medicinal character until it mellows out. My tastes have changed and if I make a ginger-flavored mead, it is usually a sweet mead with 2½ pounds of honey per U.S. gallon.

Wish I had a Kinky Slinker,
THE PROFESSOR, HB.D.

This column originally appeared in Zymurgy, Spring 1990 (vol. 13, no. 1)

## IN THE DISTILLNESS OF THE NIGHT

Dear Professor,
I have been brewing for 65 gallons worth of some very good beers, some great. I have a few questions. I have called the

Chicago water district to get information on the water my area gets from Lake Michigan; obviously it is very similar to Milwaukee's. The one thing that surprised me was that the chloride is often as high as ten parts per million. Do you think this high concentration could cause an astringent flavor in the finished beers? The best beers I have brewed so far were made with distilled water. Would the carbon filtering devices on the market today solve this problem? At $6 per 5 gallons of distilled water, I would like a better solution.

My last questions involve filtering. Is there a realistic way for a homebrewer to filter beer? Are there any beer-filtering devices on the market for the homebrewer? Would filtration increase the risk of contamination? Thanks in advance for your answers.

Studiously yours,
JOE JANET
Glenview, Illinois


Dear Joe,

*Ten parts per million is not enough to worry your sweet little wort about. Relax. And be aware that carbon filters will not take out chloride. Distilled water is fine for extract brews, but if you brew an all-grain beer with distilled water, you leach oxylates from your ingredients. They will in turn promote gushing in your beer. You should read up on adding minerals to prevent this if you ever decide to brew all-grain beer with distilled water.*

*Filtering? Take a gander at the Winter 1989 Zymurgy (vol. 12, no. 5) issue to get the latest scoop on filtering. There is some really nice equipment especially designed for the homebrewer. Sure, anytime you process beer there's a chance of contamination. You must take careful precautions to minimize contamination by being thoroughly familiar with sanitizing procedures.*

*Distilled crazy after all these beers,*
THE PROFESSOR, HB.D.


This column originally appeared in Zymurgy, Winter 1989 (vol. 12, no. 5)

## SOMETIMES A GREAT NOTION

Dear Professor,

I am a homebrewer only on my tenth or so batch of beer. I bought a book and jumped right into all-grain brewing—with great success.

In fact, I have never followed a recipe and have come up with a beer that makes the "king of beers" more like a squire.

My point is that I have brewed instinctively and by taste, so not only do I have little idea what to call it, I am wondering if it's important. I would like to know your thoughts on the purpose of titles like pale ale, Pilsener, stout, etc. Is the title's function to give the consumer an approximate idea of what the beer is like, or to indicate that the beer is attempting to approach some ideal taste? On the one hand, I dread labeling my beer because I don't want to compete with the preconceived notion of a "pale ale," but on the other hand, calling it "beer" doesn't quite say it all.

    Semantically muddled,
    LARS B. SPILLERS
Schenectady, New York

Dear Lars,

*You are certainly unique in your approach. I admire your "gutzpah." There are so many variations of beer within a "classic" style that I would tend to agree with your first statement: the title's function is to give the consumer an approximate idea of what the beer is like.*

*The AHA has a national competition where styles are defined probably more than you'd care to define them, but when you're staging a competition, standards must be set. It's kind of a special game some of us homebrewers play. Our definitions aren't the be all and end all, but a very good approximation.*

    *Call me approximately* Professor,
    THE PROFESSOR, HB.D.

## UNSTICKING FERMENTATION

Dear Professor,

Regarding the stuck fermentation problem posed by Peter Caddoo in the Summer 1989 issue of *Zymurgy* (vol. 12, no. 2). My experience with stuck fermentations in high-gravity materials (wort and musts) is that, excepting too high temperatures and toxic materials, the problem is usually lack of oxygen and nutrients. Yeasts need oxygen during the first 24 hours to produce adequate cell population to complete the fermentation. The oxygen also allows the yeast to produce lipids in the cell wall that protect the yeast from the alcohol toward the end of fermentation when the alcohol content is high enough to be toxic to the cells.

Yeasts require a rich supply of nutrients to perform well at high gravity. High levels of protein within the yeast cell also protect it from higher levels of alcohol.

A new product called "Yeast Hulls" should be considered as a nutrient supplement. It is rich in amino acids and lipids and can absorb any toxic by-products that may be produced during fermentation.

>CLAYTON CONE
>Technical Consultant
>Lallemand Inc.
>Montreal, Quebec
>Canada

Dear Clayton,
*Thanks, Clayton. We much appreciate your info.*
*No problem,*
THE PROFESSOR, HB.D.

*This column originally appeared in* Zymurgy, *Spring 1989 (vol. 12, no. 1)*

# GAS ATTACK

Dear Professor,
In your Fall 1988 issue, Jay S. Hersh asks about a system for storing grains. I recalled seeing an article in the November 1988 *Yachting* magazine. Your comments would be appreciated.

### Pests on Ice
*While on long-range cruises, I often faced the problem of wee beasties getting into—and destroying—my flour. But then a friend offered this solution:*

*Take a wide-neck plastic jug with a good, tight-fitting cap and place a small piece of dry ice (frozen carbon dioxide) in the bottom. Pour in flour, corn or whatever, and allow it to sit with the cap slightly unscrewed. The dry ice evaporates in about two hours, and creates a total atmosphere of carbon dioxide in which nothing can live. Screw* down the cap and the $CO_2$ gas remains in the container long enough to discourage the hatching of the eggs previously deposited, from whence the wigglies all mysteriously appear. With this approach, flour can last for years.

—Reese Palley

JERRY MARKEY
Oak Park, Illinois

Dear Jerry,
Sounds good to me as long as eggs and insects are susceptible to $CO_2$ gas. Could get expensive, though. Be careful when handling dry ice—it "burns."
THE PROFESSOR, HB.D.

_fourteen_

# Homebrew Cooking

## HOMEBREW COOKING WITH THE BREWGAL GOURMET: A FESTIVE FALL MENU

### BY CANDY SCHERMERHORN

*This article originally appeared in Zymurgy, Fall 1994 (vol. 17, no. 3)*

**Menu**

- Pumpkin Soup and Amber Lager
- Crystal Malt Dinner Rolls and Weissbier
- Marinated Game, Spicy Roasted Potatoes, Mixed Green Salad and Scottish Ale
- Platter of Fresh Fruit

Fall's earthiness compels many of us to dust off treasured heirloom recipes and get back to our cooking "roots." Be it roasting game using Grandmother's cursory notes, fogging up the windows with a pot of soup, or baking fragrant loaves, we long to re-create the spell of cheerful hearth and kitchen.

Needless to say, beer can weave a ribbon of continuity through this culinary scenario. From soup and bread to meat and potatoes, meals prepared and served with homebrew will have their own distinctive signature.

To create a meal for the harvest months, you might want to begin with a cup of russet-hued pumpkin soup aided by the nutty flavor of amber lager. Roasting the nuts, onions, and garlic before puréeing them with the pumpkin adds a richness to this soup. To keep expectations high, pass a basket of tender rolls made with crushed crystal malt, English brown ale, and tangy buttermilk, and serve with flutes of Weissbier. This is a perfect pause before the main course—a feast of succulent roasted game (anything from rabbit and fowl to venison, antelope, elk, and commercial buffalo) and spicy potatoes served with Scottish ale.

Modern "farming" techniques have provided a revival of game in the American kitchen. Game benefits from beer, and not just from the tenderizing qualities. Because of the meat's stronger flavor, game recipes are often heavy-handed in calling for potent herbs that can overwhelm the meat's flavor. Using the full, malty flavor of bock in a moderately seasoned beer marinade is the perfect way to balance the game's intense character without destroying its essence.

Of course, it is vital to understand how to cook game, a critical factor that is too often overlooked. Most game is very lean and if overcooked, will become extremely dry and tough. It lacks the internal basting qualities found in marbleized meat. Once the juices have been cooked out, you have nothing more than varying stages of parched toughness. Game should only be cooked to a reddish pink or slightly pink stage for prime eating (an internal temperature of 120 to 130 degrees F or 49 to 54 degrees C when roasting venison, for example). Marinating in an oil-rich marinade, rubbing with additional oil, and basting also help retain full succulence.

To complement this old-world meal, spicy roasted potatoes are a must. Here the beer is used in an innovative way that also complements root vegetables (onions, carrots, turnips, beets). A full-bodied ale is simmered with spicy crab boil until deeply flavored. The liquid is strained and the potato wedges are simmered in it before being roasted to golden perfection. If desired, you can refrigerate the broth to use again.

Embellish the evening with a salad of mixed greens and a platter of ripe fruit, and you have a meal that reaches into the past for its hearty nature and leaves us with a modern appreciation for the fine beer and food that have molded our heritage.

## 🍺 *Pumpkin Soup*

Serves six

- 1 medium brown or yellow onion, unpeeled
- 6 cloves garlic, unpeeled
- ⅔ c. shelled pecans or walnuts (optional), toasted
- ¼ tsp. ginger
- ¼ tsp. allspice
- 2 c. chicken or vegetable stock*
- 1 c. amber lager
- 1½ c. canned pumpkin
- 1 c. sour cream
- ½ c. milk
- 1 extralarge egg, beaten
- salt and pepper to taste

**1.** In a microwave oven, bake the unpeeled onion for about 6 minutes on high until softened like a baked potato. Or bake at 350 degrees F (177 degrees C) for 40 minutes. Cool thoroughly before peeling.

**2.** In a 350-degree F (177-degree C) oven, bake the unpeeled garlic and nuts on a cookie sheet for 15 minutes. Cool and peel garlic.

**3.** Combine onion, garlic, nuts, ginger, and allspice in a food processor and process until fairly smooth.

**4.** Heat stock, lager, and pumpkin. When hot, add puréed mixture and stir until combined. Simmer slowly for 15 minutes.

**5.** Stir in sour cream. Beat egg and milk together and pour slowly into pumpkin mixture while stirring constantly. Cook for 10 minutes but do not boil. Season to taste and serve hot.

*Make a rich broth with chicken or vegetables and reduce by half.

## Crystal Malt Dinner Rolls

Makes about 24 rolls

- 1 c. brown ale
- 1 c. buttermilk
- 2 tbsp. dried baker's yeast
- ½ c. honey (or unhopped malt extract)
- 2 large eggs
- 1 tbsp. salt
- 2 c. crushed crystal malt
- 5 to 6 c. bread flour
- parchment
- 1 egg white beaten with 1 tbsp. warmed honey

**1.** Warm beer and milk until lukewarm (150 degrees F or 66 degrees C). In a large bowl, whisk together the yeast and 1 cup of flour. Add warmed liquid and whisk thoroughly. Allow the mixture to stand in a warm place until foamy (about 15 minutes).

**2.** Beat in the honey, eggs, salt, and crystal malt. Add 3 cups of the flour and stir until stiff.

**3.** Add the remaining flour, 1 cup at a time. When the dough is firm and no longer sticky, stop adding flour and knead for 10 minutes, adding just enough flour to keep the dough from sticking.

**4.** When the dough is elastic and smooth, place in a lightly oiled large bowl, turning once to coat with oil. Cover with plastic wrap and leave in a warm place until doubled in bulk.

**5.** Punch down dough and divide in half. Divide each half into ten to twelve rolls and place on a parchment-covered baking sheet. Cover with a cloth and allow to rise until doubled in bulk.

**6.** Brush rolls with egg white and honey mixture and bake in a preheated 350-degree F (177-degree C) oven for about 20 minutes or until golden.

## 🍺 **Marinated Game**

This recipe works well with many types of game, including venison, elk, and antelope. Makes 5 cups of marinade, enough for a 10-pound roast that will serve eight to twelve.

- 1 hind or saddle cut (6 to 10 lbs.)
- 3 c. bock
- 1 c. virgin olive oil
- ⅔ to 1 c. quality malt vinegar (imported brands are typically of high quality)
- 4 to 6 cloves garlic, crushed and mashed with 1 tbsp. salt
- 1 small yellow onion, thinly sliced
- 1 tbsp. dried parsley
- 1 tbsp. dried crushed thyme
- 2 tsp. dried crushed rosemary
- 2 tsp. orange zest, finely chopped
- 2 tsp. whole black peppercorns
- 1 tsp. tarragon
- 1 tsp. marjoram
- 1 tsp. juniper berries (optional), lightly bruised with the back of a spoon

**1.** Combine all marinade ingredients and let stand for 1 hour at room temperature.

**2.** Cover meat with marinade and refrigerate 24 to 36 hours, turning three or four times. Remove and strain marinade, reserving the liquid to baste the meat.

**3.** Roast meat at 425 degrees F (218 degrees C) for 15 minutes for boneless tied roasts or 25 minutes for large bone-in roasts. Lower heat to 350 degrees F (177 degrees C), baste, and roast until an internal thermometer reads 120 degrees F (49 degrees C) for rare or 130 degrees F (54 degrees C) for medium rare. Baste every 10 minutes.

**4.** When done, remove from oven and cover with foil. Allow to rest for 15 minutes before carving. Serve with pan juices, if desired.

## Spicy Roasted Potatoes

Serves six

    6 c. water
  1½ c. brown ale
    ¼ c. crab boil
  1½ tbsp. salt
    6 large Idaho or Russet baking potatoes
      olive oil or melted butter for brushing
      kosher salt and freshly ground black pepper
      parchment

**1.** Bring the water, ale, crab boil, and salt to a boil and simmer for 45 minutes, covered. Strain, discard spices, and bring liquid back to a boil.

**2.** Meanwhile, quarter potatoes lengthwise. Split each quarter in half, add to the simmering seasoned liquid, and cook just until tender (about 10 minutes). Drain and place on a parchment-covered baking sheet.

**3.** Brush the potatoes lightly with oil and sprinkle with salt and pepper. Bake at 450 degrees F (232 degrees C) until golden crisp outside, moist and tender inside. Serve hot.

*Candy Schermerhorn is a culinary consultant and televised cooking personality in the Phoenix, Arizona, area. Candy takes great joy in educating the public about beer and its culinary potential through her classes. She is author of the* Great American Beer Cookbook (Brewers Publications, 1993).

# fifteen

# Homebrewer of the Year Winning Recipes

Every year since 1979, the American Homebrewers Association has staged its National Homebrew Competition, drawing thousands of entries from all over the world. From the beginning, the honor of "Homebrewer of the Year" (an award sponsored by Munton and Fison of England since 1981) has been presented to the person having brewed what the judges deem to be the best of show.

There were more than 3,000 entries in the 1996 competition alone, judged by a continually improving cadre of qualified beer judges, many of whom are now certified in the Beer Judge Certification Program. These eighteen recipes are those judged the best in each year of competition.

No fewer than six of these winning homebrewers went on to become professional brewers after winning Homebrewer of the Year.

## 1979 Tim Mead Rag Time Black Ale

13 *gallons*

- 2 gal. water
- 6 lbs. dark dry malt
- 6½ lbs. John Bull dark malt
- 4 oz. hops (plus 1 oz. Cascade hops optional)
- ¼ tsp. Irish moss
- 5 tsp. gypsum

- 2 tsp. salt
- 1 lb. black patent malt
- 2 lbs. crystal malt

All the grains were mashed for 1 hour. The rest of the ingredients were added and boiled for an additional hour.

Finally, steep in wort:

- ½ oz. cinnamon bark

Add 5 lbs. sugar to primary fermentor.

Ale yeast was added at 74 degrees F for a resulting hydrometer reading of 1.048.

### 1980 Mary Beth Millard Birthday Brew Snow-High Light Lager

13 *gallons*

- 4 cans Munton & Fison plain light malt extract
- 4 oz. Cascade hops for boiling
- 3 tsp. gypsum
- 1½ oz. Cascade hops for finishing
- 5 lbs. corn sugar
- 2 packets beer yeast
- water

Boil 2 gallons of water with malt, boiling hops, and gypsum for 30 to 45 minutes. Tie finishing hops in cheesecloth and immerse during the final 10 minutes of boiling. Save these hops for finish hopping in the primary. Turn the stove off and let the wort "steep" for 60 minutes. After steeping, sparge the wort into the primary fermentor, to which 5 pounds of corn sugar has been added. Fill primary to 13-gallon level. Add finishing hops and yeast. Beginning specific gravity: 1.045. Fermentation temperature 74 degrees F. Brewed May 3, 1978. Bottled June 15, 1978.

## 1981 Dave Miller Dutch Style Lager

*5 gallons*

- 9 gal. soft, neutral water (9 qt. mash + 5 gal. sparge, rest in reserve)
- 5½ lbs. lager malt
- 1½ lbs. rice
- 1½ tsp. or more gypsum
- 1 oz. Saaz or Cascade hops (boil)
- ¼ oz. same type hops (finishing)
- 1 tsp. Irish moss
- 1 packet lager yeast
- ¼ c. Polyclar (optional)
- 1¼ c. corn sugar (priming)

Starting gravity: 1.050

**1.** Grind the malt and prepare the brewing water. Wash the rice and boil in 6 quarts mash water until gelatinized (about 45 minutes). Add to remaining 3 quarts mash water, along with 1½ teaspoons gypsum.

**2.** Bring mash water to 125 degrees F and stir in malt. Protein rest 118–125 degrees F for 45 minutes. Boost to 155 degrees F in 20 minutes. Starch conversion rest 150–155 degrees F for 45 minutes. Boost to 168 degrees F and rest 5 minutes. During mash, check temperature and stir frequently. Always stir when applying heat.

**3.** Transfer mash to lauter tub, heat sparge water to 160 F, and sparge.

**4.** Add more gypsum (if needed) to boiler and boil wort 1½ hours. After ½ hour add boiling hops; 15 minutes before end, add Irish moss.

**5.** At end of boil, turn off heat, stir in finishing hops, and rest 1 hour.

**6.** Strain wort into primary, removing 1 quart for starter. Top up to 5 gallons if necessary, cover, and force-cool.

**7.** Meanwhile, force-cool starter wort to 80 degrees F and add yeast.

**8.** When wort is cool (70 degrees F) stir up starter and pitch it in. Ferment in a cool place (50–55 degrees F ideally). When fermentation slows down (SG approximately 17) put dry hops in bottom of secondary and rack. Fit air lock.

**9.** When fermentation is over (no gravity drop for 5 days) terminal gravity should be 11–13. Rack into primary, add priming sugar, and bottle. Age two to three months.

Recipe from David Miller's *Homebrewing for Americans*, published by Amateur Winemakers Publications Ltd., England.

## 1982 Donald F. Thompson Light-Bodied Light Lager

*5 gallons*
- 7 lbs. two-row malted barley
- 1½ oz. Cascade hops
- tap water
- "Paul's Lager Yeast" (brewery yeast for cold fermentation)
- ¾ c. corn sugar

Step infusion mash:

**1.** Mash in with 2 gallons water 126 degrees F and hold for 30 minutes.

**2.** Raise temperature to 144 degrees F and hold for 30 minutes.

**3.** Raise temperaure to 158 degrees F and hold for 1½ hours.

**4.** Sparge with 160-degree F water until wort runs out at 1.010.

**5.** Bring wort to boil for 1 hour and 20 minutes. Add all hops for the last 60 minutes of boiling.

**6.** Sparge wort through spent hop "filter bed" (hop-back).

**7.** Cool wort to 61 degrees F and pitch active yeast starter. Original gravity: 1.042. Main fermentation was carried out at about 54 degrees F.

**8.** When fermentation was complete the beer was primed with about the equivalent of ¾ cup corn sugar and bottled.

Donald emphasized to *Zymurgy* that the single most important improvement in technique that he has made as a homebrewer (and a potential microbrewer) is the discovery of the importance of yeast. "It has *all* made the difference in the world," he said. Donald uses liquid cultures of brewery-grade strains (and follows instructions very carefully) from sources advertised in *Zymurgy*.

## 1983 Nancy Vineyard

*5 gallons*

- 9½ lbs. U.S. pale malted barley
- 1½ lbs. U.S. crystal malted barley
- 1¾ oz. Northern Brewer hop pellets (for bittering)
- 1 oz. Cascade hop pellets (aromatic)
- 2 tsp. gypsum
- 1 tsp. noniodized salt
- 2 packets Red Star lager yeast

**1.** Three-step infusion mashing: 95 degrees, 125 degrees, 148 degrees F.

**2.** Use 1½ quarts water per pound of grain—mashing liquor. Add gypsum and salt.

**3.** Mash 15 minutes at 95 degrees to pH 4.8; 30 minutes at 125 degrees; and 1½ hours at 148 degrees F.

**4.** Transfer to a picnic cooler fitted with a copper strainer coil. [Editor's Note: Or set up your own sparging system.]

**5.** Heat 4 gallons of water to 175 degrees F for sparging and run off in 15 minutes to collect 6 gallons.

6. Boil with Northern Brewer hops. Add 1 ounce of hops at beginning of boil. After 45 minutes add remaining ¾ ounce and continue to boil for 20 minutes.

7. Add the Cascade hops during the final 3 minutes of boiling.

8. Strain hops by siphon transfer from boiler to primary.

9. Cool wort and pitch yeast.

   Original specific gravity: 1.042
   Terminal specific gravity: 1.007
   Age when entered (after bottling): six months
   Approximate temperature at fermentation: 58 to 63 degrees F.


## 1984 Dewayne Lee Saxton Du Bru Ale

20 *gallons*
- 6 lbs. Bavarian Gold unhopped malt extract
- 6 lbs. Munton & Fison light dry malt
- 3 tsp. citric acid
- 3.3 oz. Cascade hop pellets
- 2.5 oz. Cascade whole hops
- 2 oz. Brewers Gold whole hops
- 1 lb. wheat malt
- 12 lbs. corn sugar
- 4 packets Muntona yeast
- 1 tsp. yeast nutrient
- ½ c. Munton & Fison light dry malt
- 3 c. priming sugar
- Distilled spring water

1. Start yeast 48 hours early in ½ gallon distilled water with ½ cup M&F malt and yeast nutrient.

2. In 3 gallons of distilled water boil citric acid and Bavarian extract for 20 minutes.

3. Add M&F dry malt and continue boil for 30 minutes.

4. Add Cascade pellets and boil another 30 minutes.

**5.** Immerse a strainer bag filled with Cascade and Brewers Gold loose hops and wheat malt into the wort and continue to boil for another hour.

**6.** Sparge into a primary fermenter, add corn sugar, and stir until dissolved.

**7.** Add distilled water to make 20 gallons, cool to 60 degrees F, and add yeast starter.

**8.** Rack into secondary after two weeks and bottle with priming sugar after 1½ months.

> Original specific gravity: 1.045–48
> Approximate temperature of fermentation: 60 to 64 degrees F
> Age when judged (since bottling): three months
> Standard used for judging: Spaten Club-Weisse

### 1985 Russell Schehrer Tovene Porter

*5 gallons*
- 6½ lbs. dark malt extract
- 20 oz. Munich malt
- 24 oz. crystal malt
- 12 oz. dextrine (or CaraPils) malt
- 8 oz. black patent malt
- 1½ oz. Cascade boiling hops (1 hour)
- ¾ oz. Cascade hops (last minute)
- 2 tsp. gypsum
- pinch of Irish moss
- Great Dane ale yeast
- 1 tsp. yeast energizer
- ½ c. corn sugar to prime

Original specific gravity: 1.053
Terminal specific gravity: 1.011
Age when judged (since bottling): 8½ months

(In fond memory of Russ Schehrer.)

## 1986 Byron Burch Jerry Lee Lewis Russian Imperial Stout

5 *gallons*
- 5 lbs. Munton & Fison extradark dry malt extract
- 3.3 lbs. John Bull plain dark malt syrup
- 1 lb. crystal malt
- 4 oz. Munich malt (mashed)
- 8 oz. black patent malt (finely ground and boiled 30 min.)
- 5 lbs. Great Fermentations white rice syrup
- 1 lb. corn sugar
- 2¾ oz. Northern Brewer pellets and ¼ oz. Cluster pellets (60 min.)
- 1¾ oz. Nugget pellets and ¼ oz. Eroica pellets (30 min.)
- 2 oz. Cascade pellets and ¼ oz. Saaz pellets (dry-hopped for aroma)
- 10 g. Red Star Pasteur champagne wine yeast
- ½ tsp. Great Fermentations nutrient
- ¾ c. corn sugar (syrup) and 5 oz. lactose to prime

Original specific gravity: 1.099
Terminal specific gravity: 1.040
Age when judged (since bottling): 6½ months

## 1987 Ray Spangler Toadex/Bloatarian Ale

5 *gallons*
- 9½ lbs. pale ale malt
- 1½ lbs. Munich malt
- ¾ lb. crystal malt
- ½ lb. wheat malt
- Irish moss (end of boil)
- 2 oz. Hallertauer hops (60 min.)
- 1 oz. Tettnanger hops (60 min.)
- ¾ oz. Hallertauer hops (5 min.)
- ½ oz. Cascade hops (5 min.)
- Cultured Chimay "Rouge" ale yeast
- 2 oz. light dry malt extract
- 2 oz. dextrose boiled with 1 quart wort to prime

*Brewer's Specifics*
Decoction mash: one third mash raised to boil, added back, raising mash to 105 degrees F; one third mash raised to boil for 10

minutes, added back, raising mash to 148 degrees F; one third mash raised to boil for 5 minutes, added back, raising mash to 158 degrees F; mash raised to 170 degrees F for 10 minutes. Note: Use 10-minute rest between steps.

Original specific gravity: 1.062
Terminal specific gravity: 1.020
Age when judged (since bottling): six months

## 1988 John C. Maier Oregon Special

5 *gallons*
- 11 lbs. Williams Australian dry malt extract
- 3 lbs. Klages malt
- 5 oz. Nugget hops (45 min.)
- 1½ oz. Willamette hops (10 min.)
- 8 oz. yeast starter of Sierra Nevada culture
- ¾ c. dextrose to prime

Original specific gravity: 1.075
Terminal specific gravity: 1.025
Age when judged (since bottling): twenty-three months

*Brewer's Specifics*
Mash grains at 120 degrees F for 30 minutes. Raise heat to 130 degrees F. Infuse boiling water, raise to 152 degrees for 15 minutes. Raise heat to 158 degrees F for 10 minutes. Raise to 170 degrees F. Sparge with 2 gallons 170-degree water.

## 1989 Paul Prozeller Dubbel Queensberry Framboise

5 *gallons*
- 6½ lbs. two-row pale ale malt
- 2 lbs. wheat malt
- 1 oz. Fuggles hops (60 min.)
- ⅓ oz. Challenger hops (20 min.)
- 11 pt. raspberries
- 1 tsp. gypsum
- Williams liquid German alt yeast

Original specific gravity: 1.055
Terminal specific gravity: 1.009
Age when judged (since bottling): two months
Boiling time: 60 minutes
Duration during fermentation: three to four weeks
Approximate temperature during fermentation: 70 degrees F (21.1 degrees C)
Secondary fermentation: yes
Type of fermenter: stainless steel

*Brewer's Specifics*
Half-hour protein rest at 120 degrees F (48.8 degrees C). One-hour saccharification rest at 155 degrees F (68.3 degrees C). Boil 1 hour. Primed using 1 quart 1.052-gravity wort.

## 1990 Richard Schmit Arlington Ale No. 33

*5 gallons*
- 3⅓ lbs. John Bull light hopped malt extract
- 2 lbs. light dry malt extract
- 4 oz. toasted pale malt
- 3 oz. crystal malt
- ¼ oz. Cascade hops (10 min.)
- ¼ oz. Willamette hops (10 min.)
- ¼ oz. Cascade hops (2 min.)
- ¼ oz. Willamette hops (2 min.)
- 1 tsp. Irish moss
- 1 tsp. ascorbic acid
- Wyeast No. 1056 American ale liquid yeast
- ¾ c. corn sugar for priming

Original specific gravity: 1.042
Terminal specific gravity: 1.013
Primary fermentation: 8 days at 65 degrees F (18.5 degrees C) in plastic
Secondary fermentation: 2 days at 65 degrees F (18.5 degrees C) in plastic
Age when judged (since bottling): 6½ months

*Brewer's Specifics*
Steep grains, bring to 200 degrees F (93.5 degrees C), and sparge. Hops added at end of boil.

## 1991 Jim Post Jamie Beer

*5 gallons*
- 12½ lbs. Munton & Fison pale two-row malt
- 2 lbs. William's caramel malt
- 3 oz. William's dark dry Australian malt extract
- 3 oz. Mt. Hood hops (60 min.)
- 2 tbsp. calcium carbonate
- 2 tbsp. Irish moss
- New England Brewing Company lager yeast culture

Original specific gravity: 1.052
Terminal specific gravity: 1.012
Boiling time: 60 minutes
Primary fermentation: 24 days at 55 degrees F (13 degrees C) in glass
Age when judged (since bottling): two months

*Brewer's Specifics*
All grains mashed in a single-step infusion at 158 degrees F (70 degrees C) until conversion was complete.

## 1992 Stu Tallman Stu Brew

*10 gallons*
- 15 lbs. pale malt
- 4 lbs. Munich malt
- 4 lbs. 40 °L crystal malt
- 2½ oz. Saaz hops (90 min.)
- Wyeast No. 2026 liquid yeast

Original specific gravity: 1.060
Terminal specific gravity: 1.018
Boiling time: 90 minutes
Primary fermentation: 21 days at 50 degrees F (10 degrees C) in glass
Secondary fermentation: 21 days at 50 degrees F (3 degrees C) in glass
Age when judged (since bottling): four months

*Brewer's Specifics*
Three-step upward infusion mash.

## 1993 Paddy Giffen Kilts on Fire

5 gallons
- 4 lbs. smoked Pilsener malt
- 4½ lbs. Belgian Pilsener malt
- 5 lbs. amber dry malt extract
- 1 lb. CaraVienne malt
- ¾ lb. Special "B" malt
- 1 lb. Munich malt
- 1 lb. British crystal malt
- ¼ oz. Chinook hops (60 min.)
- ¼ oz. Chinook hops (30 min.)
- ¼ oz. British Blend hops (30 min.)
- ½ oz. Liberty hops (30 min.)
- Wyeast No. 1084 liquid yeast culture
- forced carbonation

Original specific gravity: 1.088
Terminal specific gravity: 1.038
Boiling time: 60 minutes
Primary fermentation: 11 days at 65 degrees F (18 degrees C) in glass
Secondary fermentation: eight weeks at 65 degrees F (18 degrees C) in glass
Age when judged (since bottling): four months

*Brewer's Specifics*
Mash grains for 85 minutes at 154 degrees F (68 degrees C).

## 1994 James Liddil Wild Pseudo-Lambic Gueuze

5 gallons
- 6 lbs. Briess Wiezen malt extract
- 4 oz. old hops, variety unkown (60 min.)

 Yeast and bacteria strains:
- 2 strains of *Kloeckera apiculata* (yeast)
- 4 strains of *Dekkera bruxellensis* (yeast formerly known as *Brettanomyces bruxellnsis*)
- 4 strains of *Dekkera anomoia* (yeast formerly known as *Brettanomyces lambica*)
- 1 strain of *Saccharomyces cerevisiae* (Williams Brewing Burton ale yeast)

    1 strain of *Pediococcus damnosus* (bacteria)
    1 c. corn sugar (to prime)

Original specific gravity: 1.045
Terminal specific gravity: 1.008
Boiling time: 60 minutes
Primary fermentation: four months in plastic
Secondary fermentation: four months in plastic
Age when judged (since bottling): seven months

## 1995 Rhett Rebold Central European Pils

12 *gallons* (45.5 *l.*)
    11 lbs. German Pils malt (4.99 kg.)
    8 lbs. Belgian Pils malt (3.63 kg.)
    1½ lbs. German Vienna malt (0.68 kg.)
    1½ lbs. CaraPils malt (.068 kg.)
    3 oz. Saaz hops, 4.2% alpha acid (46 g.) (50 min.)
    1⅔ oz. Saaz hops, 4.2% alpha acid (46 g.) (25 min.)
    4½ oz. Saaz hops, 4.2% alpha acid (128 g.) (5 min.)
    Wyeast No. 2124 liquid yeast culture
    force carbonated in keg

Original specific gravity: 1.048
Terminal specific gravity: 1.013
Boiling time: 120 minutes
Primary fermentation: 18 days at 54 degrees F (12 degrees C) in glass
Secondary fermentation: 30 days at 38 degrees F (3 degrees C) in glass
Age when judged (since bottling): ten months

*Brewer's Specifics*
Mash grain at 152 degrees F (67 degrees C) for 3 hours.

## 1996 John R. Fahrer Muddy Mo Amber Ale

5 *gallons* (19 *l.*)
    8½ lbs. Schrier two-row malt (3.8 kg.)
    10 oz. 120 °L crystal malt (283 g.)
    1 oz. chocolate malt (28 g.)
    ½ oz. Nugget whole hops, 12% alpha acid (14 g.) (60 min.)

½ oz. Fuggles hop plugs, 4.2% alpha acid (14 g.) (30 min.)
½ oz. Fuggles hop plugs, 4.2% alpha acid (14 g.) (10 min.)
1 oz. Tettnanger hop pellets, 4.7% alpha acid (28 g.) (finish)
   Wyeast No. 1028 London ale liquid yeast culture
66 oz. wort (21.95 l.) (to prime)

Original specific gravity: 1.050
Terminal specific gravity: 1.012
Boiling time: 75 minutes
Primary fermentation: 7 days at 68 degrees F (20 degrees C) in stainless steel
Secondary fermentation: 16 days at 66 degrees F (19 degrees C) in stainless steel
Age when judged (since bottling): seven months

*Brewer's Specifics*
Mash grains at 122 degrees F (50 degrees C) for 25 minutes, 152 degrees F (67 degrees C) for 60 minutes, 157 degrees F (69 degrees C) for 20 minutes, 169 degrees F (76 degrees C) for 10 minutes.

# sixteen

# Jackson on Beer

*Michael Jackson is internationally the best-known writer on beer. His Beer Companion (Running Press, 1994) was awarded the 1994 Glen Fiddich Trophy, an honor never before bestowed on a book on beer. His articles, books, and documentary videos have introduced beer styles to countless drinkers and brewers outside their native lands. Jackson's use of taste descriptions and accounts of his travels introduced a new genre of writing on beer.*

## DRINKABILITY: WHAT IS IT?

### BY MICHAEL JACKSON

*This article originally appeared in* Zymurgy, *Special 1987 (vol. 10, no. 4)*

It was a hot day in San Francisco and I had foolishly chosen to climb Russian Hill on foot, instead of taking a cab to my rendezvous there. Then I had schlepped the baggage of a six-week tour in and out of a shuttle bus onto the scale at the airport check-in (and I had a lazy driver, for whose lack of pains I offered no tip, and there was not a skycap to be seen). When I made landfall in Denver, it was only to change planes for further journeying. I began my layover by heading for a beer.

This being Denver, the beer was Coors. "Large or small, sir?"

# Drinkability: What Is It?

Michael Jackson
MICHAEL LICHTER

The "large" successfully beckoned. It slipped down like a Rocky Mountain stream. "Another, sir?" My thirst had been slaked sufficiently for me to think twice about another paper cup, but I complied. A regular, everyday American beer had never tasted so good. Nor has it since.

For me, the cleanness of Coors has always mitigated the corn taste that so often makes me feel slightly nauseous after the first two or three regular American beers. To be pleasantly drinkable, a light Pilsener-style beer has to be crystal-clean; that is not necessarily true of other styles. This was Colorado, and the beer was as fresh as a virgin's kiss.

Freshness is critical to the drinkability of all beers except the few that are designed to mature in the bottle. A beer that is less than fresh is less than truly drinkable; that does not mean it is necessarily undrinkable. In particular, I believe that hop aroma fades fast in most beers. So, to a lesser extent, does hop flavor. Without these elements, most styles of beer are less inviting—and lack balance.

The beer that day was on draft, and not bursting angrily with that trapped-in-the-bottle, soda-pop carbonation that is so often evident in packaged brews. The only drinkable beers with a high carbonation are the wheat brews that naturally have that character. Something to do with the way in which the carbonation is absorbed, I would guess.

This memorable beer was (as might be expected) approximately well chilled, but it was not Popsicle-cold. I have only once yielded to one of those signs that says: "Coldest beer in town." I had spent the day in a car that had no air-conditioning, crossing from Tennessee into Kentucky. The sign also said: "Last beer before dry county."

I had made the elementary mistake of asking which beers the bar had stocked. The owner of the bar (actually, it was a wooden hut) glowered at this intrusive question, and his neck suffused even further. "Beer! You want a beer?" That's what they had—beer. I was clearly a communist agent. The beer turned out to be the localish label of a national giant. Perhaps it was poisoned, though I suspect that my feeling sick to the stomach for the rest of the day derived simply from the intense cold of the beverage.

To be drinkable, a beer must not be aggressively chilled, served in a frozen glass, nor treated with any other manifestation of brutality. That includes careless or overintensive pasteurization. Even the most "industrial" of beers is a natural product of some delicacy. Don't bore me with simplistic twaddle about personal (or "American") taste. If you like your beer out of a frozen mug, you don't have any taste. Neither will the beer. If they filled that mug with Dr. Pepper, you would not be able to tell the difference.

As for "American" tastes, for every time I have enjoyed my Coors on a hot day, I have known an American to appreciate a Sam Smith's. Some beers are undoubtedly more drinkable than others, and it is often hard to say just why, but it has nothing to do with country of origin. Nor can anyone with a sense of taste be gastronomically monogamous.

Like a good partner, a fine beer responds to being cherished. Like a good partner, it should be chosen with care. In these matters, there are magic moments.

Monogamy, though, is something else. I cannot be monogamous in respect to countries, either. I love so many of them. After a long absence from Ireland, a country that I love hopelessly, I was delighted to find myself one day in Dublin, and specifically in the Hibernia, in those days the finest hotel in town. I made myself comfortable in the restaurant, and was presented with a menu that was bloated with untruths written in French. My eyes glazed, and I entertained a reverie concerning prawns from Dublin Bay, lobster from Kinsale, and oysters from Galway.

"I'll have a dozen oysters, please, to start. Then I'll have the mixed grill." The frock-coated waiter noted this down and inquired as to which vintage I would like from a cavernous wine list. "With the oysters, I'll have a pint of Guinness. With the mixed grill, I'll take a Smithwick's Ale."

To his credit, he smilingly acceded to this request, with only a barely perceptible twitch of his coattails. Guinness is rarely anything less than drinkable, and on this occasion it was magnificent. Smithwick's is never magnificent, but on this occasion it was very drinkable. It was the right beer for the dish, and for the moment.

When I make one of my occasional inspection visits to my home in London, the second thing I do is walk down the road to my local for a pint of Young's Bitter. The very sight of the full glass renders me silent in contemplation. I look around and see many other prospective drinkers caught in the same moment of reflection. They look at their pint and relish the pleasure they are about to experience. They look at their pint, raise the glass, and look again, longingly. Then they take a sip, and put the glass down again for a further pause, a moment of satisfaction.

This is a slow method of drinking. Americans generally drink much more quickly. The United States is a land of instant gratification: "I want it now; I do not have the time (or patience?) to wait. I probably do not really have the time to enjoy this."

The tortoises of Britain drink more beer per head than the hares of the United States. To be drinkable, a beer must inspire reflection. If it offers nothing upon which to reflect, it is not truly drinkable but merely easy to swallow. Like a good partner, a drinkable beer is worth waiting for, seeking out, learning to understand. It offers more than one urgent, rushed moment of pleasure.

Short of hearing the patter of rain on the window of a shared bedroom on a Sunday morning, there is nothing more pleasurable than the sight of drenched people shaking out their raincoats as they enter the haven of a pub. They say it always rains in Seattle and London. Though I have not found this to be true in either case, it would help explain the extent of civilized drinking in these two cities.

Hot weather makes for fevered drinking (and for skunky beer). In the heat of a barbecue, beer never seems quite as good as it should, but it can be very drinkable on a spring day by the Thames. Sip your pint and shed a bead of sweat for the scullers and eights rowing on the river. I find this fine exercise: sipping my pint, I mean.

Ritual makes us all feel safer, and therefore more relaxed. I watched an elderly man silently enjoying a glass of lambic in a café in the Senne Valley of Belgium. After a time, a neighbor entered the café. The first man looked up: "If it's time for you to arrive, it must be time for me to go home." You would be surprised how drinkable a glass of lambic can be in the Senne Valley. It is the right beer for the place.

When the beer gardens open each spring, the people of Bavaria stop in for a midmorning snack of veal sausage, washed down with a liter of bock. At lunch it might be "liver cheese" and half a liter. In the early evening there will be endless curls of black-skinned radish, a prelude to a picnic, perhaps with a regular pale beer.

On summer weekends there will be religious services in the mornings to bless festivals that last all afternoon and evening, and at which beer is consumed by the liter or half for the duration by a cast of thousands. A band plays and every 15 minutes or so strikes up the chorus, "Ein Prosit," upon which the assembled company toasts. A brewer once told me that in his village festival a hectoliter of beer was consumed every time the band played "Ein Prosit." He knew; he provided, and sold, the beer.

I find more over which to reflect in a pint of British ale or a glass of lambic, but I have consumed more Bavarian beer at a sitting. Far more. It is not the most complex of lager beers, though it is not lacking in subtlety. It is, in its "Export" and bock forms, quite big-bodied and strong, and would not seem to be suitable for consumption in quantity. Yet I have often drunk it in great quantity.

The beer is always as fresh as the new afternoon. Take a mouthful and taste the clean sweetness of the malt. Let the beer slip down, and be untroubled by its gentle carbonation, well absorbed during weeks of lagering. Where else would you find these qualities, except perhaps in one of the better country breweries in Czechoslovakia?

Beers of this type are, in my experience, the most drinkable of all. I am most definitely not electing them as my favorites—merely saying that they are the most drinkable. I do not know why. Not the whole story, anyway. I will continue to work on it until I have the answers. I shall not hurry. This may take some time.


# WHEN A BEER GOES TO SLEEP

## by Michael Jackson

*This article originally appeared in* **Zymurgy**, *Spring 1988 (vol. 11, no. 1)*

No knowledgeable beer lover would readily "lay down" a brew unless it was both strong and bottle-conditioned, but I have twice been invited to disregard this rule. Contrary to all of my instincts, it has been suggested that I lay down a couple of beers that are filtered and pasteurized.

I would have brushed aside these invitations had they not been made by brewers in respect of their own products. The brewers were Lees of Manchester and Pripps of Sweden. Each was referring to a specific beer.

I have heard of people keeping for years beers they found in some faraway land. The assumption is that beer of exotic origin will somehow benefit from being allowed to gather a few cobwebs. This is true only if the bottle is to be kept unopened as a memento.

Most beers are made to be drunk immediately, and are vulnerable to deterioration from the moment they are packaged. This does not mean they will deteriorate quickly, but they might. A conventional beer that invites consumption should be taken up on its offer without hesitation.

On such a beer, a long "best consumed by" date indicates a dose of preservatives, or a case of incurable optimism or terminal fecklessness on the part of the brewer.

The fuller a beer's flavor, the longer it may take to erode; the darker a beer, the more it seems able to protect itself against the effects of supermarket lighting; the stronger it is, the more chance that the alcohol will act as a preservative. But these circumstances do not attend many everyday brews.

In their case, the least damage that time will do is to diminish hop character, especially in the aroma of the beer. Worse is oxidation, causing a stale, "damp cardboard" taste. Worse still is supermarket lighting, which can impart a cabbagey character.

When a beer is made specifically for the purpose of laying down, it will usually have sufficient gravity to provide the residual sugars necessary for fermentation to continue, albeit very slowly. Also, it will usually be bottle-conditioned, thus containing the yeast to do the job.

What is not clear is how long the yeast can keep up some form of activity, however limited. Yeast gets tired. It is also overwhelmed by the alcohol it has created. When the yeast is still working, the beer is "alive"—perhaps "awake" might be a happier term. If the yeast is sent to sleep by the alcohol, then the latter protects the slumbering beer.

It can be argued that the yeast is overpowered within weeks, yet it also has been demonstrated in experiments that the flavor of a beer can evolve quite dramatically for at least five years. So is the yeast still dreamily making the changes, or are other reactions at work among the hundreds of chemical compounds that are naturally formed in the course of malting, mashing, brewing, fermentation, and conditioning?

It might be expected that research would by now have answered this question. It hasn't, even where laying down is fundamental—in the wine industry—at least, of the kind that would cause further fermentation.

If there is no fermentation, what does happen? It seems that yeast may continue to instigate many tiny microbiological activities. They may work, for example, on trace elements in wine, too. Additionally, there may be some slight, slow oxidation; too gentle to create a stale note, but sufficient to add a hint of Madeira.

Some developments of this sort may also take place in a strong beer in which there is no longer a secondary fermentation. That is, anyway, the argument of Lees and Pripps.

The Lees product that is brewed for laying down is Harvest Ale. This was first produced last year. The idea is to make one brew each year, from the new season's best malt and hops.

The brew is made, to a gravity of 1.100, at the beginning of November. It has a long boil, more than two hours, and is hopped only once, but heavily. Fermentation is conventional, though with some rousing. There is then a period of two to three weeks' warm conditioning. The beer produces 10.5 percent alcohol by volume.

While a bottle-conditioned strong ale is usually undrinkably sweet when served young, Lees Harvest Ale can be consumed immediately. When I tasted it last year, I found it rich but in no way cloying, exceptionally smooth, with a gradual dryness.

It has an immense hop character as well as warming alcohol. I felt I was being bathed in soothing hop oils. I could have erotic dreams after a nightcap of Harvest Ale.

A year later, I again sampled the 1986 vintage, and found it even smoother, more rounded and complex, as though the flavors had mingled.

No sooner had Lees surprised me with their pasteurized "laying down" beer than I received something similar from Pripps. Like the Lees product, Pripps is vintage-dated. Both represent a recognition that there is a market for a taste of tradition.

Pripps is by far the biggest brewing company in Sweden, and is owned by the state enterprise board. For years Pripps has had within its portfolio a product called Carnegie Porter.

Carnegie was a Scottish brewer who emigrated to Sweden during the heyday of porter as a style. The porter that still bears his name is the only top-fermenting brew to survive in Sweden. It is brewed in two of the three alcohol bands permitted by Swedish law, and expresses its character most truly in the top one (Class III, 5.6 percent alcohol by volume; around 1.056).

When I visited Pripps brewery at Bromma, near Stockholm, a couple of years ago, I tasted a Class III Carnegie Porter blindfold alongside a European-export version of Guinness. I was surprised to discover that the Carnegie Porter had by far the more intense palate. It is a roasty-tasting brew, with great fullness of flavor.

My tasting in Sweden was not of the vintage-dated version. The date indicates when the beer was brewed—but the product is kept in the bottle for six months before it is released.

While I was in Sweden, a brew was maturing, and I have now had the opportunity to sample it. I found it remarkably soft and drinkable for a strongish porter. It, too, makes a good nightcap—but I am happy to report there was nothing Strindbergian about my dreams.

*Reprinted with permission from* What's Brewing, *newspaper of the Campaign for Real Ale.*

# seventeen

# Homebrewing Lore

## THE LOST ART OF HOMEBREWING
### BY KARL F. ZIESLER

*This article appeared in Zymurgy, December 1978 (vol. 1, no. 1)*

While rummaging through the basement the other day I came unexpectedly upon a curious and at first unrecognizable bit of mechanism; on examination it proved to be a device I once had purchased hopefully for filtering homebrew. The discovery took me right back to pre–New Deal times, the days before the respectable art of homebrewing faded into a poignant past along with candle-molding, lard-rendering, and curry-combing. Memories returned of agonizing experiments with patent filters, cappers, siphons, bottle washers, and yeast—all those devices which characterized my humble beginning as a homebrewer—and I recalled vividly the last batch I ever concocted, the one I spent an hour wiping off the kitchen ceiling. That catastrophe, on an evening when my wife was entertaining, cured me even without her final ultimatum. Foolishly, I had allowed several bottles to warm up: the second one let go as I pried off the cap and sprayed the whole kitchen, including the supper over which my wife had lovingly labored. I had made the usual mistake of putting my thumb over the bottle, and so the suds lathered my bosom. After the bottle had finished fizzing, the room looked like the scene of a hatchet murder, and there was a good half inch of beer on the floor.

This explosion terminated years of painful, groping experiments, as far as I was concerned, experiments that often resulted in disaster, occasionally in a fluid that was actually drinkable, and once or twice, as in any hazardous pursuit, in a marvelously delectable, amber-clear, ivory-collared treasure, to be fondled, held up to the light, and sipped delicately—a gift from the gods as rare as truth from a barrister's lips and as palatable as manna. It had been made, for all I could ever determine, exactly like the other batches, which turned sour, deliquesced into suds, or outdid the Missouri in muddiness; yet there she stood, so help me, like a fan dancer at a ladies' aid meeting, a masterpiece deserving, if any beer ever did, only one name—Pilsner! Ah me, if I could but pass on to posterity the secret of those ineffable brews, I would face the prospect of another arid era with fortitude, even anticipation.

Unfortunately, however, all I can do is to record the technique of an art whose beginnings are already lost in antiquity, awaiting exhumation by some doctoral candidate. Professional *bierbrauers*, made jobless by the Volsteadian ukase, concocted the first wort for neighborhood consumption: it was several years before big business realized the profits inherent in purveying all the ingredients of beer but the water, the collar, and the kick. Even then the dehydrated materials were at first sold surreptitiously, and in ultra-dry territory malt and hop stores were subject to frequent rude visits from the police. Before the Sahara was crossed, however, malt was sold under glorified brands, and even had radio programs dedicated to it. Batch-laying became a recognized profession like piano-tuning, with reputable practitioners making the rounds of the boulevard districts to serve bankers and chiropractors rich enough to escape the drudgery. And in more modest homes, men whose domesticity, under uxorial duress, encompassed nothing more complicated than drying teaspoons became authorities on sterilizing bottles and dissolving yeast and sugar.

Homebrewing was practiced in upstairs halls, bedrooms, broom closets, telephone booths, and dumb-waiters, but my own technique required an entire basement. Mere dilettantes brewed only five gallons at a time—the quantity made from one can of malt—producing about forty-five 12-ounce bottles. But more sophisticated fermenters like myself made a double batch, netting approximately eighty-five bottles at a single ordeal. Purloining the bottles was one of the sobering elements in the whole business, for only plutocrats laid out good money for them, and many a nocturnal scavenging expedition up alleys was undertaken to meet the needs of a confirmed brewer. Next, you acquired several gross of caps and a capper—and if anyone thinks invention ceased with Edison, he should have seen some of the contraptions investors fondly imagined would affix stoppers to homebrew bottles. There were varieties that stood upright, that nailed on the doorjamb, that worked with levers and foot pedals and thumbscrews, that had universal joints, free wheeling, and knee action; but they all broke bottles, pinched fingers, and jammed, making the capping job one of the most profane in the entire business. Patent siphons exercised all the ingenuity of a hydrostatic engineer, evolving from a simple rubber tube into elaborate affairs replete with valves, hooks, strainers, bulbs, and four-wheel brakes. Strainers and filters exhausted the gamut of resourcefulness, utilizing everything from pumice to the family Bible as media for separating malt from its ineradicable sediment; but none of them ever functioned. Sooner or later most of these accessories were thrown violently behind the furnace and you got down to the fundamentals.

Aside from collecting an assortment of measuring cups, spoons, kettles, pans, dish towels, and the like, you were ready now to lay a batch. A quart tin of malt was emptied into the big crock, and 5 gallons, or thereabouts, of water added, along with enough sugar to impart the desired kick. Last, a half cake of yeast was dissolved with water and stirred in vigorously by means of a broom handle or furnace poker. Then you adjourned to the laundry tubs to wash up, and called it a night.

The occult phase came next—fermentation. Ordinarily the process took 72 hours more or less, depending on the temperature, the phase of the moon, and the number of times the furnace was shaken. The first morning produced a fine welter of coarse white suds pushing over the rim of the crock; a little later the foam became flecked with brown, like a scorched meringue; carefully these extrusions were skimmed off and the surface of the repulsive fluid scrutinized. Men have been known to crack during these anxious hours, beating their wives, frightening their children, and sending stenographers into convulsions, for it was all-important that the brew be incarcerated in the bottles just as it reached the right turn. Specific gravity meters, litmus paper, and any number of scientific devices were sold to help detect this critical point, but they proved of little value. There were a few gifted individuals who, by looking at the rioting malt, could tell infallibly when it was ready for bottling. But most brewers agreed that when the malt ceased exuding suds and attained a mottled appearance with tiny bubbles and clear brown patches, the fateful moment had arrived, usually about the third night and the one which the local dominie invariably called.

Decorum tempts me to pass over the final process of bottling; after all, it was purely mechanical. But posterity must know the difficulties as well as the pleasures of homebrewing, and I may as well be frank. To bottle, one donned a bathing suit, waders, fireman's helmet, and a grim but determined countenance. The bottles were lined up closely around the box whereon sat the crock of brew, high enough to let the siphon function. A half teaspoon, more or less, of sugar was dropped into each bottle; at least you hit as many as you could, with some getting double or triple doses and some getting none at all. (Those overcharged would later taste like maple syrup produced in a kerosene refinery, if they did not blow up; those without any sugar could easily be confused with a solution of green soap.) Then you started to toy with the siphon. Prone, with one cheek to the cold floor, you held one end of the tube deep enough in the solution to draw and not so far as to reach the sediment, put the other end in your mouth, and sucked. Alchemy never produced anything viler than the gush of tepid beer that immediately drenched your tonsils; but you managed, simultaneously, to retch and to divert the geyser from the tube into the nearest bottle. Only a few squirts could be put in at a time, as the foam ran over and made the bottles sticky; so you circulated the hose from bottle to bottle till they were all full, or the siphon ran out with an obscene gurgle, and you had to ladle out the last few bottles, half full of yeasty dregs, which were carefully put aside for your wife's brother.

Capping, too, was a purely mechanical task. If you survived without six blood blisters and a dozen broken bottles, you considered yourself lucky. If you were fastidious, you wiped the outside of the bottles and put them neatly on a shelf; but most of us just sluiced out the basement with a hose and went wearily to bed. It was considered the mark of an exceptionally masterful spouse if he could boast of making his wife clean the crock.

Commercial brewers speak today with unseemly pride of the age of their beer; this grates on the ear of a homebrewer, for it was universally believed that the homemade stuff would spoil if kept over a month. It required about 3 days for the clouds of sediment to precipitate, and hardier brewers took their first taste then. It was better, however, to let the shot of sugar complete its work and put a little life in the brew before sampling. There was a delicious instant of expectancy when you finally held a bottle to the light, found it clear to within an inch or so of the bottom, and carefully pried off the cap. If the contents didn't detonate in your face, the first crisis was passed. Then you tipped the bottle and allowed the contents to glide ever so slowly into the glass, otherwise you would pour nothing but suds; and you always had to guard against complete decantation, too, lest the dregs leave the bottom of the bottle. Finally, you tasted it. Copartners in a batch anxiously scanned the features of the first taster. If the grimace was not too demonical, sighs of relief were heaved—it was then evident that the stuff could be drunk.

Sermons were preached against homebrewing, and editorials in the dry organs pilloried it as the first station on the road to the Drink Habit, as an evil influence undermining the American home, and as insidiously breaking down moral fiber. Well, I will grant that the manufacture of homebrew was potentially destructive of character, but there was nothing insidious or underhanded about its effects. True, I have heard high school principals, after two bottles, tell stories in mixed company that would bring blushes to the cheeks of a weekend tour stewardess, but I always attributed this potency more to desire than actuality. Prohibition drinking generally was animated by expectancy, so that the effect of whatever quantity of alcohol was imbibed was heightened by a receptive conviction that intoxication must be immediate and inevitable. As for the claim that homebrewing led to chronic inebriation, I submit that exactly the opposite was true.

Unquestionably the cellar art was the greatest force for temperance in the whole insufferable Prohibition era. Practicing it, as I have revealed, was just too damn much trouble.

*Reprinted from the November 1935 issue of* The American Mercury. *Karl Ziesler was assistant* Zymurgy *editor Kathy McClurg's father.*

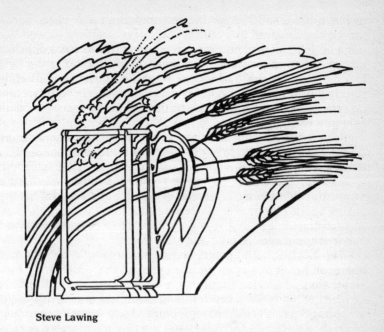

Steve Lawing

# SECRET SATISFACTION OF BREWING

## BY JOHN GOLDFINE

*This article originally appeared in Zymurgy, Winter 1984 (vol. 7, no. 4)*

We all know the pleasure of sitting down to supper and staring at the foam of the beer we have created. How well the stout goes with the rarebit, the Pilsener with the curried lamb. How inexpensive, wholesome, and tasty our beer is. We all have these satisfactions.

But they alone would not make the avid brewer I am. As I think of a brewing day, I recognize secret satisfactions that have little to do with the advertised "reasons to brew at home."

When I get out of bed on a Saturday morning in the winter and I see heavy snow falling or a temperature of 20 below, I feel the pleasure of certainty. The weather conditions have given me a chunk of time and cold air, both of which I can use. I am glad to be working with nature today, instead of fighting against it. Because of the weather, I will begin the beer-making process that will pattern my life for several hours and whose logic, once begun, I can serve, but not stop or modify.

I rattle the plates of my Star Kineo woodstove that has pumped out heat for three quarters of a century. Today it will do the heavy work of heating mash and sparge water and wort. Its real job is heating my house, and this heavy work is only a serendipity. I feel the same satisfaction for Star Kineo that I do for anyone who offers me a free lunch.

After the fire is going and the water heating, I disassemble my Corona hand mill and clean the coffee out. I love to handle this mill. Its design is obvious when it's all apart in the sink, but some genius had to think of it. As I scrub I honor the mind that saw the simplicities that eluded everyone else. A lot of the metal casting on this South American mill is crude (not like the beautiful surfaces of Star Kineo), but it is only fair that someone's backyard foundry should service my backyard brewery.

Later, as I begin my sparge, I have to taste a little of the sweet wort. I can never demonstrate often enough that I have truly converted starch to sugar. I feel like a magician whose audience is stunned by his tricks. Even I, the magician, am stunned! I could have bought a can. I could have depended on a foreign expert. But no, this sugar is mine and it is better than anything sealed in cans and sold in shops. In a small way, making this sweet wort is like repairing a balky engine, Sheetrocking a room, or getting in the winter's wood. I've done it by myself, and that makes all the difference in the world.

After the boil, the Maine winter makes its first direct contribution. Out goes the primary onto my picnic table. While I ski for an hour or two, the wind, snow, and arctic air will provide a cold break with no wort chiller to clean up later. Again I love the free-lunch feeling that, while I'm skiing, the weather is working on my brew for me.

When I get back from the trails, the liquid yeast is showing its fancy pedigree and foaming away on a cool corner of Star Kineo. Within two hours after pitching, the beer has a mat of froth and I have a sense of relief that the day's work is over. No more worries of the bad old days of dry yeast: Is that a bubble? Has the fermentation begun? Will this beer eventually be drunk from a mug or poured from a vinegar cruet onto a tomato, conquered by the evil, airborne empire?

The primary goes into the cellar with two or three lagering carboys, a Cornelius keg, ten or fifteen cases of empties, and the same number of full ones. Everything is at 40 degrees—another bonus of the Maine winter. On the derelict freezer supporting my carboys is written "Brew City" in maroon paint, relic of a fully loaded paintbrush with nothing left to paint. Tonight in Brew City everything is ticking along as I want.

There are so many things in the world, and most are too complicated for me to understand. I have to learn their mysteries in the hands of the specialist. But there is one thing in this whirlpool of puzzlement that I have plucked out and examined. I know how the experts do it. In my own little way, down in the cold cellar, down in Brew City, I am one of the experts, a master of arts and mysteries in his own domain.

And that is the secret satisfaction of brewing.


# JOY OF BREWING

## by Fritz Maytag

*This article originally appeared in* Zymurgy, *Special 1985 (vol. 8, no. 4)*

Brewing as a profession has brought me all the joys normally associated with a successful business venture, plus one.

Nearly any day (or night—I love the nights) of the year when I stroll through the brewery, I find something that has never failed to bring joy and a special feeling of awe. I am speaking of a vat filled with actively fermenting beer. Always when I see it I stop and look and wonder. Always I lean over and take a gentle sniff. Sometimes I blow the foam away—as I learned to do long ago from some forgotten brewer—to see the appearance of the wort. I watch the foam come slowly back, covering the little round window I made.

I look out over the whole batch, gradually adjusting my sense of time to that of the fermenting beer; watching for changes, watching for movement, watching foam bubbles form and burst, watching the dark swirls slowly accumulate and move toward the outer edges.

What in the world is going on here? Who am I to have caused this to happen? How many billions of billions of yeast cells are living here? How marvelous that in this vat is something so far beyond the human brain's ability to number or comprehend. And most of all, how wonderful that this will become beer!

It will move out into the world of people. Each glass will have a little foam on top, just like this fermenter. This beer is alive, and will bring a special mysterious happiness into many people's lives.

I am sure I have spent many more hours looking through a microscope at fermenting beer than any other brewery owner in the world. I have watched and watched and watched. I have looked and looked at yeast cells dividing and budding, at bacteria in all their horrible splendor and multiple shapes, at wild yeast, at hop resins and all that protein, at all those tiny things so small that they jump around from being bumped by Mr. Brown's moving molecules.

The first time I did this, early in my brewing career, I was terrified by all the bacteria I saw. How could we ever get rid of them? How to clean up this beer? How could we get a pure and healthy yeast to thrive here? Before I learned about putting a little caustic soda in the drop of beer to clean out the protein, I thought everything that moved was bacteria. God knows, even after adding the caustic soda in those days, there was plenty of movement! Have you seen this? The unbelievable life force in one single drop of "old-fashioned" beer?

Gradually over time our beer became cleaner and cleaner. Nowadays looking for bacteria is like looking for a needle in a haystack. How proud I am of that! How well I remember the long rods, the short rods, the fat rods, the cocci in all their many groupings. How close I felt to Pasteur late one night in the lab when I realized that the drawings in his great work are absolutely accurate. How proud I was as I began to understand from personal experience something of the courage and discipline and patience and effort required by the great pioneers like Koch and Pasteur.

In the real world of human beings, weather, and all the accidents of fate, an experiment is never pure. Things don't behave, nothing is ever quite proven. It can take years just to learn how to thoroughly sterilize a wort cooler. And more years to be sure you are sure it is sterile.

I have spent all night waiting for signs of budding on the yeast, wondering if there is enough oxygen, if the wort is too cold, if the yeast is viable. Will this brew take off? Will all the things we are trying to learn from this particular brew be lost because something new has gone wrong? How will we ever get it all together? Will we ever make a beer we are proud of, that people will want to buy?

From the first time I ever contemplated a batch of fermenting beer, I have been thrilled by the thought that all those billions of yeast cells are making a magical drink for people. In my early days, our one fermenter held the equivalent of about 20,000 glasses of beer. I would stand there wondering who the people were who would drink it. Where would they be? What would they be doing?

Nowadays our fermenters hold twice that much beer. And as I stand beside one, watching the living changes on the surface, I try to imagine 40,000 people each drinking a glass. Think of all the lives, all the families, all the old friends and new friends. Think of the full moons, the hard work finished proudly, the sadness and happiness.

I am very proud to know that this will be good beer. It is clean, the yeast is thriving. All those people will drink it because they choose to. And with any luck it will bring a bit of happiness into each life. I am very proud to think that this brew will be a part of all those lives, a magical part. I can barely begin to get my mind around this joy.

I leave the fermenting room. If it's daytime, there is a bustle of people working—someone here to see me, or I have somewhere to go. The brewery is all around me, and all the people who work for me and with me. The phone is ringing, there is a tour in the taproom.

If it is nighttime, the copper kettles are glowing under the lights, the brewery is quiet, I think of all the many, many people working together who have built this company, all the years, all the agony and joy. How proud we all are.

But I have a special joy that I have tried to describe, but which I can never really share with anyone. It is waiting for me any time I go into that fermenting room. The beer is alive there in the vat, we are old friends, and we know what we are up to. That is the joy of brewing!

*Fritz Maytag is President and Brewmaster of Anchor Brewing Company, San Fransisco, California. He is a Director of the Brewers' Association of America, as well as The Beer Institute, and President of the California Small Brewers' Association.*

# TRAVELING WITH HOMEBREW

## by Charles Matzen and Charlie Papazian

*This article was originally published in* Zymurgy, Summer 1979 (vol. 2, no. 2)

No one needs to tell the homebrewer about the advantages of homebrew. A cold, freshly decanted glass of your favorite recipe is unmatched by anything in the commercial universe. How about sipping mead in your favorite mountain hot springs, or a light lager while viewing a superb desert sunset? But aren't cans and bottles, six-packs and cases, so much easier to throw in your pack or car? You can buy them anywhere and don't have to worry about stirring up sediment or keeping the bottles with you for the whole trip. To this, the editors of *Zymurgy* can only say, Hog bile! After years of exhausting research into the effects of homebrew on vacations, we assert that it is not only EASY but practical to take homebrew along—even to remote areas.

There are distinct advantages. Most of us need some time for our nervous systems to leave the working-world worries behind. Homebrew is the fastest method we know of. Homebrew is also very effective in keeping your mind off those worries until the moment you return. A delicious brew will make instant friends—a much better way to say hello than a stick of gum. And of course, with homebrew, who can tell what will happen?

If you plan to backpack, remember that alcohol is lighter than water. It's an excellent mosquito repellent, or at least it helps you forget about them. Homebrew is also terrific for repelling/forgetting centipedes, scorpions, rattlesnakes, cockroaches, ants, wind, rain, snow, and flat tires. In short, homebrew makes the good parts of traveling better, and the lousy times tolerable.

Most of us have had days when nothing seems to go right. This is most intolerable while on a vacation. The editors of *Zymurgy* have taken many research and development trips with various ales, meads, and beers, but during our latest expedition, homebrew was put to one of its greatest tests.

We'd left three days ago looking for sunshine, but without success. Certainly our luck would change tomorrow, as it was a new moon today. The next morning we woke up in the snow. After hastily packing wet equipment, we headed down the road towards a hot springs we'd been hoping to make the night before. Three days in a car is too long, and all we wanted to do was find a good place to spend the day on homebrew research. The snow had stopped, but the radio warned of more.

First came a stop for gas—only 80 cents a gallon in the mountains. Three miles later, the car stopped. "It could be something simple, like water in the gas. But then again . . . " Fortunately, we were on top of a hill and able to coast a few miles (while the engine sputtered) to a garage. When the carburetor was in several pieces on the ground, the mechanic announced, "Water in the gas." We emptied the full, 80-cents-a-gallon tank, and refilled it for only 79 cents a gallon. The mechanic pointed the way to the hot springs, and we gave him a bottle of ginger mead. An hour and a half later, over homebrew-testing dirt roads, we found the springs. The only trouble was that they were on the other side of a rain-swollen river! A guy with a raft, watching our dismay, offered to take us across for a buck apiece. What would it cost to get back? We no-thanked him anyway and drove an hour and a half back up the mountainside. The car, still with some water residue in the tank, threatened to stop any second.

As the day was progressing, we needed homebrew more than ever. We decided to stop and camp at the first place that even remotely looked like a relaxing spot. The storm was still looming above us, so we decided to get out of the mountains and head for the desert again. Another hot springs was within an hour on good roads. A few miles outside a small town, we came to a gate with a tent by its side. The hot springs were apparently on private property. When a turbaned man emerged from the tent to greet us, I couldn't help but remember a conversation we'd had earlier in the day about murder-suicide cults. Well, we'd made it this far; it was the middle of the afternoon and we'd had almost another full day in the car. "No way to know unless we check it out." We were led across Oriental carpets on the tent floor to a small trailer on the other side. In the trailer, a group of people were discussing a rash of rip-offs that had been occurring at the cooperative, which was where the hot springs were located. We were invited to spend the night at the cooperative and eat with them in exchange for work or pay. We decided to pay a few dollars for use of the hot springs only.

My companion left for the money and the unlocked car containing much camera equipment. As soon as he left, a guy walked in who informed the group that everything he owned had been taken. They comforted him with thoughts on the emptiness of the material plane. I left the trailer soon afterwards, and was advised to prepare for a spiritual experience at the hot springs. They were the best we'd found—a perfect temperature in a great spot. For

CHARLIE PAPAZIAN

some reason, I found it difficult to relax, thinking about the part of my material plane located in the car.

It was becoming late afternoon. The storm was following us and there was no way we wanted to spend the night there. As it turned out, we left the hot springs minus only a few dollars and plus some anxieties. An hour or so later, we were driving up a dirt road in the Superstition Mountains, wondering if we'd ever find a place to relax that day. As the sun was setting, the road turned into a wash. "Boy, it'd be awfully bad to get stuck in here!" We checked the sand. It seemed firm enough, and there were plenty of other tire tracks leading up the wash. Pretty stupid, eh? Well, we hadn't had any homebrew yet. Ten minutes up the wash, we were stuck in deep sand. You know the sound—urruhuururruruh-uh. Thunder was getting closer, it was getting dark, and a ranger had warned us about flash floods in these canyons.

We decided that now was the time to have a homebrew. Then we tried digging down and placing boulders under the wheels. It was a real workout, but the homebrew made us stronger. After much sweat, we managed to dig the car into the sand up to the back bumper. Sand was slipping into the exhaust pipe, and it was beginning to rain. I was trying to imagine what the inside of my car would look like filled with mud. My mood had changed with the day—from pissed off, to anxiety, then to hysteria, and now, comedy. We opened up another bottle of homebrew, which helped our night vision consid-

CHARLIE PAPAZIAN

erably. My companion was able to find a suitable piece of wood to put under one of the back wheels. There was nothing we could find for the other wheel. We opened another bottle of homebrew.

You've no doubt heard those stories about people achieving superhuman strength in times of crisis—all those mothers who have lifted automobiles, stopped buses and the like to save their endangered babies. We needed something for the other side of my car to get out of the wash. Now, I admit, there are some who may believe this picture is staged. I ask only that you use your knowledge of the power of homebrew to erase any doubts about the implications of this photo.

At any rate, we were able to escape the wash, and our bad luck seemed to be all used up in one day. We were able to return to the serious research on traveling with homebrew and the effects of intense relaxation on one's outlook towards life.

Through several years of experimentation in different climates and conditions, we were able to compile this list of tips for taking homebrew with you on your next adventure:

1. Decant, either before you leave home or begin a hike. Sediment adds weight and a yeasty flavor when stirred up. Homebrew

decanted into a jar or bottle with a tight lid will stay plenty carbonated for at least a week. If you decant and then recap the brew in another bottle, it will stay carbonated longer. Don't worry about spoilage for trips up to a month (or perhaps longer). Decanting is especially important for meads, as they are usually more sensitive to being stirred up.

**2. How to carry.** If possible, carry glass. As you probably know, plastic imparts some flavor and decreases the carbonation. Champagne bottles are remarkably strong. Mason jars work very well. A few bottles should prove no more difficult to carry than water. Be sure to firmly strap in, or pack in. For larger amounts, we experimented with the device pictured below over all types of terrain, including sand and rocks. It was remarkably stable, could be pulled with either hand, and it did the job. Under impossible conditions, you can always carry it over your shoulder (though we've never had to).

**3. How to keep it cold.** When you get up in the morning, attend to first things first. Put the homebrew into the shade. In the mountains, it's usually no problem. Snow or a cold stream are ideal (be careful not to freeze). The desert or beach are more challenging. At the beach, pouring seawater on sand around the homebrew in shade will help. In the desert, one must carefully rotate homebrew with the shade of a cactus, and count on cool nights to maintain a decent temperature. If using the official AHA cross-country carrier, you can pack ice in the bottom.

**4. Type of beer.** It depends on where you're going. Dark beer in the desert absorbs heat—light ales and lagers don't get hot as quickly. A variety is usually best so you can choose the right brew to accompany your mood or food.

**5. In public places.** We've experimented on trains and planes with no problem. If anyone questions you, just offer a glass. If you're self-conscious or in an area off-limits to alcoholic beverages (a church social?), remember that dark ale looks like cola. Things truly go better with homebrew.

**6. Be sure to carry enough.** This, of course, is the most important tip. You'll want to share, and you won't want to run out. The AHA offers awards for missionary work proselytizing the merits of homebrew. Write *Zymurgy* for details.

Wherever you're going, homebrew makes it better. We've tested transporting quart bottles over the roughest of roads, walking into deep canyons, hot deserts, mountains, and tropical jungles. It's always been worth the effort!

*Charles Matzen was cofounder of the American Homebrewers Association and Zymurgy along with Charlie Papazian. Both still homebrew.*

# eighteen

# The Last Drop

## LAST DROP—A BIT OVERCARBONATED

### by John Isenhour

*This column originally appeared in* Zymurgy, Spring *1995 (vol. 18, no. 1)*

I always attend the annual Minnesota Brewfest at Sherlock's Home in Minnetonka, which is organized by the Minnesota (home) Brewers Association (MhBA). It has been one of my favorite events because it was the first place I had experienced Bill Burdick's hand-pulled ales served in an authentic English environment. (They even accept English currency.)

Last year while judging lambics I opened a 10-ounce bottle of Kriek and was surprised by a huge pink geyser that shot out of the bottle, coming down all over me, the judging forms, table, and everything else in the vicinity. The sound alone momentarily stopped the entire judging event. Following my own rule of "possible single bottle infection," I asked for the second bottle.

When it was presented, the cap was obviously bulging. Wishing I had some tongs with which to hold the bottle, I decided it might be prudent to open it outside. I asked MhBA member John Desharnais if he would take a picture at the moment I opened the bottle. The results were spectacular, and once again there was not enough brew

left in the bottle a split second after opening to give it a taste. A bit overcarbonated, down to the *last drop*.

*John Isenhour is a National BJCP Judge and hop-aroma aficionado who is involved in growing hops, culturing yeast, and managing an Internet BJCP exam study group while pursuing a graduate degree in information science. Current passions include ginger and beersenhour.*

# LAST DROP—YOU KNOW YOU'RE A HOMEBREWER IF . . .

### BY DEAN BOOTH

*This column originally appeared in Zymurgy, Fall 1995 (vol. 18, no. 3)*

The Greater Attleboro Suds Suckers were inspired by Jeff Foxworthy's comedy routine "You know you're a redneck if . . ." (for example, if fewer than half the cars in your yard actually run or if your father walks you to school because he's in the same grade).

**If** you've ever tried to improve a Budweiser by stirring in a crumbled hop pellet.

**If** you have a hose adapter permanently attached to your kitchen faucet.

**If** you wanted to name the puppy "Fuggles."

**If** you own a sterile trash can.

**If** you have more than 10 gallons of beer in your home right now.

**If** you measure beer in gallons.

**If** you don't think that 10 gallons of beer is a lot.

**If** you've ever used a mop on a ceiling.

**If** all party invitations you receive say "bring a keg."

**If** you have a large stove pot that no one else is allowed to use.

**If** you've ever driven your car in winter with the windows down and the heat off because you were afraid the cooler in the backseat was getting warm.

**If** you've ever stumped the tour guide on a megabrewery tour, deliberately.

**If** you have a glass that you wash by hand instead of in the dishwasher.

**If** you've ever said any of these phrases:
"In a not-frosted glass, please."
"Probably dirty hoses."
"What kind of beer is it supposed to be?"
"By weight or volume?"
"My yeast is ready."
"Aw, crap, twist-offs."

**If** there is a bottle in your refrigerator with an air lock on it.

**If** you've ever butted into the conversation of total strangers because you overheard the word "sparge."

**If** you can't remember the last time you popped open a flip-top beer can.

**If** your favorite character on *Bonanza* reruns is Hop Sing.

## Last Drop—You Know You're a Homebrewer If . . .

**If** you've ever cut a hole in a refrigerator door.

**If** you've ever gone to a redemption center to buy bottles.

**If** your 10-year-old critiques the clarity and head retention of her root beer.

**If** the owner of the beer store doesn't remember your name anymore.

**If** a waitress has said you're the first person to ever send a beer back.

**If** your kid entered the science fair with a demonstration of fermentation.

**If** you've ever bought a six-pack of beer just because you liked the empties.

**If** walking across the kitchen floor sounds like Velcro.

**If** you've even *thought* about adding hop oil to unscented love oil.

**If** you've ever pulled bottles out of other people's recycling bins.

**If** you've ever put the kids to bed dirty because the tub was full of soaking bottles.

**If** every T-shirt you own is from a brewpub or microbrewery.

*Dean Booth, a homebrewer for 6 years, is President of the Attleboro (Massachusetts) Suds Suckers. His cartoon "Yeast Culture" won the 1993 AHA cartoon contest (Zymurgy, Winter 1993, vol. 16, no. 5).*

# APPENDIX

# ZYMURGY

1978–1996 Abbreviated Index
Selected Feature Articles Published in Zymurgy
19 Years of the Finest Writing on Homebrewing and Beer

## SPECIAL ISSUES 1985–1996

### SPECIAL 1996 (VOL. 19, NO. 4)
### WHY WE BREW
### 47 Homebrew Recipes and 19 Profiles of 22 World-Class Homebrewers

The Best of 1996
28 Gold-Medal-Winning Recipes from the 1996 AHA National Homebrew Competition

### SPECIAL 1995 (VOL. 18, NO. 4)
### THE GREAT GRAIN ISSUE
### 45 recipes and 21 articles by 22 authors

Grain: The Heart of Beer
Principles and Practice
Gear and Systems
The Best of 1995
29 First-Place Recipes from the 1995 AHA National Homebrew Competition

### SPECIAL 1994 (VOL. 17, NO. 4)
### SPECIAL INGREDIENTS AND INDIGENOUS BEER
### 76 recipes and 35 articles from 31 authors

Around the World in 76 Brews
Indigenous Beers
Herbs and Spices
Fruits and Vegetables
Grains
The Best of 1994

29 First-Place Recipes from the 1994
AHA National Homebrew Competition

## SPECIAL 1993 (VOL. 16, NO. 4) TRADITIONAL GERMAN, BRITISH AND AMERICAN BREWING METHODS

### Three sections with 29 articles from 22 authors

German Section
German Brewing History
Malting Techniques
Lager Beer: A Brief History
The Art and Science of Decoction Mashing
Basic Techniques for Formulating Lactic Acid
Rests

British Section
The British Brewing Scene
Cask-Conditioned Ale
British Malting and Brewing Practices
Early British Ale: Bittering, Flavoring and Aroma Ingredients
Infusion Mashing

American Section
Extract Brewing: American Style
The Lagering of Lagers
Brewing with Adjuncts
Cereal Mashing
Mashing Methods and Malt Compared

The Best of 1993
29 First-Place Recipes from the 1993 AHA National Homebrew Competition

## SPECIAL 1992 (VOL. 15, NO. 4) THE BREWING PROCESS, GADGETS AND EQUIPMENT
30 authors fill 120 pages with articles on these topics:

Equipment Utilization
Tips and Gadgets
Sanitation
Ingredients
Mashing
Wort Preparation
Fermentation
Bottling and Kegging
The Best of 1992
28 First-Place Recipes from the 1992 AHA National Homebrew Competition

## SPECIAL 1991 (VOL. 14, NO. 4) TRADITIONAL BEER STYLES

The Moods of Beer
*Michael Jackson*

The Beer Judge Certification Program

Treasury of Beer Styles
72 Beer and Mead Styles Discussed in Detail

Bibliography of Beer Style References

The Best of 1991
26 First-Place Recipes from the 1991 AHA National Homebrew Competition

## SPECIAL 1990 (VOL. 13, NO. 4) HOPS AND BEER

Hops Through the Years: A Brief History
*Kihm Winship*

It's Funny What Hops Into Your Memory
*Stephen Foster*

Development of Hop Varieties
*Alfred Haunold*

Hop Varieties and Qualities
*Bert Grant*

Assessing Hop Quality
*Dave Wills*

Processing Hops Into Bales and Pellets
*Ralph Olson*

# Appendix

Hop Products in 1990
*Lloyd Rigby, Ph.D.*

Hop Oil Equals Aroma and Flavor
*Gus Guzinski*

Growing Hops at Home
*Pierre Rajotte*

The Best of 1990
23 First-Place Recipes from the 1990 AHA National Homebrew Competition

## SPECIAL 1989 (VOL. 12, NO. 4) YEAST AND BEER

Yeast—The Sorcerer's Apprentice
*Charlie Papazian*

Healthy Homebrew Starter Cultures
*Paul Farnsworth*

Yeast Biology and Beer Fermentation
*Jean-Xavier Guinard, Mary Miranda and Michael Lewis, Ph.D.*

Wild Yeast
*George Fix, Ph.D.*

Yeast Nutrients in Brewing
*Paul Monk, Ph.D.*

Commercial Production of Dried Yeast
*Clayton Cone*

Yeast Stock Maintenance and Starter Culture Production
*Paul Farnsworth*

Collecting Yeast While Traveling
*Pierre Rajotte*

## SPECIAL 1988 (VOL. 11, NO. 4) BREWERS AND THEIR GADGETS

Profiles of Five Homebrewers
Kathy Ireland, Eric Furry, Mark Hillestad, Michael Bosold and Rodney Morris

Gadgets Section
16 Ideas and Homemade Gadgets
Illustrated Dictionary of Homebrewing Equipment
*Diane Keay*

The Best of the Best of 1988
*Compiled by Wayne Waananen*
22 First-Place Recipes from the 1988 AHA National Homebrew Competition

## SPECIAL 1987 (VOL. 10, NO. 4) TROUBLESHOOTING

Drinkability: What Is It?
*Michael Jackson*
Different Beers for Different Occasions

A Brewer's Detective Story
*Dave Miller*
Tracking Down Off-Flavors

Tasting Techniques
*William Pfeiffer, Ph.D.*
Effectively Evaluating Beer

Flavor Profiling
*Charlie Papazian*
Standardized Flavor Profiles

## SPECIAL 1986 (VOL. 9, NO. 4) Malt Extract

What Makes Your Beer Special
*Alan Tobey*
Designing Your Own Brew

Tips and Hints
*Edited by Christine Schouten*
Guidelines for Using the Recipes

The Low-down on Malt Extracts
*Compiled by Jill Singleton*
AHA's Definitive Guide

Beer Styles and Recipes
*Beer Style Descriptions by Charles Hiigel and Edited by Christine Schouten*
A Homebrew Cookbook

## SPECIAL 1985 (VOL. 8, NO. 4)
## ALL-GRAIN BREWING

A Primer on Malt
*Staff of Great Western Malt*
Malting Definitions in Layman's Terms

Home Malting for Homebrewers
*R.C. Dale*
Advantages of Doing It Yourself

Yeast Cycles
*George Fix, Ph.D.*
Ethanol, $CO_2$ and By-products

Mashing Systems and Lautering Vessels
*Al Andrews*
A Wide Range of Proven Systems

## 1996

### SPRING (VOL. 19, NO. 1)
The Science of the Art of Beer
*Randy Mosher*

From Hot to Cold: A Cool Brew Cruise
A Counterflow Wort Chiller Road Test
*Daniel S. McConnell Ph.D. and Kenneth D. Schramm*

The Competitive Edge
*Steve Dempsey*

It's Been a Pleasure Meading You—An International Meadery Tour
*Jim Martella*

Jackson on Beer
*Michael Jackson*
Giving Good Beer the IPA Name

Homebrew Cooking
*Candy Schermerhorn*
Scrumptious Spring Supper

Tips and Gadgets
*Our Members*
Water Conservation, Racking Tube Rinser, Well-Guarded Wort, Double Bottle Rinser

For the Beginner
*Jeff Shurts*
Easy Starter Steps

World of Worts
*Charlie Papazian*
Felicitous Belgian Stout

Best From Kits
*Jack Hagens*
Add a Little Hoppiness to Your Life

### SUMMER (VOL. 19, NO. 2)

Extracting the Essentials
*Bill Metzger*

A Peek Into Porter's Past
*Keith Thomas*

Perfect Your Porter
*Terry Foster*

Adventures in Brewing—A Profile Starring Wendy Aaronson and Bill Ridgely
*Scott Bickham*

Jackson on Beer
*Michael Jackson*
Elemental Homebrewing: Midnight Sun, Saunas and Sahti

Homebrew Cooking
*Joseph Ascoli*
The Rebirth of Summer and Brewing

Tips and Gadgets
*Our Members*
Boilover Abatement, CO$_2$ Tank Stands, Quick Fill, Increased Pourability, No More Lumps

For the Beginner
*Dave Bone*
Hop Basics

World of Worts
*Charlie Papazian*
Klibbety Jibbit Lager

Best From Kits
*Larry Johnson*
Quaffable German Ales

## FALL (VOL. 19, NO. 3)

Light and Beer
Peter A. Ensminger, Ph.D

Brown Ale in a Can
*Jefferey A. Seeley, Ph.D., and Todd L. Mansfield, Ph.D.*

Cool Coils—Immersion Chiller Road Test
*Daniel S. McConnell, Ph.D., and Kenneth D. Schramm*

Homebrew Bayou Conference Report
*Ed Greenlee*
AHA 1996 National Homebrew Competition Winners

AHA 1996 National Homebrew Competition Second-Round Brewers

Jackson on Beer
*Michael Jackson*
State of Confusion

Homebrew Cooking
*Brian Glover*
Celebrating Creativity in the Kitchen

Tips and Gadgets
*Our Members*
Avoid Heart/Carboy Break, Dry Yeast Tips, Nitrogen Dispense for Minikegs, Trub Removal

For the Beginner
*Jeff Shurts*
Secondary Importance

World of Worts
*Charlie Papazian*
Islandic Vellosdricka

Best from Kits
*Dennis Fisher and Joseph Fisher*
Descending Into the Bock Underworld

## WINTER (VOL. 19, NO. 5)

Inside Berlin's Own Beer
*Dennis Davison*

Unlock the Secrets of Old Ale
*Ray Daniels*

Lagering in Louisiana—The Brewing of Paddlewheel Pilsener
*Ralph Latapie*

Jackson on Beer
*Michael Jackson*
Brew Like an Egyptian

Homebrew Cooking
*Joseph Ascoli*
A Brewer's Holiday

Tips and Gadgets
*Our Members*
Carboy Tote, Bottle Tree on Wheels, Hopper Helper, Heat Shield, Carboy Drying Rack

For the Beginner
*Martin P. Manning*
Airing Things Out: Aeration vs. Oxygenation vs. Oxidation

World of Worts
*Charlie Papazian*
Cuba's Havana Gold

Best from Kits
*Mark Moylan*
Quick and Easy Ales

## 1995

### SPRING (VOL. 18, NO. 1)

Mead Success: Ingredients, Processes and Techniques
*Daniel S. McConnell, Ph.D., and Kenneth D. Schramm*

Belgium: A Land of Endless Riches
*Martin Lodahl*

Is Belgium's Brewing Culture at Risk?
*Jan Maes*

Brewing Better Belgian Ales
*Phillip Seitz*

Enhance Your Beer-Drinking Pleasure With the Proper Glass
*Mark R. Anderson*

Jackson on Beer
*Michael Jackson*
Will Euro Drinkers Have to Face a Red Barrel Threat?

Tips and Gadgets
*Bob Jones*
Pseudo Beer Engine

For the Beginner
*David A. Weisberg*
Holy Hydrometer, Batman!

World of Worts
*Charlie Papazian*
Sure as Tootin' Pumpernickel Stout

AHA 1995 National Homebrewers Conference—Planet Beer

### SUMMER (VOL. 18, NO. 2)

Confessions of Two Bitter Men
*Tony Babinec and Steve Hamburg*

A Bottler's Guide to Kegging
*Ed Westemeier*

Brew by the Numbers: Add Up What's in Your Beer
*Michael L. Hall, Ph.D.*

The Experimental Homebrewers: Creators of a Revolution
*Jim Neighbors*

AHA Registered Club List

Jackson on Beer
*Michael Jackson*
Be on Your Guard for Fine French Beers

Tips and Gadgets
Clean Kettle Bottoms, Draining Bottles, Dry-Hopper, Easy Siphon Starting, Turbo Carboy Rinsing, E-Z Bottling Chart and Simple Single Infusion

For the Beginner
*Fred Hardy*
Can We Talk? A Beginner's Vocabulary

World of Worts
*Charlie Papazian*
Ira's Teeter Tottler Altbier

### FALL (VOL. 18, NO. 3)

Ward Off the Wild Things: A Complete Guide to Cleaning and Sanitation
*James Liddil and John Palmer*

Oktoberfest: German Engineering in a Glass
*Brad Kraus*

The Counterpressure Connection
Where Bottles and Kegs Unite
*David Ruggerio, Jonathan Spillane and Doug Snyder*

Taking Off for Planet Beer
*Jim Dorsch*

AHA 1995 National Homebrew Competition Winners

AHA 1995 National Homebrew Competition Second-Round Winners

Jackson on Beer
*Michael Jackson*
A Brewpub Blooms in Brooklyn

For the Beginner
*Bill McKinless*
Brewing as a Hobby

Tips and Gadgets
Carbonating in PET Bottles, 30 Cent Beer Engine and High-Pressure Sprayer

World of Worts
*Charlie Papazian*
Ginger Dinger—Ahead of Its Time

## WINTER (VOL. 18, NO. 5)
Eisbock: The Original Ice Beer
*Dennis Davison*

The Haze Maze: Fine Your Way to Clear Beer
*Jeff Mellem*

Rhythm and Brews: A Profile of Homebrewer Paul Sullivan
*Bill Wald*

The Spirit of Homebrewing
1995 Commemorative Belgian Strong Ale
*Greg Kitsock*

1995 Great American Beer Festival® Winners

Jackson on Beer
*Michael Jackson*
In Search of a Ruby Plus an India Pale Ale

For the Beginner
*Ray Ballard*
High-Gravity Brewing: Homebrewing 1.101

World of Worts
*Charlie Papazian*
Switch and Toggles Preposterous Porter

Tips and Gadgets
*Dan Hall*
Homebrewery Improvements: Beverage Line Rinser

## 1994

## SPRING (VOL. 17, NO. 1)

Heat Capacity Calculations for Mashing
*Kurt Froning*

Brew Indigenous Beers: Chicha and Chang
*Wendy Aaronson and Bill Ridgely*

How Sweet It Is—Brewing with Sugar
*Jeff Frane*

Groovy Ways to Remove Trub
*Kinney Baughman*

Scroll Through Brewing Software
*Ray Daniels and Steve Hamburg*

Brew-on-Premises
*Bruce Brode*

Jackson on Beer
*Michael Jackson*
Egging Them on in the Land of Midnight O1

For the Beginner
*Fred Hardy*
Oh Those Bottles!

World of Worts
*Charlie Papazian*
Gregor's Violet Ray No Drip Stout

## SUMMER (VOL. 17, NO. 2)

Dutch Brewing in the 17th Century
*Dick Cantwell*

Yeast Management Techniques
*Michael Ligas*

Become Saccharomyces Savvy
*Patrick Weix*

Gas Gossip—Carbon Dioxide vs. Nitrogen in Brewing
*Cliff Tanner*

Conversion Chart
*Philip Fleming and Joachim Schuring*

Jackson on Brewing
*Michael Jackson*
Bock Brings Germans Rushing to the Beer Garden

For the Beginner
*Bruce Susel*
Primed to Perfection

World of Worts
*Charlie Papazian*
Slumgullion Amber Ale

Tips and Gadgets
*Randy Mosher*
Power Sparger

## FALL (VOL. 17, NO. 3)

Roll Out the Mills
*Bob Gorman, Steve Stroud and Mike Fertsch*

Grain Crushing Criteria
*George Fix*

Open Fermentation
*Jim Busch*

Brewers Lobby for Beer Freedom in Japan
*Ellen Custer*

Brewstorm '94 AHA Conference Report
*Tom Dalldorf*

AHA 1994 National Homebrew Competition Winners

Jackson on Beer
*Michael Jackson*
Hints of Europe Return to Canada

For the Beginner
*Fred Hardy*
Equipping Your Kitchen Brewery

World of Worts
*Charlie Papazian*
Tower of Truth Dunkel

Tips and Gadgets
*Charlie Stackhouse*
Aroma Hop Back

## WINTER (VOL. 17, NO. 5)

The Enchanting World of Malt Extract
*Norman Farrell*

Extract Magic—From Field to Kettle
*Carol O'Neil*

The Regal Altbiers of Düsseldorf
*Roger Deschner*

In Search of American Altbier
*Ben Jankowski*

Lightning and Hail—The 1994 Commemorative Brews
*Mark Groshek*

1994 Great American Beer Festival Winners

Jackson on Beer
*Michael Jackson*
Finland's Top Brewer in a Brown Study

For the Beginner
*Fred Hardy and Al Korzonas*
Judging Your Brew

World of Worts
*Charlie Papazian*
Thistle Do It Export Pilsener

Tips and Gadgets
*Peter Van Zile*
Constant Temperature Bath

## 1993

### SPRING (VOL. 16, NO. 1)

Basic Organic Chemistry for Brewers
*Eddie Brian*

For Every Season There Is a Beer
*Ray Daniels*

pH and the Brewing Process
*Eric Warner*

A Brew Dream Fulfilled
*Steve Klover*

Jackson on Beer
*Michael Jackson*
Now Japan Will Tackle Litter Louts

For the Beginner
*Alberta Rager*
Yeast and the Beginner

World of Worts
*Charlie Papazian*
Tennessee Waltzer Dunkelweizenbock

### SUMMER (VOL. 16, NO. 2)

Summer Brewin'
*Ray Daniels*

Boost Hop Bouquet by Dry Hopping
*Mark Garetz*

Stalking the Wild Meads
*Ralph Bucca*

Institute for Brewing Studies: Serving the Industry's Needs for 10 Years
*Jeff Mendel*

Jackson on Beer
*Michael Jackson*
How Black Beer Survived Behind the Wall

For the Beginner
*Rusty McCrady*
Kegging—The Easier Alternative

World of Worts
*Charlie Papazian*
Autumnal Special Reserve

### FALL (VOL. 16, NO. 3)

1993 National Competition Winners

Fusel Alcohols
*George Fix*

Take Pride in Your Label
*Dana Rowe*

Barley Wine—The King Kong of Beers
*Tom Bachman*

Chili Beer
*Karl Bremer*

Jackson on Beer
*Michael Jackson*
A Brewery With Its Own Abbey—It Must Be Ireland

For the Beginner
*Seth Schneider*
Five Crisis Points for the New Brewer

World of Worts
*Charlie Papazian*
Nomadic Kölsch

## WINTER (VOL. 16, NO. 5)

1993 Great American Beer Festival Winners

Saving Time by Saving Beer
*Douglas Serrill*

Hop Schedule Record
*Richard Larsen*

The Making of a Microbrewer
*Stephen Snyder*

Special Malts for Greater Beer Type Variety
*Ludwig Narziss, Ph.D.*

Dear Professor Turns 15
*Elizabeth Gold*

The Oregon Nut Brown Ale Trail—Brewing the '93 Commemorative Beer
*Dena Nishek*

Jackson on Beer
*Michael Jackson*
Beer Archaeologists

For the Beginner
*Fred Hardy*
Care and Feeding of Your Carboy

World of Worts
*Charlie Papazian*
Here to Heaven Oktoberfestwine Ale

## 1992

## SPRING (VOL. 15, NO. 1)

Step into the CompuServe Beer Forum Libraries
*Bill Crisafulli and Brad Krohn*

Simple Math and Your Homebrew
*Jim Hilton*

Wort Chillers: Three Styles to Improve Your Brew
*Ross and Mindy Goeres*

Brewing the 17th Century Way
*Rich Wagner*

Jackson on Beer
*Michael Jackson*
A Seasonal Search for the Phantom of Brewing

For the Beginner
*Rusty McCrady*
Hopping and Straining

World of Worts
*Charlie Papazian*
Turtles Wheat Beer

## SUMMER (VOL. 15, NO. 2)

Good Cider Starts in the Summer
*Paul Correnty*

From Carboy to Beer Glass: A Note on Froth
*Michael Tierney*

Fun with Fruit Beer, an Essay
*Russ Schehrer*

Brewing with Wild Hops
*Dan Fink*

# Appendix

Jackson on Beer
*Michael Jackson*
How Porter and Stout Went Global

For the Beginner
*Rusty McCrady*
First Brews that Are Quick and Easy

World of Worts
*Charlie Papazian*
Gnarly Roots Lambic-Style Barley Wine Ale

Sanctioned Competition Program
*James Spence*
More Than 10 Years of Homebrew Competition Support

## FALL (VOL. 15, NO. 3)

Just Brew It! 1992 AHA Conference Report
*Greg Giorgio*
1992 National Competition Winners

Stimulate Your Senses With Mead
*Susanne Price*

Sulfur Flavors in Beer
*George Fix*

Hitting Target Gravities
*Ray Daniels*

The Oldest Brewery in America
*Charlie Papazian*

Jackson on Beer
*Michael Jackson*
How Scot's Yeast Made a Belgian Classic Ale

For the Beginner
*Rusty McCrady*
An Easy Lager

World of Worts
*Charlie Papazian*
Slanting Annie's Chocolate Porter

## WINTER (VOL. 15, NO. 5)

1992 Great American Beer Festival Winners

Flaming Stone: Brewing Traditional Steinbiere
*Phil Rahn and Chuck Skypeck*

The Detriments of Hot-Side Aeration
*George Fix, Ph.D.*

Beer Stability
*Micah Millspaw and Bob Jones*

The Sensory Aspects of Zymological Evaluation
*David W. Eby*

Jackson on Beer
*Michael Jackson*
Supping at Santa's Knee

For the Beginner
*Rusty McCrady*
What Else Should Go into My Beer?

World of Worts
*Charlie Papazian*
Quarterbock

## 1991

## SPRING (VOL. 14, NO. 1)

Beers/Breweries of Philadelphia
*Rich Wagner*

Packing Your Beers
*Russ Wigglesworth*

Dry Yeast for Homebrewers
*Rodney Morris*

The Trouble with Trub
*Dan Gordon*

Jackson on Beer
*Michael Jackson*
Viennese Beer

For the Beginner
*Alberta Rager*
Oxygen and Beer

World of Worts
*Charlie Papazian*
Vicarious Gueuze Lambic Sour Mash

**SUMMER (VOL. 14, NO. 2)**

Great American Beer Festival: 10 Years
*Dan Fink*

There's a War Going On
*George O'Brien*

Pilsner Urquell: The Brewery
*Darryl Richman*

Counterpressure Transfer
*Dan Fink*

For the Beginner
*Rob Brook*
Evolution of a Beginning Brewer/Beginners Luck

Jackson on Beer
*Michael Jackson*
Russian Stout

For the Beginner
*Rusty McCrady*
From Haze to Clarity

World of Worts
*Charlie Papazian*
Colorado Cowgirl Brown Ale

**FALL (VOL. 14, NO. 3)**

1991 AHA Conference Report
*Robin Garr*
1991 AHA National Competition Winners

Sanitation
*Quentin Smith*

In Mead We Trust
*Russell Schehrer*

The Jockeybox Draft Chiller
*Teri Fahrendorf*

Homebrewing in Leningrad
*M. Todd Breslow*

Jackson on Beer
*Michael Jackson*
Belgium's Rodenbach

For the Beginner
*Alberta Rager*
Brewing Gadgets

World of Worts
*Charlie Papazian*
Roll Over Mr. Rogers

**WINTER (VOL. 14, NO. 5)**

Beer From Water: Minerals and Beer Styles
*Jon Rodin and Glenn Colon-Bonet*

Institute for Brewing Studies for Small Brewers
*Jeff Mendel*

Record Keeping for Better Beer
*Ray Daniels*

1991 Association Survey Results

Aass Brewery Visit
*Darryl Richman*

Jackson on Beer
*Michael Jackson*
Belgian Trappist Beers

For the Beginner
*Rusty McCrady*
Kitchen Management

World of Worts
*Charlie Papazian*
Shikata Ga Nai American Light Lager

## 1990

### SPRING (VOL. 13, NO. 1)

Germany: A Visit to Beer Heaven
*Charlie Papazian*
Hundreds of Years of Beer History

Amber Waves of Wheat: America's Wheat Beers
*Don Hoag and John Judd*
A Survey of Commercial Wheat Beer Recipes

How to Build a Simple Counterpressure Bottle Filler
*Steve Daniels*
Successfully Bottle Homebrew from Your Kegs

Jackson on Beer
*Michael Jackson*
Your Finest Mesopot Ale

World of Worts
*Charlie Papazian*
Five of Spades Schwartzbier

### SUMMER (VOL. 13, NO. 2)

Brewing at the Bottom of the World
*Glen E. Eickman*
Homebrewing in Antarctica

A Visit to Pilsner Urquell
*Jonathan B. Frey*
A Brewery Tour of the Czech Original

Love or Money? Why We Brew
*Rusty McCrady*
A Cost Analysis of Prohibition Beer Versus All-Malt

Beer Filtrations for the Homebrewer
*Rodney Morris*
Easy, Inexpensive Ways to Build a Homebrew Filter System

World of Worts
*Charlie Papazian*
Unspoken Passion Raspberry Imperial Stout

### FALL (VOL. 13, NO. 3)

Homebrew Expo
*Greg Giorgio*

AHA Competition Grows More than 50 Percent

1990 National Homebrew Competition Winners

I'm a Mild Man Myself, But . . .
*Howard Browne*
Reflections on Mellow Mild Ale

Teaching a Homebrew Course
*Don Hoag and John Judd*
How to Spread the Word to New Homebrewers

Beer and Rocks
*Stephen Cribb, Ph.D.*
The Geology Behind the Word's Classic Brewing Waters

World of Worts
*Charlie Papazian*
A Trace of Autumn Oktoberfest

### WINTER (VOL. 13, NO. 5)

The CompuServe Beer Forum and How It Works
*Russ Wigglesworth*
How to Use Your Computer to Improve Your Brewing and Networking

Evaluating Beer
*Charlie Papazian*
Quality Analysis and Off-Flavors in Homebrew

Homebrew—What It Really Contains
*Fred M. Scheer*
A Detailed Chemical Analysis of 13 Homebrews

New Tales of Old Ales
*Don Hoag*
An Overview of Commercial Strong Ale Recipes in America

World of Worts
*Charlie Papazian*
What the Helles

# 1989

## SPRING (VOL. 12, NO. 1)

Carbonation Techniques and Safety
*Don Hoag*
How to Get the Right Amount of Fizz in Your Beer

Bottling Beer? Test Your Sugar
*Nancy Vineyard*
How to Adapt a Sugar—Testing Kit for Use With Beer

Beer Bitterness and Boiling
*Terry Foster, Ph.D.*
How Boiling Your Wort Influences Hop Utilization

Wheat Beers
*Gary Bauer*
Brew Them Now for the Warm Months Ahead

World of Worts
*Charlie Papazian*
Limnian Doppelbock II

Jackson on Beer
North vs. South in the Netherlands

## SUMMER (VOL. 12, NO. 2)

Lakefront Brewery
*Marty Nachel*
The Homebrewer's Dream Is Alive and Well in Milwaukee

Scratch Brewing the Belgian Browns
*Michael Matucheski*
A Wisconsin Farmer Gets Back to the Basics

Colonial Brewing and Malting
*Rich Wagner*
Modern Demonstrations of Colonial Brewing Practices

Trends in Water Treatment
*Dave Miller*
Changes in City Water Regulations May Affect Your Beer

Reverse Osmosis Water Purifiers
*Marty LaBeny*
How They Work and How You Can Build One for Home Use

Mashing Made Easy
*Diane Keay*
The Benefits of Mashing

World of Worts
*Charlie Papazian*
Saunders' Nut Brown Ale

## FALL (VOL. 12, NO. 3)

Claude of Zeply
*Ray Spangler*

1989 National Homebrew Competition Winners

How to Keep the Skunks out of Your Homebrew
*Michael Tierny, Ph.D.*
Any Off-Odors in Your Beer?

Low-Alcohol Beers: Brewing Fad or Future
*Ken Kane*
Both Commercial and Homebrewers Can Brew Them

Taming the Wild Fridge
*Kurt Denke*
How to Install a Reliable Thermostat on Your Brewing Fridge

Toward Greater Purity
*Rusty McCrady*
Keeping the Arch-Enemies of Pure Beer at Bay

World of Worts
*Charlie Papazian*
Chief Niwot's Mead

## WINTER (VOL. 12, NO. 5)

GABF℠ VIII a Success
Professional Panel Tasting and Consumer Poll Winners

I'm a Bitter Man (Reflections on Real Ale)
*Philip Davia*

Soda Keg Draft Systems—Better than Bottles?
*Jackie Rager*

Racking from Carboys to Soda Kegs
*Cy Martin*

Kraeusening and Cold-Hopping Soda Kegs
*Cy Martin*

Water Treatment: How to Calculate Salt Adjustment
*Darryl Richman*
What Do These Chemicals Do for Your Beer?

World of Worts
*Charlie Papazian*
Tempestuous Pilsener

## 1988

## SPRING (VOL. 11, NO. 1)

Bockin' in the U.S.A.
*Don Hoag*
A Survey of American-Made Bocks

Brewing Competitions of the Past
*Kihm Winship*
Evolution of Commercial Beer Judging

Testing Light Malt Extracts
*Jeff Frane with Peter Jelinek*
Analysis of 12 Malt Extracts

World of Worts
*Charlie Papazian*
Possum Porter

When a Beer Goes to Sleep
*Michael Jackson*

## SUMMER (VOL. 11, NO. 2)

Beer 101
*John C. Berry*
Brewing in the Classroom

Pilgrimage to Chico
*Paul Farnsworth*
A Visit to Sierra Nevada Brewing Co.

Roast, Roast, Roast Your Grains
*Randy Mosher*
Excerpt from *The Brewer's Companion*

Principles of Beer Dispensing
*L. Z. Creley*

World of Worts
*Charlie Papazian*
Contentful Horizon Pilsener

## FALL (VOL. 11, NO. 3)

Best of the Better Beer Slogans
*Greg Jenson*
Beer Advertising in Days Gone By

Dare to Design Your Own Recipes
*Kurt Denke*

Beer Color Evaluation
*George Fix, Ph.D.*
A New Way to Measure Beer Color

Getting a Lift from Your Yeast
*Dave Miller*
Make and Use Your Own Cultures

World of Worts
*Charlie Papazian*
Dunkel in the Dark Lager

**WINTER (VOL. 11, NO. 5)**
Miracles and Milestones
*Charlie Papazian*
The AHA Is 10 Years Old

Computerbrew
*Steve Conklin*
Use Your Computer to Learn About Brewing

The Stouts of America
*Don Hoag and John Judd*

How to Make Maple Sap Beer
*Morgan Wright*

World of Worts
*Charlie Papazian*
Who's In the Garden Grancrew?

## 1987

**SPRING (VOL. 10, NO. 1)**
What Makes a Champion Beer?
*Greg Noonan*
An Analysis of Competition-Winning Brews

The Cellos and Piccolos of Beer
*Paul Freedman*
A Tour of Coors' Pilot Brewery

Dealing with the Great Unwashed
*John V. Hedtke*
Acquisition and Cleaning of Bottles

Brewing Stalworts
*Scott Walker*
The Button-Down Brews of Gary Bauer

Michael Jackson
*Michael Jackson*
Thomas Hardy Ale

World of Worts
*Charlie Papazian*
Vorogomo Märzen

**SUMMER (VOL. 10, NO. 2)**
Ripley's Beerlieve It or Not
*Michael Zalkind*
Adventures of a New York Brewer

Black Patent Malt
*Kihm Winship*
The Evolution of Porter

Silkscreening Permanent Labels
*Steven Grossnickle*
How to Build a Silkscreen Press

A Brewer's Herbal
*Gary Carlin*

The "New Medical Prohibition"
*Colleen D. Clements, Ph.D.*
A Response to Views on Alcoholism

Prickly Pear Cactus Mead
*Charlie Papazian*
The Ultimate Mead Is Discovered

World of Worts
*Charlie Papazian*
The Devil Made Me Do It Brown Ale

**FALL (VOL. 10, NO. 3)**
Special Yeast Section
*Charlie Thompson, Gary Bauer, AHA staff*
All About Yeast, Yeast Culture Directions, Liquid Yeast Sources

Beer Trek Through China
*Joseph Weeres*
A Tour of 15 Breweries

Reaching the Summit
*Grant Gengel*
Profile of Summit Brewery

Beer Strength: A Matter of Degree
Gareth John, Ph.D.
A Look at Hydrometers

Winners Circle
Wayne Waananen
Six Favorite Fall Beer Styles

World of Worts
Charlie Papazian
Elementary Penguin London Ale

## WINTER (VOL. 10, NO. 5)

The Holiday Beer Tradition
Kihm Winship
Strong Beers for Long Nights

The Art of Making Mead
Brother Adam
Meadmaking at Buckfast Abbey

The Porters of the United States
Don Hoag

Starch Testing with Iodine
Paul Farnsworth
A Look at the Mashing Process

Boiling Methods and Techniques
Greg Walz
How to Boil Your Extract Water

All About Plastic Bottles
William Montague
An Alternative to Glass

World of Worts
Charlie Papazian
Vernal Weizenbock

## 1986

## SPRING (VOL. 9, NO. 1)

Stalworts of Brewing
Scott Walker
Byron Burch

'Ale-y's Comet Brew
Phil Angerhofer
Brew for a Once-in-a-Lifetime Event

A Hops How-To
Diane Keay
A Guide to Hops Characteristics

Beer Flavor Evaluation
Charlie Papazian
Evaluating Beer Flavor Is an Art and a Science

World of Worts
Charlie Papazian
Colonel Coffin Barleywine

## SUMMER (VOL. 9, NO. 2)

In Search of Greater Beer
Donald Curtis
Seeking the Good Life and Great Brews Around the World

Homebrewer Turned Microbrewer
Jill Singleton
Dewayne Saxton Takes the Plunge

The Fine Art of Carbonation
John R. Scanlon

Kvass
Jill Singleton
Homebrew in the U.S.S.R.

World of Worts
Charlie Papazian
Rain Forest Light Lager

## FALL (VOL. 9, NO. 3)

Tropical Toddy
Michael Jackson
The Coconut Brew of Sri Lanka

Tuak—Toddy of the Rice Farmer
Melissa Ballard

Using the Bruheat Boiler
*Ron Valenti*
Trouble-Free Mashing

A Sterile Siphon Starter
*Patrick T. Pickett*
Using an Aspirator to Start a Siphon

Munich's Oktoberfest
*Tom and Bill Bauer*

World of Worts
*Charlie Papazian*
Colinbock—Special Occasion Beer

## WINTER (VOL. 9, NO. 5)

Better Water for Your Home
*Richard Leviton*
Filter Systems for Tap Water

Brewing Water
*Diane Keay*
A Guide for Beginners

World of Worts
*Charlie Papazian*
Blitzweizen Barley Wine Lager—1986
National Conference Beer

Monster Mash
*Phil Angerhofer and Friends*
Thomas Hardy Ale

For the Beginner
*Diane Keay*
Brewing Jargon

## 1985

### SPRING (VOL. 8, NO. 1)
Why Not Blend Beers?
*Gerard Sparrow*
A Competition Winner Tells How

Judging Is Good for You
*Phil Angerhofer and Paul Freedman*
Homebrewers Benefit From Judging

Improving Malt Extract Beers
*Pat Anderson*
Brewing Great Beer in One or Two Hours

Your Mouth and Beer Flavors
*David J. Welker, D.D.S.*
Overlooked Aspects in Tasting

The August Schell Brewing Company
*Donald G. Crenshaw*
Six Generations of Tradition

World of Worts
*Charlie Papazian*
Jupiter's Return Märzen and Europe Oktoberfest

### SUMMER (VOL. 8, NO. 2)

Where's the Porter?
*Don Hoag*
Eulogy for a Fine Beer

The Mystery of Malt Extract
*Pat O'Neil*
Edme Proves a Point

Beer Bang Theory
*Charlie Papazian*
Brewing in Outer Space?

Coffee Pot Beer
*Clifford T. Newman, Jr.*

The Foam at the Top
*Phil Angerhofer*
Judging with the Best of the Best

Especially for the Beginner
*Byron Burch*
Going for Greatness

World of Worts
*Charlie Papazian*
Armenian Imperial Stout

# Appendix

## FALL (VOL. 8, NO. 3)

Flanders Top Ten
*Michael Jackson*
Belgium's Idiosyncratic Beers

Temperature-Controlled Brewbox
*Patrick T. Pickett*
Homemade Brewbox Maintains Good Brewing Temperatures

Full Wort Boil Improves Extracts
*Charles J. Brem*
Brew All-Malt-Extract Beers that Taste Like All-Grain

Especially for the Beginner
*Byron Burch*
Quickbeer Revisited

Wonderful World of Worts
*Charlie Papazian*
Gillygaloo Pale Ale

## WINTER (VOL. 8, NO. 5)

Brewing Techniques Influence Bitterness
*Charlie Papazian*
How to Vary Hop Utilization

'Simmon Beer
*Gary Carlin*
An American Classic

H. L. Mencken, Homebrewer
*Kihm Winship*
He Had a Passion for His Craft

What Is the Quintessential Beer?
*Vickie Simms*
The AHA Asks the Stars

Brewing History of Pittsburgh
*Rich Dochter and Rich Wagner*
A Microcosm of the U.S. Brewing Industry

World of Worts
*Charlie Papazian*
Unkleduckfay Oatmeal Stout

# 1984

## SPRING (VOL. 7, NO. 1)

Virgin Brew
*Phil Angerhofer*
That First Batch

Homebrew's Mr. Wizard
*Al Andrews*
An Inexpensive Sparging System

John Barley-Corn Is Dead but Finnegan's Awake
*Edie Stone*
A Drunken Dream of the Fall and Reawakening of Humanity

The African Beer Gardens of Bulawayo
*Harry F. Walcott*
Brewing in the City of Bulawayo

World of Worts
*Charlie Papazian*
Sinfully Red Cherry Ale

## SUMMER (VOL. 7, NO. 2)

A Potent Religion
*Michael Jackson*
Monastery Beers of Belgium

Beer Bugs
*Brian Hunt*
Hints for Preventing Spoilage

Stone Beer Brewing
*Verlag W. Sachon*
An Old Procedure Rediscovered

World of Worts
*Charlie Papazian*
Summer Solace Pilsener

Especially for the Beginner
*Terence Foster*
Hops for the Beginner

## FALL (VOL. 7, NO. 3)

Czechoslovakian Beer
*Ian Priddley*
A Tour of Traditional Breweries

Start Mashing!
*Charlie Papazian*
An All-Grain Primer

Beer in the Wood
*John Alexander*
Traditional English Ales

I'm Mad
*Lee V. Giles*
Busted for Homebrew in Utah

Arizona Breweries of the Past
*Shelby Meyers*
Oases in the Desert

World of Worts
*Charlie Papazian*
On Deck India Pale Ale

## WINTER (VOL. 7, NO. 4)

Beer and Nutrition
*Doralie Denenberg Segal*
Try Beer for Running Marathons

Secret Satisfactions of Brewing
*John Goldfine*

Michael Jackson
*Michael Jackson*
Ireland's Other Stout Fellows

New Zealand Honey Mead
*Charlie Papazian*
Leon Havill's Mazer Meads

Apache Beer
*Johnny C. Clack*
Brew of American Indians

World of Worts
*Charlie Papazian*
Winds of Endeavor Sparkling Wheat Beer

## 1983

## SPRING (VOL. 6, NO. 1)

Bock Beer Mystique and Tradition
*Alan S. Dikty*
American Styles, German Origins

The Living Lager
*Brent Warren*
A Twilight Foam Adventure

Kiss of the Hops
*Dave Wills*
The Goodness of Hops

Homebrew's Mr. Wizard
*Al Andrews*
Sediment-Free Draft Beer

Especially for the Beginner
*Charlie Thompson*
Mashing for Fun and Flavor

World of Worts
*Charlie Papazian*
Masterbrewers Dopplebock

## SUMMER (VOL. 6, NO. 2)

Yeasts, Beasts and Disinfection
*George H. Millet*
A Primer on Sterilization Procedures

Beer—What's in a Name
*A. Ugur Akinci*
Brewer, Maine and Other Geographical Points of Interest

Wonderful World of Worts
*Charlie Papazian*
The Best Beer Ever Canned by Anheuser-Busch?

Especially for the Beginner
*Charlie Thompson*
Equipment for the Masher

Quickbeer
*Alan Tobey*
From Brewpot to Bottle in 10 Days

**FALL (VOL. 6, NO. 3)**

The Noble Experiment
*H. L. Mencken*
Prohibition's Geological Epoch Remembered

The Complete Sparger
*R.C. Dale*
Theory and Practice of Wort Separation

Guerrilla Brewing in Saudi Arabia
*Sediqui*
Prohibition in the Middle East

Yeast Cycles and Fermentation
*Don Crenshaw*
About Single-Celled Fungi

World of Worts
*Charlie Papazian*
Hassled? Brew Some Up-the-Wall Stout

**WINTER (VOL. 6, NO. 4)**

Beer and Mysticism
*Shelby Meyer*
Better Beer Through Magic

Get Cultured—With Yeast
*Jay Conner*
Procedures for Advanced Brewers

Teaching a Homebrew Class
*Charlie Papazian*
Inspiring Better Beer

Lick It, Stick It and Drink It
*Jim Kincaid*
Designing Homebrew Labels

Better Beer from Your Malt Extract
*Fred Eckhardt*
Quality Beer Is Easy to Make

World of Worts
*Charlie Papazian*
Sparrow Hawk Black Porter

# 1982

**SPRING (VOL. 5, NO. 1)**

The Secret of My Success
*Dave Miller*
1981 Homebrewer of the Year

Especially for the Beginner
*Charlie Thompson*
Understanding Hops

German Biers and Brewing Styles
*Paul Freedman and Joe Ritchie*
Beers of Germany Offer Surprising Regional Variety

German Beer Coasters
*Gregory O. Jones*
Everything You Wanted to Know . . .

500 Bottles of Beer on the Wall
*Charles Matzen*
Tales of Horror Force the Question of Serious Stockpiling

World of Worts
*Charlie Papazian*
Smoky the Beer—German Style

## SUMMER (VOL. 5, NO. 2)

**Beer Flavors**
*Richard Severo*
A Flavor Chemist Isolates 850 Chemical Compounds

**Secrets of Porter**
*Fred Berry*
Insights From 210-Year-Old Treatise

**Breweries Then and Now**
*Barbralu C. Manning*
Breweries of the Old West

**Irish Farmhouse Cheese**
*Charlie Papazian*
Farmhouse Cheese and Homebrew Shape a Lifestyle in Southern Ireland

**Munton & Fison Ltd.**
*Charlie Papazian*
Maltsters Earn Worldwide Respect

**World of Worts**
*Charlie Papazian*
Tumultuous Porter

## FALL (VOL. 5, NO. 3)

**Beer from Down Under**
*Ernie Melville*
Notes from an Australian Brewer

**Cider**
*Michael Jackson*
There's More to Apples than Pie

**Facts About Alcohol**
*Carolyn Reuben*
What's Your Limit?

**Sake—Japanese Rice Wine**
*Fred Eckhardt*

**Beer Is Born**
*Robert L. Palmer*
Fermentation in the Fertile Crescent

**How to Balance Your Beer**
*Alan Tobey*
Designing Beers from Scratch

**Word of Worts**
*Charlie Papazian*
Oktobersbest Golden Malt Lager

## WINTER (VOL. 5, NO. 4)

**Belgian Brewery—250 Years**
*Harlan Feder*
Village Brewery Specializes in Kriek and Gueuze

**Beers of Belgium**
*Michael Jackson*
Specialty Beers Provide Astounding Variety

**Specialty Malts**
*Peter Bowles*
Malt Companies Describe the Use of Their Products

**World of Worts**
*Charlie Papazian*
Barkshack Gingermead

**Especially for the Beginner**
*Charlie Thompson*
You Too Can Be a Masher

## 1981

## SPRING (VOL. 4, NO. 1)

**Special Report**
*Bill Petrij*
Report from a UC—Davis Class

**Morkimer's Missing Mug**
*Sir Philip Anthony Arnolds*
A Sherlock Foams Classic

**Readers Forum**
*Harlan Feder with Bill Franks*
High-Altitude Homebrewing

# Appendix

World of Worts
*Charlie Papazian*
Danger Knows No Favorites Black Premium Lager

## SUMMER (VOL. 4, NO. 2)

Things Your Homebrew Kit Never Told You
*Dan McCoubrey, Paul Freedman and Bob Siner*
Avoiding Homebrew Frustrations

World of Worts
*Charlie Papazian*
Things Are Looking Up Red Bitter

## FALL (VOL. 4, NO. 3)

Sahuaro Cactus Wine
*Bill Litzinger*
A Native American Indian Tradition

Bloopers
*Zymurgy Readers Reveal Their Blunderings and Miscalculations*

Cider—A New England Specialty
*Dr. Sanborn C. Brown*
Apples Are Refreshing in a Pinch

Beer Design
*Alan Tobey*
Analyzing the Components of Flavor

World of Worts
*Charlie Papazian*
Cheeks-to-the-Wind Brown Lager

## WINTER (VOL. 4, NO. 4)

Pennsylvania Lager
*George J. Fix, Ph.D.*

Especially for the Beginner
*Charlie Thompson*
About the Hydrometer

Tasting: 1-2-3
*Bill Petrij*
Factors Affecting Beer Taste

A British Beer Festival
*Charlie Papazian*
*Zymurgy* Surveys the Ale Zone

World of Worts
*Charlie Papazian*
Righteous Real Ale

## BONUS ISSUE (VOL. 4, NO. 5)
## POST-CONFERENCE ISSUE

The Taste of Beer
*Michael Jackson*
Report from a Panel Discussion

Mr. Wizard
*Steve Callio*
Al Andrews' Homemade Brewery

From the Winners Circle
Recipes from Homebrewer of the Year and Meadmaker of the Year

World of Worts
*Charlie Papazian*
Feelicks the Cat's Cherry Lager

## 1980

## SPRING (VOL. 3, NO. 1)
Readers Forum
*Bill Petrij*
Malting Your Own Barley

The L.Draught
*Phil Arnolds*
People in Dead End Canyon Still Talk About It . . .

World of Worts
*Charlie Papazian*
Bruce and Kay's Black Honey Spruce Lager

## SUMMER (VOL. 3, NO. 2)

The Ins and Outs of What You're Drinking
*Rebecca Greenwood*
No More Nitrosamine Blues

A Bedtime Story
*Barbra Wakshul*
Sleep-Inducing Qualities of Hops

Hops in Your Brew
An Introduction to Hops

World of Worts
*Charlie Papazian*
Whitey's No-Show Amber Ale

## FALL (VOL. 3, NO. 3)

England's Campaign for Real Ale
*Paul Freedman*
The Revolt Away from Tasteless Beer

Especially for the Beginner
*Tim Mead*
Equipment to Get Started

"Native" Brewing in America
*William Litzinger*
Native American Beers and Wines

World of Worts
*Charlie Papazian*
Whimmy Diddle Brown Lager

## WINTER (VOL. 3, NO. 4)

Stout Is Good for You
*Paul Freedman and Dan McCoubrey*
History of the Guinness Brewery

Mead Making: The Most Ancient Art of Brewing
*Bill Litzinger*

From the Winners Circle
Kelly Irish Stout Ale
Jubilee Stout

World of Worts
*Charlie Papazian*
Whitehead's Stout

## 1979/1978

### SPRING (VOL. 2, NO. 1)

I'll Take Mine Without Hog Bile, Please
Just What's in That Beer You've Been Drinking?

Treatise on Siphoning
How to Set Up a Gravitational Pump

Home Brew Bread Making
*Roger Bassett*
You Might Get What You Knead

World of Worts
*Charlie Papazian*
Barkshack Gingermead

### SUMMER (VOL. 2, NO. 2)

The Low-down on U.S. Beers
*Mike Royko*
Commercial Beer Additives

World of Worts
*Charlie Papazian*
Elbro Nerkte Brown Ale

Backwoods Chainsaw Beer
*Doug Daugert*
One Brewer Who Never Says "Impossible"

Traveling with Homebrew
*C. Matzen and C. Papazian*
It's Easy and Practical

### FALL (VOL. 2, NO. 3)

Advertising vs. Flavor
*Mike Royko*
It Wasn't Always This Way . . .

From the Winners Circle
Recipes from the First Annual
National Homebrew Competition

Gettin' Useless in Eustis
Encounter with a Maine Homebrewer

World of Worts
*Charlie Papazian*
Rocky Raccoon's Light Honey Lager

## WINTER (VOL. 2, NO. 4)

Homebrew on Tap
Kegging Your Own

Homebrewed American History
*John Gerstle*
Homebrew Angle on History

Vale Vakaviti
*Charlie Papazian*
Fiji Homebrew

World of Worts
*Charlie Papazian*
Joda's Jolly Lager

## DECEMBER (VOL. 1, NO. 1)

The Lost Art of Homebrewing
*Karl F. Zeisler*
Memories of Agonizing Experiments

Black Lava Ale
*Charlie Papazian*
Homebrew in Hawaii

Feast Food and Foam
Stuffed Whole Lobster A-la-mazing

World of Worts
*Charlie Papazian*
Vagabond Black "Gingered" Ale

# Index

## A

Aeration, 41, 81, 131–34, 209–10, 216, 302–09
AHA (American Homebrewers Association), xvi–xviii, 30, 46, 60, 63– 64, 67–68, 143–44, 169, 314–15, 364
Alcohol, 78–79, 154–55, 156–58, 215, 218
Alcohol, fusel, 37
Altbiers, 18, 28–34, 190, 257, 258
Association of Brewers, xvi, 49
Attenuation, 53, 156, 164, 166, 215, 217

## B

Bacteria, 3, 5, 32, 40, 81
Bacterial infection, 174, 216, 292
Barleywine, 42, 146, 155, 207, 218
Beer brands/names
    Affligem Dubbel, 35
    Affligem Tripel, 37
    Alaskan Amber, 33
    Anchor Steam, 34
    Artevelde Grand Cru strong ale, 38
    Badger Best Bitter, 68
    Bass Ale, 63
    Bateman's XXXB, 70
    Best Burton Bitter, 74
    Big Lamp, 64
    Bishop's Bitter, 74
    Blackjack Porter, 58
    Blanche de Bruges, 39
    Boddington's Bitter, 63
    Brakspear's Ordinary, 70
    Brakspear's Special Bitter, 70
    Brigand strong ale, 38
    Brugse Tripel, 37
    Carnegie Porter, 384, 385
    Celis Grand Cru strong ale, 38
    Celis White, 39
    Chouffe strong ale, 38
    Club Premium lager, 9
    Coors, 378–80
    Corsendonk strong ale, 38
    Courage Directors Bitter, 63
    Dortmunder, 257, 258
    Draught Bass, 68
    Düsseldorf Ale, 34
    Duvel strong ale, 38

Beer brands/names (*continued*)
  Everard's Tiger, 68
  Exmoor Ale/Dark, 16
  Fuller's Chiswick Bitter, 63
  Fuller's ESB, 63, 64, 70
  Fuller's London Pride, 70
  Gotlandsdricka, 21–23
  Gouden Carolus strong ale, 38
  Grimbergen Dubbel, 35
  Grimbergen Tripel, 37
  Guinness, 14, 318, 381
  Leann Fraoch, 23–28
  Lee's Harvest Ale, 384
  Liberty Ale, 71
  Maize kernel beer, 3–5
  Marston Moor, 64
  Mateen strong ale, 38
  Mitchell's of Lancaster, 64
  Moondog Ale, 74
  Moscova Beer, 336
  Old Detroit Ale, 34
  Pauwel Kwak strong ale, 38
  Pinkus Münster Alt, 30
  Pitfield, 64
  Rauchenfels Steinbier, 18–20
  Red Star Beer, 336
  Riva Blanche, 39
  Saison DuPont strong ale, 38
  Samuel Adams Boston Lager, 314
  Samuel Smith, 69, 249, 380
  Scaldis strong ale, 38
  Schlsser Alt, 33
  Schmaltz' AltMalt, 34
  Sierra Nevada Pale Ale, 71
  Sierra Nevada Porter, 49
  Smithwick's Ale, 381
  St. Stan's Alt, 34
  Steenbrugge Dubbel, 35–36
  Steenbrugge Tripel, 37
  *Sticke* Altbier, 30, 32
  Victory Ale, 70
  Westmalle Dubbel, 35
  Westmalle Tripel, 37
  Widmer Alt, 34
  Yorkshire Bitter, 74
  Young's Bitter, 381
  Young's Ramrod, 70
  Young's Special, 63
  Young's Special London Ale, 70
  Zum Uerige Altbier, 30, 31, 34
Belgian ale, 35–48, 207, 372
Belgian lambics, 21, 381–82
Bitter, 13–14, 60–74, 102, 188, 207, 257, 258
BJCP (Beer Judge Certification Program), 21, 48, 315, 326, 364
Bock, 191, 207, 257, 258, 382
Bottling, 110–31
Brewers
  Alexander, John, 81
  Babinec, Tony, 60, 74
  Bauer, Gary, 255
  Bergmeister, Joe, 179
  Booth, Dean, 403, 405
  Bosch, Shawn, 274, 279
  Brem, Charlie, 7
  Brosious, Dan, 7
  Browne, Howard, 13, 16–17
  Bucca, Ralph, 271
  Burch, Byron, 87, 148, 150, 371
  Cavicchi, Clare Lise, 6–8
  Colon-Bonet, Glenn, 252, 257
  Conner, Jay, 149
  Corran, H. S., 50
  Dahlgren, Rob, 179
  Daniels, Ray, 54
  Deschner, Roger, 28, 35
  Eby, David W., 314, 325
  Eckhardt, Fred, 17, 340
  Fahrendorf, Teri, 109, 135, 138
  Fahrer, John R., 376
  Farnsworth, Paul, 219
  Farrell, Norman, 168, 182
  Favre, Monica, 142
  Fix, George, 82, 110, 113, 175, 177, 301, 310
  Foster, Terry, 48, 60, 63, 70, 340
  Giffen, Paddy, 375
  Goeres, Mindy, 81, 84
  Goeres, Ross, 81, 84
  Goldfine, John, 391

# INDEX

Grant, Bert, 198
Grossman, Robert, 52
Gudmestad, Neil C., 182
Hait, Layne, 179
Hall, Michael L., 150, 167
Hamburg, Steve, 60, 74
Harper, Tim, 179
Harrison, John, 52
Harwood, Ralph, 53
Jackson, Michael, 17, 24, 28, 30, 34, 49, 63, 70, 378, 382
Klisch, Russ, 131, 134
Kobylinski, Adrienne, 179
Kobylinski, Lee, 179
Konis, Ted, 175
Korzonas, Al, 259
Liddil, Jim, 375
Line, Dave, 248–50, 341
Loysen, Tracy, 142
Lundgren, Hakan, 21
Maier, John C., 372
Martella, Jim, 58
Matzen, Charles, xv, 396
Maytag, Fritz, xv, 393, 395
Mead, Tim, 364
Millard, Mary Beth, 365
Miller, Dave, 82, 87, 108, 366
Morris, Rodney, 209, 213
Moscoso, Guillermo, 10
Mosher, Randy, 31, 139–41, 261, 266
Moylan, Mark, 84
Noonan, Gregory J., 87, 172, 248, 250, 283
Papazian, Charlie, 11, 86, 172, 248, 279, 315, 327, 332, 396
Post, Jim, 374
Protz, Roger, 53, 64, 70
Prozeller, Paul, 372
Rahn, Phil, 17, 21
Rajotte, Pierre, 42
Rebold, Rhett, 376
Richman, Darryl, 247, 255
Rodin, Jon, 252, 257
Ruggiero, David, 110, 131
Sailer, Franz-Joseph, 20
Sarlandt, Oscar, 179
Saxton, Dewayne Lee, 369
Schehrer, Russell, 16, 370
Schmit, Richard, 373
Seitz, Phillip, 35, 48
Skypeck, Chuck, 17, 21
Snyder, Doug, 110, 131
Spangler, Ray, 371
Spillane, Jonathan, 110, 131
Tallman, Stu, 374
Taylor, Raymond J., 182
Taylor, Tod, 179
Thompson, Donale F., 367
Wagner, Rich, 5, 8
Warner, Eric, 310, 313
Watkins, Laurie, 179
Watkins, Mike, 179
Weisberg, David, 75, 80
Weix, Patrick, 214, 219
Wenzel, Fred, 179
Westemeier, Ed, 89, 109
Wheeler, Graham, 50, 52, 70
Williams, Bruce, 23
Brewing companies
  Alaskan Brewing, 33, 57
  Anchor Brewing, xv, 34, 395
  Anheuser-Busch, 10
  August Schell Brewing, 34
  Berliner Brewery, 336
  Boston Beer Co., 314
  Commonwealth Brewing, 74
  Coors Brewery, 301–02
  Frankenmuth Brewery, 34
  Golden Hill Brewery, 16
  Goose Island Brewing, 74
  Great Lakes Brewing, 74
  Hirsch Brewery-Hotel, 28
  Indianapolis Brewing, 34
  La Victoria Malting/ Brewing, 11
  Lakefront Brewery, 134
  Lees of Manchester, 383, 384
  Left Hand Brewing, 58
  Moscow Brewery, 336
  Old Brewery, Tadcaster, 249
  Pripps of Sweden, 383–84
  Rauchenfels, 18–20
  San Francisco Monastery, 10–12

Brewing companies (*continued*)
  St. Stan's Brewery, 34
  Steelhead Brewery, 138
  Tabernash Brewing, 313
  Thistle Brewery, 27
  Travre Brewing, 145
  Vermont Pub & Brewery, 57
  West Highland Brewery, 27
  Whitbread brewery, 49, 58
  Widmer Brewing, 34
  Wynkoop Brewing, 16
  Yuenglings, Pennsylvania, 49
  Zum Schlussel brewery, 30
  Zum Uerige brewery, 29–30, 32
British mild ale, 16, 257, 258
British pubs, 13–15
Brown ale, 146, 207, 257, 258
BURP (Brewers United for Real Potables), 46, 48

## C

Calorie content, 158–60, 164
CAMRA (Campaign for Real Ale), 14–15, 52–53, 62, 64
Carbonation
  and air lock, 79–80
  calculating level of, 104–05, 161–66
  forced, 33, 55–56, 101–08
$CO_2$ tanks, 95–96, 101–02
Counterpressure bottle filler (CPBF), 110–31

## D

Degrees Plato, 76, 154, 171–72, 264
Diacetyl, 69, 177
Doppelbock, 257, 258, 282

## E

Extract brewing, 36, 38, 39, 51, 55, 72, 82, 142–43, 327

## F

FAN (free amino nitrogen), 176–77, 178
Flemish strong ale, 38, 48
Fruit, 2, 259–61
Fruit character, 35, 36–37, 46, 69

## G

Gelatinization, 195–96
Grain mills, 39, 86
Grains, unmalted, 38, 39, 45, 73, 187, 193
Great American Beer Festival, 314
Great British Beer Festival, 58, 64

## H

Hecklerfest, 7
Home Wine and Beer Trade Association, 44, 149, 315
Homebrew suppliers
  Benjamin Machine Products, 117–19, 128
  The Beverage People, 116–17, 128, 150
  Braukunst Homebrewer's Systems, 123–25, 129
  Brewer's Resource, 73
  DeWolf-Cosyns, 71, 72
  Foxx Equipment, 120–22, 128
  Hugh Baird, 71, 72
  J. E. Seibel & Sons, 241
  malt companies, 183–84, 192
  Maris Otter, 71, 72
  surplus outlets, 139–41
  Vinotheque, 122–23
  Yeast Culture Kit Co., 73
  Zahm & Nagel, 114–15, 128
Hop bitterness, 144, 146, 198, 201–08
Hops, 198–208
  alpha acid, 28

Hops (*continued*)
  Aquila, 200, 203
  aroma, 28, 53, 55, 198
  Banner, 200, 203
  Brewers Gold, 198, 200, 203, 369
  British Blend, 375
  British Columbian Golding, 45
  Bullion, 198, 200, 203
  Cascade, 16, 53, 198, 200, 202, 203, 278, 300, 334, 364–71, 373
  Centennial, 333
  Challenger, 53, 372
  Chinook, 53, 198, 199, 200, 203, 375
  Cluster, 198, 200, 202, 203
  Columbia, 198, 200, 203
  Comet, 198, 200, 203
  East Kent Goldings, 40
  English classic, 51
  English Goldings, 53
  Eroica, 200, 371
  fresh, 346–47
  Fuggles, 51, 53, 56, 71, 73, 200, 202–03, 372, 377
  Galena, 198, 199, 200, 203
  German-type, 32
  Goldings, 200
  Hallertauer, 32, 37, 40, 44, 149, 198, 200, 203, 331, 371
  Hersbrucker, 44, 198, 201, 329
  Kent Goldings, 58, 59, 71–73, 200, 265
  Liberty, 375
  Mittelfrüh, 198
  Mt. Hood, 45, 198, 199, 201, 203, 278, 374
  Northern Brewer, 16, 19, 53, 57, 71, 72, 73, 201, 265, 277, 278, 329, 331, 368, 371
  Nugget, 53, 198, 199, 201, 203, 371, 372, 376
  Olympic, 198, 201, 203
  Perle, 23, 53, 57, 201, 202, 203
  Pride of Ringwood, 201
  Progress, 54
  Saaz, 31, 32, 37, 40, 44, 45, 59, 198, 201, 278, 300, 329, 366, 371, 374, 376
  Spalt, 31, 32, 201
  Styrian Goldings, 40, 45, 71–73, 201, 265
  Talisman, 201, 203
  Tettnanger, 19, 31, 32, 43, 199, 201, 203, 300, 329, 334, 377
  Willamette, 16, 51, 53, 59, 198, 201, 202–03, 334, 372, 373
  Wye Target, 201
Hydrometer, 75–81, 151, 170, 172, 303
Hydrometer Correction Table, 80–81

# I

IBU (International bitterness units), 64–68, 207
Internet, Judge's Digest, 46
Iodine test, 19, 87, 174, 296
Iodophor, 101, 170
Irish moss, 27, 44, 45, 55, 87, 329, 334, 364, 366, 370, 371, 373, 374
Isinglass, 10, 19, 69, 218

# J

Jockeybox, 108–09, 135–38
Juniper, 20–23

# K

Kegging, 89–109, 126–27

# L

Lactic acid, 40, 173, 174, 254, 292
Lager, 207
  components of, 191, 198, 257, 258, 365–68

Lager (*continued*)
  pH of, 174, 312
  processes for, 81–82, 108, 148, 177, 300–301

## M

Malt Available in the U.S., 185–87
Malt, 182–97
  amber, 27, 54, 174, 186, 189, 190
  aromatic, 36, 45, 186, 189, 190, 278
  biscuit, 36, 43, 52, 57, 186, 189, 278
  black, 11, 31, 50, 52–54, 57, 58, 72, 146, 147, 173, 190, 192, 274, 282
  black patent, 32, 143, 144, 145, 187, 192, 274, 365, 370, 371
  brown, 50, 52–53, 56–58, 186, 189, 190
  Brumalt, 192
  Carafa, 187, 192
  Carahell, 187
  CaraMalt, 186
  caramel, 187
  caramel Pils, 186, 191
  CaraMunich, 36, 43, 45, 73, 187, 191
  Carastan, 186, 191
  CaraVienne, 187, 191, 375
  carmelized, 19, 20, 52, 186, 190–93, 195
  chocolate, 187
  crystal, 31, 195
  honey, 186, 192–93
  Klages, 372
  lager, 185, 366
  Munich, 186
  pale, 185
  pale ale, 185
  peated, 52–53, 57, 185, 189
  Pilsener, 185
  Rauch, 52, 57
  rye, 185, 189
  Sauer (acid), 185, 192
  Scottish ale, 24, 27
  smoked, 57, 185, 189, 190, 375
  Special B, 36, 43, 73, 187, 191, 278, 375
  Special Roast, 186
  toasted, 36, 39, 52, 57, 186, 282, 373
  two-row, 51, 56, 57, 185, 288
  Victory, 183, 189
  Vienna, 31, 32, 185, 189, 331, 373, 376
  wheat, 19, 31, 56, 176, 185, 188–89, 333, 369, 371, 372
  white, 192
Malt companies, 183–84
Malt extract, 44, 163, 168–82, 185–86, 338–39
  amber, 16, 277, 375
  Bavarian Gold, 369
  Briess Wiezen, 375
  dark, 51
  dry, 16, 72, 373
  John Bull light hopped, 373
  Munton & Fison, 365
  Northwest Gold liquid, 44
  pale, 36, 38, 51, 57–59
  Williams Australian, 372, 374
Malt Extract Tables, 173, 180–81
Maltodextrin, 51, 55, 57–59
Mashing techniques
  decoction, 41, 280, 283–300
  infusion, 40–41, 280–81, 283
  step, 40–41, 280–81
  step infusion, 329, 334, 367–68, 368–69, 374
Master Brewers Association of the Americas, 114
Mead, 21, 155, 270, 353

## N

National Homebrewers Competition, 364
National Homebrewers Conference, xv, 67

## O

Oxidation, 82, 101, 112, 303–04, 308

## P

Pale ale
  as bitter, 60, 61, 70, 71
  characteristics of, 49, 146, 207
  components of, 148, 195, 252, 257, 258
Phosphoric acid, 173
Pilseners, 327–30
  characteristics of, 31
  components of, 149–50, 188, 192, 252, 257, 258
  processes for, 82, 177
  South American, 9
Porters, 48–60
  characteristics of, 82, 147, 207, 384–85
  components of, 147, 277 malt, 190, 192 water, 257, 258
  Victorian, 49, 53, 58
Protein rest, 39–41
Pulque, 2–3, 5

## R

Recipes
  American Pilsener, 149–50
  Andy Anderson's Aaron's Abbey Ale, 43–44
  Bitter, 73
  British mild ale, 16
  Byron Burch Jerry Lee Lewis Russian Imperial Stout, 371
  Chocolate Decadence, 58–59
  Christmas Ale, 265–66
  Damikola Beer, 273
  Dandelion Beer, 272
  Dave Miller Dutch Style Lager, 366
  Delano Dugarm's Batch #28 Triple, 44
  Dewayne Lee Saxton Du Bru Ale, 369–70
  Donald F. Thompson Light-Bodied Light Lager, 367–68
  Double Dipper (Belgian Double), 278
  Düsseldorfer Altbier, 32–33
  Flaming Stone Steinbiere, 18–19
  Flossmoor Best Bitter, 73
  Georgia Stream Common Beer, 278
  ginger beer, 16
  Herb Beer, 271–72
  James Liddil Wild Pseudo-Lambic Gueuze, 375–76
  Jeff Frane's Strong Ale, 45
  Jim Post Jamie Beer, 374
  John C. Maier Oregon Special, 372
  John R. Fahrer Muddy Mo Amber Ale, 376–77
  Kentish Porter, 59
  King Gorm's Gotlandsdricka, 23
  London Porter (malt extract), 58
  Mary Beth Millard Birthday Brew Snow-High Light Lager, 365–66
  Never-Let-Me-Down-Porter, 277–78
  Not So Brown Porter, 56
  Oktoberfestwine Ale, 331–32
  1.049 Lager, 300–301
  Ordinary Bitter, 72
  Paddy Giffen Kilts on Fire, 375
  Paul Prozeller Dubbel Queensberry Framboise, 372–73
  Popeye Porter, 57
  porter, 54–56
  Ray Spangler Toadex/Bloatarian Ale, 371
  Rhett Rebold Central European Pils, 376
  Richard Schmit Arlington ale no. 33, 373–74

Recipes (*continued*)
 Rick Garvin's Cherry Blossom Wit, 45–46
 Scottish Heather Ale, 27–28
 Slumgullion Amber Ale, 333–35
 Smoky the Beer, 57
 Starter Wort, 217
 Stu Tallman Stu Brew, 374
 Thistle Do-It Export Pilsener, 328–29
 Tim Mead Rag Time Black Ale, 364–65
Regulator, $CO_2$, 96–98, 107

## S

Saccharification, 41, 44, 196, 283–88, 292–94, 297, 312
Saison ale, 48
Sanitation, 100–101, 133, 136–37, 216
Scottish ale, 257, 258
Scottish heather ale, 23–28
Shandy, 14
Specific gravity, 172–73
 and alcohol content, 156–57
 equations for, 76, 150–54, 172
 and hydrometer, 76–78
Spirit of Belgium Homebrew Competition, 46
Standard Research Method (SRM), 174
Steinbier, 17–21
Stouts, 49–50
 characteristics of, 49–50, 82, 147, 174, 207, 318
 components of, 371 malt, 147, 178, 192 water, 252, 257, 258
 imperial, 149, 218, 371
Sugars
 brown, 11
 Candi, 36, 42, 43, 45
 Cane, 72, 82
 carmelized, 17, 19
 dextrose, 44, 277, 278, 371, 372
 glucose, 166, 175–76, 178, 292
 honey, 2, 23, 163
 maltose, 175, 209
 molasses, 209–10
 priming, 149, 162–65, 366, 369, 370
 sucrose, 172

## T

Trappist ale, 48
Trub removal, 82, 176

## V

Versuchsanstalt für Bierbrauerei, 20

## W

Water treatment, 247–58
 "Burtonizing," 71–73
 mineral additives, 55, 72
White beers, 39–40, 42–43, 48
Wine, Native American, 2–3
Wort
 oxygenation of, 131–34
 pH of, 173–74
 specific gravity of, 76–78, 150–55
 Starter recipe, 217
Wort chiller, 81–84

## Y

Yeast, 209–19
 ale, *S. cervisiae*, 5, 214, 220–27
 bakers, 23
 Belgian, 36, 37, 40, 44
 champagne/wine, 218, 264, 371
 Cultured Chimay "Rouge," 371
 *Dekkera anomoia*, 375
 *Dekkera bruxellensis*, 375
 German wheat beer, 40
 *Kloeckera apiculata*, 375

lager, S. uvarum, 34, 177, 218, 228–33, 300, 329, 366–68, 374
 single-strain, 31
 white beer, 442
 wild, 24, 212–13
Yeast starter, 211–12, 219, 238–46
 quantity used, 19, 41
 in recipes, 44
Yeast Strain Table, 220–37

# HOMEBREWING.

**Ground Control to Major Homebrewer** Making great beer doesn't have to be as difficult as putting a man on the moon. You decide how much time and effort you want to spend on your homebrew. And whether you brew an easy-as-1-2-3 pale ale or experiment with a Dark Side of the Moon Stout that you plan to enter in one of our hundreds of sanctioned homebrew competitions nationwide, the American Homebrewers Association® is your Ground Control. We're there to help you each step of the way.

For more information
**CALL TOLL FREE
1-888-U-CAN-BREW**

1-888-822-6273 (U.S. and Canada only)

American Homebrewers Association PO Box 1510, Boulder, Colorado 80306-1510 USA. (303) 546-6514, FAX (303) 447-2825, AHA web address http://beertown.org

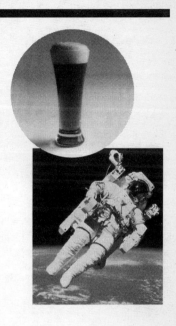

*It isn't Rocket Science.*
Unless you want it to be.
Join the **American Homebrewers Association**®

BOZ

# Thirsty for an Exotic Home Brew?
# Anxious to Get Started on Your Own Home Brewery?

### Let Brewing Expert
# Charlie Papazian
### Show You How With:

### THE NEW COMPLETE JOY OF HOME BREWING
76366-4/$12.00 US/$16.00 Can
Learn first-hand about the history of beer;
discover the secrets of brewing world class styles of beer;
and find out how to add "spice" to your favorite blend

### THE HOME BREWER'S COMPANION
77287-6/$12.00 US/$16.00 Can
Special advice on fermenting, yeast-culturing and
stovetop boiling; helpful trouble-shooting tips;
and answers to the most often asked questions

---

Buy these books at your local bookstore or use this coupon for ordering:

Mail to: Avon Books, Dept BP, Box 767, Rte 2, Dresden, TN 38225         G
Please send me the book(s) I have checked above.
❏ My check or money order—no cash or CODs please—for $_____ is enclosed (please add $1.50 per order to cover postage and handling—Canadian residents add 7% GST). U.S. residents make checks payable to Avon Books; Canada residents make checks payable to Hearst Book Group of Canada.
❏ Charge my VISA/MC Acct#_____Exp Date_____
Minimum credit card order is two books or $7.50 (please add postage and handling charge of $1.50 per order—Canadian residents add 7% GST). For faster service, call 1-800-762-0779. Prices and numbers are subject to change without notice. Please allow six to eight weeks for delivery.
Name_____
Address_____
City_____State/Zip_____
Telephone No._____        BER 0997